THE PHYSICS OF NANOELECTRONICS

OXFORD MASTER SERIES IN PHYSICS

The Oxford Master Series is designed for final year undergraduate and beginning graduate students in physics and related disciplines. It has been driven by a perceived gap in the literature today. While basic undergraduate physics texts often show little or no connection with the huge explosion of research over the last two decades, more advanced and specialized texts tend to be rather daunting for students. In this series, all topics and their consequences are treated at a simple level, while pointers to recent developments are provided at various stages. The emphasis is on clear physical principles like symmetry, quantum mechanics, and electromagnetism which underlie the whole of physics. At the same time, the subjects are related to real measurements and to the experimental techniques and devices currently used by physicists in academe and industry. Books in this series are written as course books, and include ample tutorial material, examples, illustrations, revision points, and problem sets. They can likewise be used as preparation for students starting a doctorate in physics and related fields, or for recent graduates starting research in one of these fields in industry.

CONDENSED MATTER PHYSICS

1. M.T. Dove: *Structure and dynamics: an atomic view of materials*
2. J. Singleton: *Band theory and electronic properties of solids*
3. A.M. Fox: *Optical properties of solids, second edition*
4. S.J. Blundell: *Magnetism in condensed matter*
5. J.F. Annett: *Superconductivity, superfluids, and condensates*
6. R.A.L. Jones: *Soft condensed matter*
17. S. Tautz: *Surfaces of condensed matter*
18. H. Bruus: *Theoretical microfluidics*
19. C.L. Dennis, J.F. Gregg: *The art of spintronics: an introduction*
21. T.T. Heikkilä: *The physics of nanoelectronics: transport and fluctuation phenomena at low temperatures*
22. M. Geoghegan, G. Hadziioannou: *Polymer electronics*

ATOMIC, OPTICAL, AND LASER PHYSICS

7. C.J. Foot: *Atomic physics*
8. G.A. Brooker: *Modern classical optics*
9. S.M. Hooker, C.E. Webb: *Laser physics*
15. A.M. Fox: *Quantum optics: an introduction*
16. S.M. Barnett: *Quantum information*

PARTICLE PHYSICS, ASTROPHYSICS, AND COSMOLOGY

10. D.H. Perkins: *Particle astrophysics, second edition*
11. Ta-Pei Cheng: *Relativity, gravitation and cosmology, second edition*

STATISTICAL, COMPUTATIONAL, AND THEORETICAL PHYSICS

12. M. Maggiore: *A modern introduction to quantum field theory*
13. W. Krauth: *Statistical mechanics: algorithms and computations*
14. J.P. Sethna: *Statistical mechanics: entropy, order parameters, and complexity*
20. S.N. Dorogovtsev: *Lectures on complex networks*

The Physics of Nanoelectronics

Transport and Fluctuation Phenomena at Low Temperatures

Tero T. Heikkilä

Low Temperature Laboratory, Aalto University, Finland

OXFORD

UNIVERSITY PRESS

Great Clarendon Street, Oxford, OX2 6DP,
United Kingdom

Oxford University Press is a department of the University of Oxford.
It furthers the University's objective of excellence in research, scholarship,
and education by publishing worldwide. Oxford is a registered trade mark of
Oxford University Press in the UK and in certain other countries

First Edition published in 2013
Impression: 1

British Library Cataloguing in Publication Data

Data available

ISBN 978–0–19–959244–9 (hbk.)
        978–0–19–967349–0 (pbk.)

Printed and bound by
CPI Group (UK) Ltd, Croydon, CR0 4YY

For Mari, Silja, Elisa, Kaisla and my parents, with love.

Preface

I started to write this book in autumn 2005 when I presented a course of lectures on nanoelectronics at Helsinki University of Technology. I had presented the course once before with three other people, and could not find a book with which I would have been satisfied. Initially, I intended to write only lecture notes, but after running the course again in 2007 and 2009 I thought that the 250 pages of lecture material could easily be published as a book. Little did I know. After being late for my initial deadline by two years, I now finally dare to submit the text to the publisher.

There are other books on electron transport and nanoelectronics. Nevertheless, I found them too narrow in their topic, outdated, too deep for a textbook, containing too much formalism, or having other deficiencies. In other words, I wanted to write my own book that would contain a wide selection of topics and explain them without too difficult mathematics or formalism, and which would tell both about experiments and theory. So here it is.

Because of my aim of avoiding too heavy formalism I have chosen not to describe non-equilibrium Green's function approaches to transport phenomena. Courses detailing these approaches usually spend half the time on the formalism, finding the poles of the various Green's functions, and figuring out analytic continuations and so on. Whereas this is a necessary tool for theorists in the field, most of the transport phenomena as such can be explained without using these tools. Because of this my book concentrates on different types of transport phenomena, without forgetting about the main theoretical approaches (but at a lower level than non-equilibrium Green's functions) to describe them: Boltzmann equations, scattering theory, master equations and so on. On the other hand, I also do not discuss the different fabrication or measurement techniques that are the bread and butter of nanoelectronics experimentalists. However, I try to tell what and how things are measured on a general level. Whereas many of the basic theories presented in the book date back to the 1980s or 1990s (or even before), I have tried to include references and especially pictures of more recent experiments.

Some may complain about the selection of phenomena discussed in this book. Nanoelectronics is a vast field and I could have made different choices. In particular, I have decided to omit the quantum Hall effect altogether. Neither do I discuss, for example, Luttinger liquids or other strongly correlated phenomena, or topological insulators. Spintronics is only a section in the chapter on semiclassical theory, whereas a whole book could be written about the topic. My choice is obviously strongly affected by my own background. Because of this, I have subtitled the book 'Transport and fluctuation phenomena at low temperatures'. Low temperatures are not a necessity for everything I describe, but many of the effects are in practice accessible only at cryogenic temperatures (below a few Kelvin). That is where the interesting physics lies.

Use as a textbook

This book is intended for advanced undergraduate and starting graduate students who have passed courses on quantum mechanics, solid state physics and some statistical physics. In Appendix A I summarize some of the main technical tools needed when reading the book, but I assume that the reader has already encountered most of them before.

I have tried to arrange the book such that as a rule of thumb every chapter could be lectured in one two-hour lecture. The exception to this is Ch. 6, which requires two lectures. It is probable that not all aspects of every chapter can be discussed during such lectures, so the lecturer should decide which details to omit. This pace means that the whole book could be lectured on a course containing 12–14 lectures, which is, at least in my university, quite a typical length of a single-semester course having one lecture each week. Many of the chapters have been written such that they form almost independent entities, so that some of them can be bypassed without too big problems. I would not advise skipping Chs. 2, 3 and 7, which are needed in many other chapters as well, although Secs. 2.7, 2.8, 3.6 and 7.6 are not entirely necessary. Moreover, the concepts of noise and the fluctuation–dissipation theorem discussed in Sec. 6.1 are also needed in other chapters. Chapter 5 is a precursor to Ch. 9, but it can be skipped if the students know the basics of the theory of superconductivity. Part of the text has also been organized as examples and complements, which aim to give some deeper insight on the topic at hand. The difference between the two is that the examples are intended for all readers, whereas the complements are more for the advanced students. The appendices contain first a set of technical tools that students usually learn in basic courses on quantum mechanics or solid state theory, but which can be used as reminders. Also, some of the long derivations are presented in the appendices to improve the flow of the main text.

Each chapter contains several exercises, that are directly related to the text or offer ideas for further development. A solutions manual will be available upon request once the book is published. In addition, every chapter contains one or two questions on recent scientific papers, by which the students are expected to take a particular recent experimental paper and to understand what was measured.

While every effort has been made to carefully check and edit this book, corrections to any errors that may have crept in will be posted on the book website at http://www.thephysicsofnanoelectronics.info.

Happy reading!

Acknowledgements

Over the course of writing this book I have benefited from comments from many people. Foremost, I would like to thank the assistants of my course, Pauli Virtanen, Matti Laakso, Janne Viljas and Francesco Massel, for various helpful remarks and suggestions. I also thank all the students for enduring the draft versions of this text. Special thanks are extended to Philip Jones, Ville Kauppila, Raphaël Khan and Andreas Uppstu for detailed comments on the text.

Parts of this text have also been read and commented upon by other researchers in nanoelectronics. I would like to thank Norman Birge, Guido Burkard, Sebastiaan van Dijken, Francesco Giazotto, Andreas Isacsson, Nikolai Kopnin, Peter Liljeroth, Sorin Paraoanu, Jukka Pekola, Peter Samuelsson, Mika Sillanpää, Janine Splettstösser and Fabio Taddei for helpful suggestions and comments. Thanks are also due to Jukka Pekola, Edouard Sonin and Pertti Hakonen, with whom we started the lectures on nanoelectronics. I also apologize to those whose names I have forgotten to include, but who contributed in different ways. Needless to say, responsibility for any errors is mine. Finally, I would like to thank Sonke Adlung of Oxford University Press for his infinite patience and encouragement over the past three years.

During recent years my work has been supported by the Academy of Finland and the European Research Council, whose help I gratefully acknowledge.

Tero Heikkilä
In Aalto University Espoo, June 2012

Contents

List of symbols

λ_F	Fermi wavelength
ℓ_{el}	elastic scattering length
$\ell = v_F \tau, \sqrt{D\tau}$	relation of a scattering length to a scattering time in the ballistic, diffusive limit
$\tau_\varphi, \ell_\varphi$	dephasing time, length
$\tau_{\text{e-ph}}, \ell_{\text{e-ph}}$	electron–phonon scattering time, length
$\tau_{\text{e-e}}, \ell_{\text{e-e}}$	electron–electron scattering time, length
$\tau_{\text{sf}}, \ell_{\text{sf}}$	spin-flip scattering time, length
$\ell_m = \sqrt{\Phi_0/B}$	length scale related to the magnetic field B
N_F	density of states at the Fermi level
$p_F, \mathbf{p}_F$	Fermi momentum (magnitude and the vector)
$v_F, \mathbf{v}_F$	Fermi velocity (magnitude and the vector)
E_F	Fermi energy
$\mu_{L/R}, T_{L/R}$	chemical potential (Fermi level) and temperature for left/right leads
$f(E), f^0(E)$	electron energy distribution function, Fermi function
$n(E), n^0(E)$	boson (phonon, photon) distribution function, Bose function
$D = v^2\tau/d$	diffusion constant for dimensionality d
T_n	transmission coefficient for channel n
$\bar{T}(E), \bar{R}(E), M(E);$ $\bar{T}(E) + \bar{R}(E) = M(E)$	total transmission probability (summed over modes), total reflection probability, number of modes
$s_{nm}^{\alpha\beta}$	scattering matrix or its element connecting mode m in lead β to mode n in lead α
E	energy or Young's modulus
I_y, I_r	bending moduli
T	temperature or tension
ρ	(mass) density
$R_K = 1/G_K = h/e^2 = 25812.8 \ \Omega$	quantum of resistance
R_T	tunnelling resistance
$\Phi_0 = h/e = 4.13567 \times 10^{-15} \ \text{Wb}$	flux quantum
$\Phi_S = h/2e$	flux quantum for Cooper pairs
$\sigma, \bar{\sigma} \in \{\uparrow, \downarrow\}$	spin and the opposite spin
$F(\vec{r}) = \langle \psi_\uparrow(\vec{r})\psi_\downarrow(\vec{r}) \rangle$	pairing amplitude
Δ	superconducting pair potential, BCS gap
λ	interaction constant in the theory of superconductivity
λ_p	magnetic field penetration depth into a superconductor
$S(\omega)$	noise power spectral density (Fourier transform of a two-operator correlator)
$P(E)$	(in the dynamical Coulomb blockade theory) probability for an environment to absorb energy E
$\partial_x = \frac{\partial}{\partial x}$	short-hand for derivative

Introduction

This book describes electron transport phenomena in small, *mesoscopic* systems. The word mesoscopic comes from the Greek word *mesos*, which means middle, indicating that the mesoscopic world resides between the microscopic and the macroscopic. Typically this means that the phenomena take place in systems consisting of a large number of atoms and electrons, but they only occur if the system is small enough, i.e., the size of the system is smaller than some length scale characterizing the transition from the microscopic to the macroscopic world.

Let us take a few examples that are discussed in the remainder of the book. In a perfect metallic crystal, electron transport is essentially dissipationless, i.e., the conductivity is infinite. But there are no everyday-scale perfect crystals. The dislocations, vacancies and other impurities distort the crystal periodicity and lead to a finite conductivity. Also, the presence of lattice vibrations, phonons, induce scattering that further decreases conductivity. These processes come with a length scale, a mean free path, that characterizes the average distance between the impurities, or between subsequent scatterings of conduction electrons from them, or from the phonons. A metallic wire which is smaller than this mean free path is said to be ballistic. In a ballistic wire, the resistance does not depend on the length of the wire, and therefore the conductivity can be said to be infinite. As explained in Ch. 3 of this book, however, even with this infinite conductivity, the measured resistance of such a wire is non-zero, as the finite number of quantum channels in the wire leads to a finite resistance. In this example the crossover from mesoscopic (where a macroscopic observable such as resistance is well defined, but it is determined by the quantum nature of the wire) to macroscopic (where resistance scales with the length of the wire) is characterized by the mean free path.

Another example concerns the definition of the electron temperature. From statistical physics we learn that temperature characterizes the width of the electron energy distribution function turning from one below the chemical potential μ to zero above μ. This temperature is a parameter in the Fermi-Dirac distribution function. In a non-equilibrium setting, an applied voltage V giving rise to a current I leads to the power $P = VI$ applied to the sample, and the sample heats up. But as discussed in Ch. 2, the resultant distribution function may not always be of the Fermi-Dirac form, and one can have many different types of definition for the electron temperature. This non-equilibrium form persists in the wire until after some distance the electrons relax into the

Length	Symbol	order of magnitude
Fermi wavelength (metals)	λ_F	0.1 . . . 1 nm
Fermi wavelength (2DEG)	λ_F	10 . . . 100 nm
Elastic scattering length (metals)	ℓ_{el}	10 . . . 100 nm
Elastic scattering length (2DEG)	ℓ_{el}	. . . hundreds μm
Energy relaxation length at $T = 1$ K	ℓ_{en}	1 . . . few tens μm
Dephasing length at $T = 1$ K	ℓ_{φ}	0.1 . . . 10 μm

Table 1.1 Orders of magnitude for some length scales in typically studied mesoscopic systems. These should be compared to the structure sizes with dimensions some tens of nanometres, fabricated with standard lithography techniques (at minimum, larger conductors can obviously also be made), which also set the scale for the potential profiles in two-dimensional electron gas (2DEG) systems. With more advanced techniques, such as mechanically controllable break junctions, electromigration, atomic layer deposition, manipulations with atomic force microscopes, etc., one may reach wires with atomic widths and thicknesses. Moreover, such ultra-thin conductors may be reached with 'bottom to top' approaches, e.g., from carbon nanotubes.

usual equilibrium shape. Such a relaxation is characterized by a relaxation length, and this again describes a crossover between mesoscopic and macroscopic regimes.

Especially the length scales concerning energy relaxation are strongly temperature dependent. For example, for a typical copper wire,[1] the relaxation length at room temperature is of the order of a few nanometres, at $T = 10$ K it is close to a micrometre, at $T = 1$ K it already reaches a few tens of micrometres, and for the minimum typically achievable electron temperature, $T = 10$ mK, it is a 'macroscopic' length, of the order of centimetres. This shows that mesoscopic effects are best seen at low temperatures. Reaching temperatures below 1 K is nowadays an everyday task in many physics laboratories. As many of the effects discussed in this book require a large enough wire so that level quantization effects are small within the phase breaking or energy relaxation lengths, finding some of those effects really requires cooling the sample to low temperatures. This also shows up in the assumptions made of the operating temperatures in the remainder of this book.

Other typical length scales relevant for mesoscopic conductors are summarized in Table 1.1.

The opposite end of mesoscopic effects, i.e., turning of microscopics into mesoscopics, is a matter of convention. Molecular electronics—measuring charge transport through single molecules—is rarely included in the strict definition of mesoscopic physics, but apart from the energy scales many of the observed effects are quite similar to those seen in quantum dot systems consisting of millions of atoms and electrons. One characteristic feature of mesoscopics is that the systems are open: the small system under interest is coupled to a larger system, and this coupling may modify the properties of the small system, for example, inducing finite lifetimes of the electronic states. For these reasons, the def-

[1] See details in Sec. II.C.2 of the review on thermal effects (Giazotto *et al.*, 2006); in this particular case the electron–phonon scattering length is proportional to $T^{-3/2}$.

inition of the mesoscopic realm can be extended down to the molecular level, when concentrating on those effects where the materials-dependent features affect only quantitative details.

1.1   Studied systems

This book concentrates on electron transport phenomena in small conducting structures, rather than the detailed (often materials-dependent) characteristics of those structures themselves. But in order to acquire the correct mindset on where these phenomena are found, let us shortly discuss those structures and their main characteristics as well.

Mesoscopic phenomena are studied in four different types of conducting system depending on the main materials used for their construction: metallic wires, semiconductor structures, molecules and graphene.[2] In addition, different kinds of nanowires typically fall within all of these categories, but often have properties characteristic to those nanowires rather than the bulk material of which they are made. These systems are described in some detail in the following.

Common to the studies of any nanostructure, one also has to fabricate the contact from this structure to the macroscopic measuring devices (see the large-scale picture in Fig. 1.1). Typically this is done by structuring much wider electrodes than the studied wires in contact with them (see Figs. 1.2, 1.6, 1.8 and 1.9 for examples). Such an electrode ideally works as a *reservoir* (heat and particle bath) of electrons: once the electrons enter the electrode, they quickly thermalize with the lattice. In practice, this means that their energy distribution function obtains the Fermi–Dirac form[3]

$$f^0(E) = \frac{1}{\exp[(E-\mu)/(k_B T)] + 1}, \quad (1.1)$$

where μ is the chemical potential and T the lattice temperature of the reservoir. In the ideal case, the reservoir is then unperturbed by what happens in the nanostructure. These wide electrodes continue for some hundreds of microns and connect to contact pads, the latter with dimension of the order of a millimetre. To these the experimentalist connects the external wires, which then connect the sample to the measurement apparatus. At this point there are often a few signal amplification stages. As the mesoscopic nanoelectronics experiments are often carried out at low temperatures, the thermal noise in the measurement apparatus (residing at room temperature) and the electronic heat current through the wires typically heat up the electrons in the sample. Therefore, one typically needs a good thermal contact of the wires with the low-temperature equipment, and several electronic filters for the noise between the sample and the room-temperature equipment. All this is of extreme importance when measuring the mesoscopic effects in small conductors. However, this is a book concentrating on phenomena, so I do not dwell on the detailed problems faced when fabricating and measuring nanoelectronic samples.

[2]Recently, quite a lot of attention has also been paid to a fifth class of mesoscopic systems, fabricated on systems called topological insulators, which have a conducting surface because of the surface states formed at the interface between a topological insulator and a conventional insulator or vacuum. For a recent review, see (Qi and Zhang, 2011).

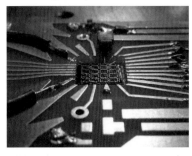

Fig. 1.1 Photograph of a GaAs chip used for measurements of spin qubits (see Sec. 8.5.4). The chip (in the centre) is about 5 mm × 5 mm. It is bonded to the external measurement equipment via the many aluminium wires, which supply the control voltages to the qubit. The (sub-nanosecond) qubit driving is carried out via the two coaxial transmission lines at the left, and the readout via the surface mount copper inductor (in the background). Courtesy of F. Kuemmeth, Marcus Lab, Harvard University.

[3]Below, when I want to emphasize that the distribution function has this form, I use the superscript 0.

Fig. 1.2 Scanning electron micrograph (SEM) of a single-junction thermometer structure, consisting of aluminium wires attached to each other via aluminium oxide tunnel contacts. This sample can be used for a very precise measurement of the temperature via the Coulomb blockade, as discussed in Ch. 7, see (Pekola *et al.*, 2008). This structure was fabricated in NEC Nanoelectronics Research Laboratories in Japan, and measured in the Low Temperature Laboratory of the Aalto University, Finland. Courtesy of Jukka Pekola and Matthias Meschke.

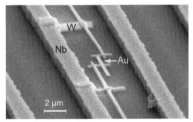

Fig. 1.3 SEM figure of a normal-metal (Au)–superconductor (W) loop attached to a superconducting resonator (made of Nb, only a small part shown). This structure was fabricated in the University of Paris Sud for the measurements of the high-frequency impedance of a superconductor–normal-metal–superconductor junction (Chiodi *et al.*, 2011). Courtesy of F. Chiodi and H. Bouchiat.

1.1.1   Metallic wires and metal-to-metal contacts

Perhaps the simplest studied systems in nanoelectronics are small metallic wires and metal-to-metal contacts fabricated with the help of a lithographically patterned mask through which the metals are evaporated. Typically used metallic materials are based on aluminum, copper, gold, niobium, silver, platinum or palladium, but also other metals are used. As aluminum and niobium become superconducting at sub-Kelvin temperatures (the critical temperature of Al is 1.1 K and for Nb it is 9.3 K), a combination of superconducting and normal-metallic (non-superconducting) effects can be studied using them (see Chs. 5 and 9). Most single-electron transistors (Ch. 7) and superconducting Josephson junctions (Ch. 9) are fabricated with metals. For an example of an ultrasmall system consisting of metallic wires and many tunnel contacts, see Figs. 1.2, and for one not so small but containing many different metals, see Fig. 1.3.

In the following I discuss very briefly the simplest approach to the electronic structure of metals. For most purposes this is a sufficient description. Besides being needed for describing the transport phenomena in metals, it is also needed for understanding the electronic transport through other types of material, as the samples made from them are eventually always contacted to metal electrodes for the measurements.

'Free' fermions in metals

Electrons interact strongly via the Coulomb interaction. However, it turns out that the low-energy properties of the conduction electrons in metals or doped semiconductors can be well described within an independent electron description. This means that for most bulk properties we need to consider the Schrödinger equation for a single particle wave function ψ,

$$i\hbar\frac{\partial\psi(\vec{r})}{\partial t} = \left(-\frac{\hbar^2}{2m^*}\nabla^2 + U(\vec{r})\right)\psi(\vec{r}), \qquad (1.2)$$

where $U(\vec{r})$ describes the possibly random local variations of a mean-field potential felt by the electron. That such a description is possible is explained by Landau's Fermi liquid theory.[4] We do not dwell on this theory here, and it suffices to state that this theory is valid only at excitation energies much smaller than the Fermi energy, and in this case the effect of electron–electron interactions is captured mostly in the effective mass m^*. In fact, the first term on the right-hand side of eqn (1.2) is the lowest-order term in the expansion of the bulk electron energy vs. the momentum around the Fermi level.

There is one remaining effect of the electron–electron interactions even in this case: the collisions between the electrons serve as a route for relaxation towards an equilibrium state. This effect is described in Ch. 2.

---

[4]See, e.g., Ch. 17 in (Ashcroft and Mermin, 1976) or Sec. 11.2.1 in (Mahan, 2000). In fact, the free electrons in this theory are replaced by *Landau quasiparticles*, which are elementary excitations of the interacting electron system. These quasiparticles are the single electrons 'dressed' with the interactions.

In restricted geometries the electron–electron interactions may again become more relevant, and a mean-field type of a description is no longer sufficient. For transport through zero-dimensional islands, this shows up in the Coulomb blockade phenomenon (see Chs. 7 and 8). In one-dimensional wires the (sufficiently weak) interactions result in the formation of a Luttinger liquid.[5] Finally, if the effective interaction between electron pairs is attractive, the electronic system may turn into a superconducting state. The consequences of this are discussed in further detail in Chs. 5 and 9.

Besides the neglect of electron–electron interactions, there is another relevant approximation usually applicable when describing the transport properties of metals. Namely, typical Fermi energies of metals are in the range of electron volts, or in temperature units, tens of thousands of Kelvin. Typical excitation energies in transport studies and the operating temperatures of the devices are much lower than this scale. In other words, the relevant electron momenta lie within $\delta \mathbf{p}$ from the Fermi momentum $\mathbf{p}_F$. In this case the electron dispersion relation (relation between the momentum and the kinetic energy of an electron) for the relevant excitations can be often approximated via

$$\epsilon_p = \frac{\mathbf{p}^2}{2m^*} = \frac{(\mathbf{p}_F + \delta \mathbf{p})^2}{2m^*} = \frac{\mathbf{p}_F^2}{2m^*} + \frac{\mathbf{p}_F}{m^*} \cdot \delta \mathbf{p} + o(\delta p^2) \approx E_F + \mathbf{v}_F \cdot \delta \mathbf{p}, \quad (1.3)$$

where $\mathbf{p}_F = \hat{p} p_F$, $p_F = \sqrt{2m^* E_F} = m^* v_F$ is the Fermi momentum and $\mathbf{v}_F$ is the Fermi velocity. This is called the semiclassical approximation, and it implies that for many purposes electrons can be viewed as classical particles moving with the Fermi velocity. Such an approach is employed in Ch. 2 to study the semiclassical limit of transport, widely applicable for metals.

Linearizing the dispersion relation around the Fermi energy has also another important consequence: it makes the density of states constant. Assume we want to calculate the value of some energy-dependent observable $F(\epsilon_p)$ given as an integral over the momentum states, i.e.,

$$\int \frac{d^3 p}{4\pi^3 \hbar^3} F(\epsilon_p) = \int_0^\infty \frac{p^2 dp}{\hbar^3 \pi^2} F(\epsilon_p) = \int_0^\infty d\epsilon N(\epsilon) F(\epsilon). \quad (1.4)$$

On transforming from the momentum integral to the energy integral we introduced the density of states

$$N(\epsilon) = \frac{1}{\pi^2 \hbar^3} \frac{p^2 dp}{d\epsilon_p} \approx \frac{2m^* E_F}{\pi^2 \hbar^3 v_F} = \frac{m^* p_F}{\pi^2 \hbar^3} \equiv N_F, \quad (1.5)$$

where the last three forms are independent of energy. This approximation is used throughout the text when dealing with metals. Moreover, for typical observables related with electron transport, $F(\epsilon)$ is non-zero only within a small energy window around the Fermi energy E_F much larger than this energy window.[6] Therefore the integral over the energies in eqn (1.4) can be extended to start from $\epsilon = -\infty$.

[5]Luttinger liquids are outside the scope of this book. For an introduction, see, for example, Sec. 19.4 in (Bruus and Flensberg, 2004).

[6]This applies for all 'transport' observables, such as average current, which vanish in the absence of a bias (voltage), or thermal noise, which vanishes at $T = 0$.

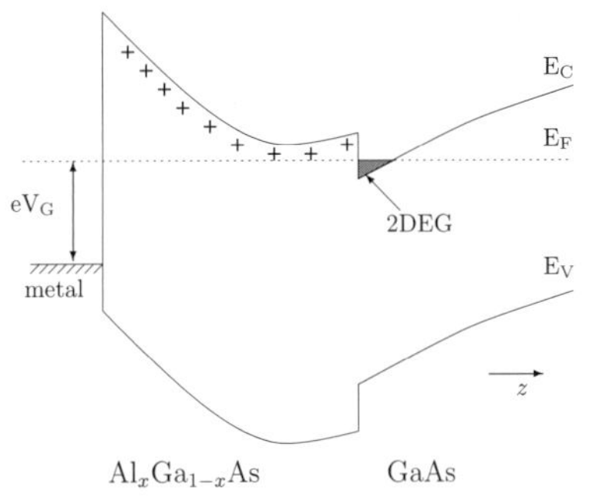

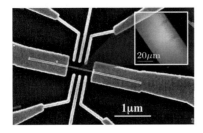

Fig. 1.4 Band-bending diagram of modulation doped GaAs/$Al_x Ga_{1-x}$As heterostructure. A 2DEG is formed in the undoped GaAs at the interface with the *p*-type doped AlGaAs.

Fig. 1.5 Single electron transistor fabricated starting from InAs/InP nanowires deposited on a SiO2/Si substrate. Devices are contacted by a Ti/Au metallization and local nanogates at the sides of the wire can be used to tune the spectrum and filling of electron orbitals. Device fabricated in Scuola Normale Superiore, Pisa, Italy. Courtesy of Stefano Roddaro and Francesco Giazotto.

1.1.2   Semiconductor systems

Fabricating semiconducting materials requires more massive and expensive techniques than with metals, but the resultant systems are also generally more controllable. Besides replacing metals with heavily doped semiconductor materials, for mesoscopic transport studies semiconductors are most often used in two distinct flavors: metal-gated two-dimensional electron gases (2DEGs) formed between semiconductors with different levels of doping, or nanowires grown out of different types of semiconductor material.

Figure 1.4 shows the energy diagram of a 2DEG forming at the interface between non-doped GaAs and Al-doped GaAs. To adjust to the variation in the doping, there is charge transfer at the interface. As a result, the band energies are bent as shown in the figure. With suitable doping, the conduction band drops to just below the Fermi level within a small region near the interface. As a result, excitations to higher-order states in the perpendicular direction to the interface require a large energy, whereas in the direction parallel to the interface electrons can move almost freely. The electron gas is thus effectively two-dimensional, and the electron density can further be controlled *in situ* by attaching metal gates on top of the heterostructure. Moreover, patterning these gates suitably, one can locally change the electron density (also to zero) and thereby produce different types of confining potential for the electrons. With such a scheme one can realize even lower- (than two-)dimensional objects, such as one-dimensional wires, quantum point contacts, or zero-dimensional quantum dots (see Fig. 1.6 for an example).[7] These systems are described in this book.

Different types of semiconductor can nowadays be quite controllably grown into thin nanowires (see an example in Fig. 1.5). Typical materials used for these wires are Si and different III-V semiconductor compounds, such as InAs, InP and GaAs. The thickness of these wires is typically a few tens of electron wavelengths, i.e., they contain a few tens of electron modes contributing to the charge transport. In effect, semiconductor nanowires have many similar properties as (multi-wall) carbon nanotubes, but their different materials parameters allow for studying a wider range of topics. Especially, some of them have interesting magnetic properties, such as strong spin-orbit scattering and large effective electron *g*-factor for Zeeman splitting, which may make them interesting for spintronics applications.

Depending on the particular semiconductor concerned, on the magnitude of the excitation energies and the observable studied, doped semi-

---

[7]Nearly all systems in condensed-matter physics are strictly speaking three-dimensional. However, often systems can be described by being effectively lower-dimensional: in this case one or more of the dimensions (width, thickness, length) is smaller than some characteristic length scale describing certain type of physics. For example, 2DEGs are thin in the *z*-direction compared to the Fermi wavelength, and therefore excitation to higher states in that direction are costly. In some cases there are also other relevant length scales, such as those related to screening of interactions, dephasing of interference effects (see Ch. 4) and energy relaxation. Therefore, a given conductor may be three-dimensional in one respect and, for example, one-dimensional in another. Moreover, often systems showing gradients of variables in only one (or two) directions are called *quasi-one-(or two-)dimensional*, even though otherwise they would be three-dimensional.

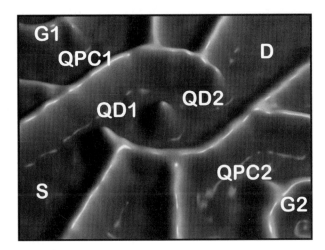

Fig. 1.6 Atomic force microscope (AFM) image of a double quantum dot fabricated on a GaAs/AlGaAs heterostructure by a local oxidation. The double quantum dot (QD1 and QD2) is formed in the centre of the structure. The radius of the QD1–QD2 system is of the order of 100 nm. The two quantum point contacts (QPC1 and QPC2) can be used to monitor the charge state of the quantum dots, and the whole system can be controlled by applying voltages to the two gates (G1 and G2) and between source S and drain D. This structure was used to measure the charge dynamics of the quantum dot, see (Gustavsson *et al.*, 2007). Courtesy of Klaus Ensslin, Swiss Federal Institute of Technology, ETH, Zürich, Switzerland.

Fig. 1.7 Two techniques used for making molecular break junctions: mechanically (b) and electronically (c) broken junctions. These techniques allow reaching ultra-narrow gaps between the electrodes, so that molecules can be deposited in the free space between them as illustrated in (a). In reality, the molecules are typically in a solution everywhere near the junction, and only a transport measurement reveals whether one of them links the two electrodes. In this setup, usually the transport takes place dominantly only through one molecule. (van der Molen and Liljeroth, *J. Phys. Condens. Matter* **22**, 133001 (2010), Fig. 6.)

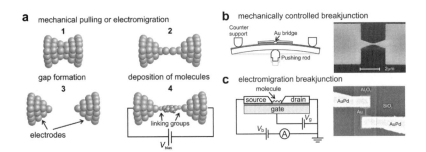

conductors and low-dimensional structures based on them may often still be described with the free-electron picture similar to that of metals—at least when constrained in the relevant number of dimensions. However, semiconductors have typically a more complicated Fermi surface than metals, and this often shows up in the applications. Besides the possibility of realizing low-dimensional structures, such materials-specific features of semiconductor nanostructures are not discussed in this book.

1.1.3 Carbon nanotubes and molecules

There are techniques to contact molecules between metallic leads and study transport through them. Two commonly used techniques are depicted in Fig. 1.7 showing a schematic way of making extremely small break junctions into which molecules can be placed; a third alternative is to use scanning tunnelling microscopy.[8] Many different types of

[8]For a review of these techniques, see (van der Molen and Liljeroth, 2010).

molecule have been studied, for example the electric current through a single Hydrogen molecule has been measured (Smit *et al.*, 2002). The transport through the molecules show behaviour reminiscent of quantum wires (Ch. 3) or quantum dots (Ch. 8). In this respect the most notable difference to semiconductor quantum dots is that the energy level spacing in molecules tends to be larger.[9]

For ordinary molecules, the bulk band structure is no longer an applicable concept. Rather, typically the relevant topic of study is the discrete electronic spectrum on the molecule, and the dependence of this spectrum on the number of electrons. Therefore, the energy scales encountered in molecular electronics are closely related to the chemical properties of these molecules.

In terms of transport studies, carbon nanotubes (CNTs) are quite different from ordinary molecules. They are typically long enough so that their studies generally do not require break-junction techniques, and they are long enough so that a description as a bulky object is often justified.[10] In them the nature of the extended electron states depends strongly on how the nanotubes are wound into the tubes. Therefore, some nanotubes are insulating whereas some are metallic, depending on the exact microscopic structure. Because of this, the transport properties of different carbon nanotubes vary considerably.[11] Moreover, nanotubes exist in multiple forms: Perhaps the best understood are the single-wall nanotubes (SWNTs), which contain only a single carbon sheet rolled into a tube. What makes the SWNTs special is that they contain two degenerate orbital states at the Fermi energy,[12] corresponding to the possible two directions of the chiral orbits around and along the tubes. In contrast, multi-wall tubes (MWNTs) contain many more states, and in some cases they can be likened to thin metal wires. In some cases the tubes form bundles through which the transport can also be studied, but the properties of the bundles are less controllable than those of the individual tubes.

1.1.4   Graphene

In 2004, physicists at the University of Manchester attached scotch tape to a piece of graphite, then removed it and placed it on top of a Si substrate. After peeling off the tape, they discovered that some flakes of graphite were left on the substrate. They also found that the optical response of these flakes depends on the number of carbon layers in the flakes. Closer inspection showed that some of these flakes had only a few carbon layers, and some only one. Such single-layer *graphene* had been studied since the 1940s (Wallace, 1947) as a mathematical model electronic system with properties very different from ordinary conductors or insulators. Namely, besides being the thinnest-known two-dimensional material, the electrons in graphene seem to behave like massless Dirac fermions. As discussed shortly below and with more detail in Ch. 10, this property is related to the graphene honeycomb lattice structure.

[9]Note that the viewpoint in molecular electronics is often different from the one in this book: there the aim is often, besides the generic picture of the transport process, to connect the chemical valence to the charge transport characteristics. That means that the exact positions of the energy levels matters rather than the consequences of the mere level quantization.

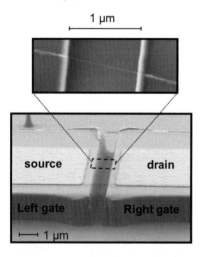

Fig. 1.8 Double-gated suspended ultra-clean single-wall carbon nanotube (SWNT) between two metal electrodes. This type of a sample was used for the measurements of the coupling of the spin and orbital motion of electrons in SWNTs (Kuemmeth *et al.*, 2008). Courtesy of Shahal Ilani, Weizmann Institute of Science, Israel.

[10]However, CNTs are often also used to make quantum dots.

[11]I do not dwell on the electronic structure of the carbon nanotubes here. For those interested, there are books written on them; see for example (Ando, 2005).

[12]When adding spin, the total degeneracy is thus four.

Since its initial finding, many laboratories have been able to start fabricating graphene samples (see examples in Figs. 1.9 and 1.10). Besides the curious behaviour of the electrons, there are many properties of graphene that are attractive for the electronics industry: for example, graphene samples can be made extremely clean, the transport can be tuned via gate voltages, the charge and heat conductivities in graphene are large and optical properties qualitatively different from other materials.

Massless Dirac fermions in graphene

Graphene is a two-dimensional layer of graphite. It is formed from carbon atoms that are arranged in a honeycomb lattice, shown in Fig. 1.10. The fact that the unit cell contains two similar atoms has an important consequence for the electronic structure: each electron is described by a Dirac spinor wave function, with components describing the amplitude of occupying one of the two atoms in the unit cell. The electronic dispersion relation in graphene is derived in Ch. 10. According to this derivation, one gets at low energies an effective Hamiltonian

$$H_g \approx \hbar v_F \begin{pmatrix} 0 & \pm k_x - ik_y \\ \pm k_x + ik_y & 0 \end{pmatrix} = \hbar v_F \left(\pm k_x \sigma_x + k_y \sigma_y \right). \quad (1.6)$$

The two signs are for two non-equivalent valleys in the reciprocal space. This Hamiltonian has the form of the Dirac Hamiltonian for massless[13] particles, but the speed of light has been replaced by $v_F \approx 10^6$ m/s.

Besides the single-layer graphene, bilayer graphene has also been studied actively. In the bilayer case, the electrons are still described by two-component spinors, but the Hamiltonian has the form

$$H = -\frac{\hbar^2}{2m^*} \begin{pmatrix} 0 & (k_x \mp ik_y)^2 \\ (k_x \pm ik_y)^2 & 0 \end{pmatrix}. \quad (1.7)$$

Hence, the excitations of bilayer graphene are still spinors with two indices for the atoms in the unit cell, but contrary to the case in single-layer graphene, they have a finite mass m^*, which is roughly 5% of the electron mass. However, the particular Hamiltonian of the bilayer depends on the orientation of the two layers with respect to each other, and in some cases it is possible to induce a finite gap in the bilayer density of states. This is discussed in more detail in Ch. 10.

The particular dispersion relations and the presence of the 'pseudospin' (corresponding to the two similar atoms in the unit cell) result in many exotic electronic properties seen only in graphene.[14] This is why the successful fabrication and measurements of single graphene sheets have resulted in massive activity in the detailed study of its properties.

1.2 Classical vs. quantum transport

The forthcoming chapters discuss quantum-mechanical transport properties of nanostructures. These show deviations from the classical Ohm's

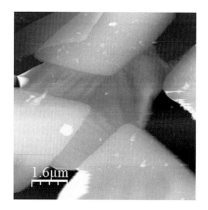

Fig. 1.9 Atomic force microscope (AFM) image of a suspended graphene sheet between four metal electrodes residing at the corners of the image, fabricated in the Low Temperature Laboratory, Aalto University, Finland. Courtesy of Jayanta Sarkar and Peter Liljeroth.

[13]Masslessness here refers to the picture of general relativity: the dispersion relation does not contain the usual mass term mc^2, which would correspond to an energy gap in the dispersion relation.

Fig. 1.10 High-resolution scanning tunnelling micrograph of graphene. The carbon atoms form a single layer honeycomb lattice, residing at the edges of the honeycomb as shown in the schematic figure. The unit cell encloses two neighbouring atoms of the honeycomb. (Stolyarova *et al.*, *PNAS* **104**, 9209 (2007), Fig. 2.) © 2007 National Academy of Sciences, USA.

[14]For a short review see (Geim and Novoselov, 2007), and for a more thorough one (Castro Neto *et al.*, 2009).

law $I = U/R$, stating that the current I through a given sample is a linear function of the voltage U, the coefficient R being the resistance. In the classical case, this resistance scales linearly with the wire length,

$$R = \frac{L}{\sigma A},\tag{1.8}$$

A being the cross-section of the wire and σ its conductivity. In Ch. 2 we derive this law using semiclassical arguments: ignoring the interference effects (Ch. 4), which often are weak, and single-electron effects (Ch. 7), which arise in strongly interacting systems.

1.2.1 Drude formula

The Boltzmann-equation analysis of Ch. 2 yields for the conductivity in the semiclassical limit

$$\sigma = e^2 N_F D = e^2 N_F v_F \ell_{\text{el}}/d,\tag{1.9}$$

where e is the electric charge, N_F is the density of states at the Fermi level, D is the diffusion constant, v_F is the Fermi velocity, ℓ_{el} is the elastic mean free path and d is the dimensionality of the wire.

Besides the semiclassical derivation of eqn (1.9), quite a similar result was found by (Drude, 1900a; Drude, 1900b), before the microscopic theories of solid state. Drude's derivation goes as follows. Let us consider an electron in a solid where one has applied an electric field $\vec{E}$. The electric field produces a Lorentz force $-e\vec{E}$, accelerating the electron. Assume that a time t has passed since the electron last collided with the lattice. The velocity of the electron has increased to $\vec{v}_0 - e\vec{E}t/m$ due to the Lorentz force, $\vec{v}_0$ being the initial velocity after the last collision. If that electron collides in random directions from the impurities, the initial velocity has no contribution to the average velocity of the electron, which must therefore be given as the average of $-e\vec{E}t/m$. Denote the average time between collisions as 2τ.[15] This implies the average electron velocity given by

$$\vec{v}_a = -\frac{e\vec{E}\tau}{m}.\tag{1.10}$$

The average current carried by the electrons is given by $\vec{j} = -en\vec{v}_a$, where n is the electron density. Therefore, we get $\vec{j} = \sigma_D \vec{E}$ with the Drude conductivity

$$\sigma_D = \frac{ne^2\tau}{m}.\tag{1.11}$$

This is the same as eqn (1.9) (up to a prefactor of order unity) after we identify $n \sim N_F v_F^2 m/2 = N_F E_F$. This is why also the result (1.9) is often referred as the Drude conductivity.

1.2.2 Quantum effects

Typical quantum effects encountered in nanoelectronic systems arise due to the energy and/or charge quantization effects, tunnelling, and interference effects. A large-scale quantum phenomenon is the transition to

[15]The factor 2 does not matter much, but is included here to get the original Drude result in eqn (1.11).

the superconducting state taking place in many materials at low temperatures. This transition is not a mesoscopic effect, but also in nano-electronic circuits it gives rise to many types of phenomena which would not be present in simple normal-metallic systems.

There are many possible reasons for violating Ohm's law, i.e., either linearity between current and voltage or the scaling of the resistance with the length of the wire. Often in these cases both of them are simultaneously violated, but there are exceptions to this. Such non-classical effects are, for example:

- **Tunnelling** through a thin insulating region or a vacuum gap between two metals. Such *tunnel junctions* are an important part of mesoscopic electronics, and they can be used to realize many types of structure. A single tunnel junction fabricated between two normal (non-superconducting) metals is typically a linear object, i.e., its resistance is independent of voltage up to very large voltages of the order of the work function difference between the tunnel barrier and the metals. However, the scaling of the resistance with the thickness of the tunnel junction is exponential, not linear. Calculating the current due to tunnelling through a single insulating barrier is illustrated in Example 1.1 below.

- In low-dimensional systems, such as quantum dots, electron **energy level quantization** within the system shows up in the current through the system. This is discussed in detail in Ch. 8.

- **Single-electron effects.** When tunnel barriers have a small capacitance C, the energy required for charging the capacitance with the charge of a single electron, $E_C = e^2/(2C)$, may become relevant. In this case, to get a finite current through the tunnel contact, the external circuit must provide this *charging energy E_C* to the electrons crossing the barrier. It turns out that in a double-junction system ('single-electron transistor' SET, see Ch. 7), this energy has to be provided by the bias voltage or the temperature. Otherwise no current can flow, and the system is in the state of a *Coulomb blockade.*

- In metal or semiconductor wires, carbon nanotube or graphene structures whose size is smaller than the *phase relaxation length* ℓ_φ, **interference effects** between different electron paths within the wire may alter the conductivity. Such effects are Aharonov–Bohm effects and persistent currents in multiply connected systems under an applied magnetic field, and localization and universal conductance fluctuations in disordered systems. These effects are discussed in Ch. 4.

- In **ballistic wires**, i.e., conductors whose size is smaller than the mean free path, the resistance no longer scales with the length of the wire, but it becomes quantized, depending on the ratio between the width of the wire and the Fermi wavelength λ_F of electrons (see Ch. 3).

- At a high magnetic field, *Landau levels* are formed within the conductor, and the resistance through the system is strongly field dependent. The resultant **quantum Hall effects** are not discussed in this book, but for a general introduction see for example (Imry, 2002) or (Datta, 1995).

- In **superconducting** wires and junctions, and superconductor–normal metal heterostructures, quantum-mechanical phase coherent effects show up, and Ohm's law ceases to be valid. Some of these phenomena are discussed in Chs. 5 and 9.

Before embarking on the multitude of transport phenomena I would like to point the reader to Appendix A, which contains a set of necessary theoretical tools, concepts and formulas employed in the book. An advanced or impatient reader may skip these appendices, and if necessary return to them when they are needed.

Example 1.1 Current through a tunnel junction

As an example of calculating the current through a low-dimensional object, consider the case of a tunnel junction depicted in Fig. 1.11. The proper microscopic calculation of the current in this case is carried out in Ch. 7, but the formula for the current can also be argued based on a simple picture. With an applied voltage V through the junction, the Fermi levels of the electron systems on the two sides of the junctions are shifted by eV, i.e., they are μ_L and $\mu_R = \mu_L - eV$. Assume that the average transmission probability for an electron on the left side of the junction to enter the right side is τ (taken for simplicity independent of energy). According to elementary quantum mechanics, this depends on the junction parameters as $\tau \sim \exp(-2d\sqrt{2mU}/\hbar)$, where d is the width and U is the height (work function difference between the metal and the tunnel barrier) of the tunnel barrier, and m is the electron mass. At a vanishing temperature, the current from the left to the right at energy E is then $eAc\tau N_L(E - \mu_L)N_R(E - \mu_R)$, where A is the area of the junction, $N_{L/R}(E)$ is the density of states in the left/right side of the junction and c is a constant fixing the dimensions of this expression to (current/energy). In what follows, we identify the prefactor as the resistance, and eliminate c. At non-zero temperature we have to include the occupation numbers (distribution functions) $f_i(E)$ of electrons in reservoir i: the initial state has to be filled and the final state has to be empty. Therefore, the total current (current from the left to the right minus that from the right to the left) is

$$I = ecA\tau \int_{-\infty}^{\infty} dE N_L(E - \mu_L)N_R(E - \mu_R)\{f_L(E)[1 - f_R(E)]$$
$$- f_R(E)[1 - f_L(E)]\} \qquad (1.12)$$

$$= ecA\tau \int_{-\infty}^{\infty} dE N_L(E - \mu_L)N_R(E - \mu_R)[f_L(E) - f_R(E)], \qquad (1.13)$$

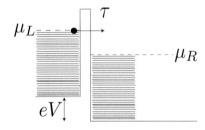

μ_L　τ

μ_R

eV

Fig. 1.11 Schematic model of a tunnel junction with transparency τ.

where $\mu_R = \mu_L - eV$. When the two sides of the tunnel junction are at a local equilibrium, $f_{L/R}(E) = f^0(E; \mu_{L/R}, T)$, where $f^0(E; \mu, T)$ is the Fermi–Dirac distribution function, eqn (1.1). The integrand is non-vanishing within the window of width $\sim \max(k_B T, eV)$ around μ_L and μ_R, i.e., around the Fermi energies E_F of the two leads.[16] If one of the reservoirs, say the one on the right, is made of a normal metal and the voltage is much smaller than E_F/e, its density of states is almost constant within this window (see

[16]In the limit $k_B T \ll E_F$, E_F is the same as the chemical potential; see Appendix A.5.

the discussion in Sec. 1.3). It can therefore be replaced by its value at E_F, $N_R(E) \approx N_R(E_F) \equiv N_F^R$. In this case, assuming Fermi–Dirac distributions and equal temperatures on both sides of the tunnel junction, and taking the voltage derivative of the current yields

$$\frac{dI}{dV} = e^2 c A \tau N_F^R \int dE \frac{N_L(E - \mu_L - eV)}{4T \cosh^2\left(\frac{\epsilon - \mu_L}{2T}\right)} \overset{T \to 0}{\to} e^2 c A \tau N_F^R N_L(-eV). \quad (1.14)$$

The latter part of the equation was taken in the limit of a low temperature. The differential conductance of the junction at a given voltage is directly proportional to the density of states at energy $E = -eV$. This way the tunnelling current can be used for measuring the energy dependence of the local density of states on one side of the tunnel junction. An example of such a measurement is presented in Fig. 5.8.

If both reservoirs are normal metals, the current is given by

$$\begin{aligned} I &= ecA\tau N_F^L N_F^R \int_{-\infty}^{\infty} dE[f^0(E; \mu_L, T_L) - f^0(E; \mu_R, T_R)] \\ &= ecA\tau N_F^L N_F^R(\mu_L - \mu_R) = e^2 cA\tau N_F^L N_F^R V \equiv V/R_T. \end{aligned} \quad (1.15)$$

Here we used the integral given in eqn (A.64a). The resistance R_T of a tunnel junction, defined in this formula, is thus independent of temperature and voltage.

Current through a tunnel junction can also be used to actually measure the distribution function or (in the equilibrium case) the temperature, provided the densities of states are energy dependent and their form is known. Such measurements are explained in Example 2.4 and in Sec. 9.2.

We may also write for the *heat current* carried by the electrons from the left reservoir

$$\dot{Q}_L = \frac{1}{e^2 R_T} \int_{-\infty}^{\infty} dE(E - \mu_L)[f^0(E; \mu_L, T_L) - f^0(E; \mu_R, T_R)] \quad (1.16)$$

and from the right reservoir

$$\dot{Q}_R = \frac{1}{e^2 R_T} \int_{-\infty}^{\infty} dE(E - \mu_R)[f^0(E; \mu_R, T_R) - f^0(E; \mu_L, T_L)]. \quad (1.17)$$

In the absence of a bias voltage, $\mu_L = \mu_R$, these heat currents are opposite, i.e., $\dot{Q}_L = -\dot{Q}_R$. In the presence of a voltage, the sum $\dot{Q}_L + \dot{Q}_R = IV$ results from the Joule power dissipated into the reservoirs.

Further reading

There are a few other books detailing some of the topics discussed in this book. For example:

- (Imry, 2002) contains a discussion on many basic mesoscopic effects, such as quantized conductance, many of the interference effects, noise and meso-

scopic superconductivity. It also includes the quantum Hall effect, which is not discussed here.

- (Nazarov and Blanter, 2009) is a recently published book on quantum transport in mesoscopic systems for advanced readers.

- (Datta, 1995) is a classic treatment of the scattering approach for quantum transport.
- (Dittrich *et al.*, 1998) is a compendium of independent chapters from six authors, detailing, for example, theory of coherent transport, conductance quantization and single electron effects.
- Many recent many-body quantum mechanics books have chapters on quantum transport, such as (Rammer, 1998) and (Bruus and Flensberg, 2004). There the topics are typically discussed with the full arsenal of the many-body Green's function theory.

Exercises

(1.1) The charging energy required to charge a capacitor with capacitance C by the charge of a single electron is $E_C = e^2/(2C)$. Let us assume a simple parallel-plate model for a capacitor, $C = \epsilon A/d$, where $\epsilon = \epsilon_r \epsilon_0$, typical $\epsilon_r = 10$ and $\epsilon_0 = 8.85$ pF/m, A is the area of the plates and d their separation. We can also set $d = 2$ nm, a typical thickness of the oxide layer ('tunnel contact') formed between two metals. Assuming a square plate, $A = w^2$, estimate a width w of the junction which would correspond to the charging energy E_C being equal to 1 K. Estimate also the corresponding capacitances. How should these scales change so that charging effects would be observable at room temperature, i.e., $E_C/k_B \approx 300$ K? Remember that $e = 1.6 \times 10^{-19}$ C and $k_B = 1.38 \times 10^{-23}$ J/K.

(1.2) At the end of Example 1.1 the density of states and the transmission probability through the tunnel barrier are assumed to be energy independent. To mimic the finite width Δ of the conduction band, assume that the density of states has a sharp cutoff at $|E - \mu_L| = \Delta/2$, i.e., it is constant for energies $|E - \mu_L| < \Delta/2$ and vanishes for $|E - \mu_L| > \Delta/2$. Show that the correction to the linear tunnel barrier conductance is proportional to $\exp[-\Delta/(2k_B T)]$ when $\Delta \gg k_B T$, i.e., in this limit the corrections to the linearity are exponentially small. In other words, show that

$$1/R_T - \frac{dI}{dV}\Big|_{V=0} \overset{\Delta \gg k_B T}{\propto} \exp[-\Delta/(2k_B T)]$$

for the current defined in eqn (1.12) and R_T defined below eqn (1.15).

(1.3) Show that the heat current through a tunnel barrier in the linear response regime obeys a Wiedemann–Franz law, i.e., $\dot{Q}/\Delta T \propto T/R_T$. Find also the prefactor of this expression. Hint: Assume a vanishing voltage and that the temperatures $T_{L/R}$ of the left/right reservoirs are $T_{L/R} = T \pm \Delta T/2$. Finally, take the linear order in ΔT.

(1.4) **Question on a scientific paper.** In nanoelectronics measurements, the most typically measured observable is either the current as a function of voltage or the (linear) conductance as a function of some control parameter, such as the gate voltage or a magnetic field. Sometimes the voltage is measured as a function of current. Consult the papers Smit *et al.*, *Nature* **419**, 906 (2002), Kuemmeth *et al.*, *Nature* **452**, 448 (2008) and Heersche *et al.*, *Nature* **446**, 56 (2007), and determine what the measured observables and the control parameters were in the reported experiments. Note that the size of the typically measured nanoelectronic structures ranges from a few nanometres to some micrometres. On the other hand, the measurement equipment is on our everyday scale (from some tens of cm to metres). Argue why and in which case the latter can show some information about the former.

Semiclassical theory

This book discusses many types of electron transport phenomena taking place in small structures at low temperatures. Characteristic for transport phenomena is *non-equilibrium*: the electrons cannot be assumed to share a mutual equilibrium, as is typical in isolated systems in the stationary limit. However, the small size of the mesoscopic samples brings one benefit: typically the studied system is connected to large electrodes, which can be treated as reservoirs of charge and heat. These electrodes are to a good approximation isolated systems, and within them the electrons are in mutual equilibrium. Therefore, inside these reservoirs electrons occupy different energy states according to the Fermi–Dirac distribution function $f^0(E; \mu, T)$, specified through the local potential μ and temperature T.

Inside the studied mesoscopic systems the situation may be different: due to the applied biases, there may be a potential and/or a temperature gradient through the sample. In some cases, even the temperature is not well-defined, as the distribution function (occupation probability of the states) takes a a non-equilibrium form. Therefore, before dwelling on the various quantum effects on electronic transport, let us study how the electron distribution function behaves in these structures as a response to an applied bias, and what we can say about the resultant charge and heat currents. This chapter introduces a semiclassical Boltzmann approach for describing these effects. Besides forming a guideline for the treatment of different types of thermal effect related to different energy relaxation mechanisms,[1] it allows us to derive the Drude conductivity, eqn (1.9), and it is useful in the evaluation of the out-of-equilibrium current noise (see Ch. 6). Moreover, as discussed in Secs. 2.7 and 2.8, the Boltzmann approach can also be used to describe spin-dependent effects and thermoelectric phenomena.

Traditionally the Boltzmann approach has been mostly used in the linear response regime (at low voltages eV compared to the thermal energy $k_B T$ and/or the intrinsic energy scales of the system). There the Boltzmann equation allows us to relate the current flowing through the system to the scattering rates describing the various scattering mechanisms. Here we take another approach, concentrating on the behaviour of the distribution function itself, and the current is calculated as a bi-product.

The aim in this chapter is to understand in which cases the electron energy distribution function can actually be defined, and to determine how it behaves in a mesoscopic sample. We also aim to find prescriptions for

---

[1] For a review of such phenomena, see (Giazotto *et al.*, 2006).

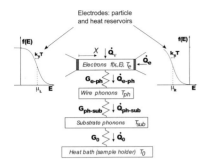

Fig. 2.1 Typical thermal balance of a small wire. The electrons may be heated or sometimes cooled via a direct current (Joule heating $\dot{Q}_e$) or by coupling the system to a radiation source $\dot{Q}_\nu$. The final temperature then depends on various couplings to the phonons of the different sub-systems: phonons in the wire, substrate and the sample holder which is held in some externally controllable temperature. Often it is enough to assume that the wire or the substrate phonons constitutes the heat bath, so that their heating can be disregarded. For further discussion, see (Giazotto *et al.*, 2006).

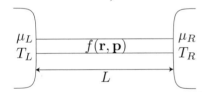

Fig. 2.2 System studied in this chapter. Reservoirs at the ends have well-defined potentials μ_i and temperatures T_i.

computing the values of different observables from this behaviour. Besides allowing us to calculate the charge and heat currents, this analysis also paves a way to studying the thermal balance of a small sample. A typical case is illustrated in Fig. 2.1. In general, the state of the electrons depends on biasing conditions and the different heat fluxes between the electron system and other sub-systems present in the setup.

In macroscopic samples the electron system under study can be described, even in the presence of an applied voltage, by an equilibrium (Fermi) distribution $f^0(E)$ with given potential profile $\mu(x)$ and temperature T equal to that of the underlying lattice T_{ph}. In mesoscopic systems, the temperature T_e of the electrons can become position dependent, and the distribution function can deviate from the Fermi function form.

Let us assume that the system under consideration can be separated into the studied, interesting sub-system S and the reservoirs (electrodes) to which it is connected (see Fig. 2.2). The most relevant assumption about the reservoirs for the study of the distribution functions is that they contain sufficiently strong energy relaxation, such that the electron distribution function in them is a Fermi function.

The aims of this description can now be stated more clearly:

- What is the distribution function $f(\mathbf{r}, \mathbf{p})$ for given potentials and temperatures $\{\mu_L, T_L\}$ and $\{\mu_R, T_R\}$ in the electrodes?
- From this $f(\mathbf{r}, \mathbf{p})$, how do we find the observables such as charge and heat current?

In general, the distribution function and the associated currents depend on the form of the excitation spectrum in the reservoirs and in the system, and on the type and strength of scattering within S. In the following, we assume normal (non-superconducting[2]) reservoirs and a system with approximately energy-independent density of states.

In general, the scattering mechanisms are described by collision integrals which we specify in Secs. 2.4 and 2.5. But to get some feeling of their strength, we may apply the relaxation-time approximation (see Sec. 2.3) and attribute a rate or time scale for the various types of scattering process. Moreover, for the given relaxation time scale, there exists a corresponding length scale which is, loosely speaking, the average distance that a particle traverses between the scattering events. The relevant scattering mechanisms and the corresponding scales are:

- Elastic scattering from impurities, dislocations, etc. The length scale for this process is the elastic mean free path ℓ_{el}.
- In magnetic heterostructures, the spin-flip scattering, ℓ_{sf}.
- Electron–electron scattering, ℓ_{e-e}.
- Electron–phonon scattering, $\ell_{e-\mathrm{ph}}$.
- Coupling of the electrons into the external electromagnetic field (electron–photon scattering), $\ell_{e-\gamma}$

The latter three are inelastic, i.e., energy is transferred in the scattering process. The electron–photon scattering is typically very weak, but its effect on the energy relaxation has been observed; see (Meschke *et al.*, 2006).[3] Such scattering mechanisms give rise to a stochastic phase shift to the particle wave function. Therefore, they also lead to dephasing, described with length ℓ_φ (see Complement 6.1). Dephasing is the mechanism that suppresses the quantum-mechanical interference effects as discussed in Ch. 4.

At low temperatures in metals the typical order of these length scales is[4]

$$\lambda_F \ll \ell_{\rm el} \ll \ell_{\rm e-e}, \ell_{\rm e-ph}. \tag{2.1}$$

The relative order between $\ell_{\rm e-e}$ and $\ell_{\rm e-ph}$ depends strongly on temperature: in metals, typically down to roughly 1 K $\ell_{\rm e-ph} < \ell_{\rm e-e}$, and below this $\ell_{\rm e-e} < \ell_{\rm e-ph}$. This is due to the strong temperature dependence of the electron–phonon scattering rate.

2.1 Semiclassical Boltzmann equation

The electron distribution function $f(\mathbf{r}, \mathbf{p})$ is defined so that the average number of electrons in the element $\{\mathbf{dr}, \mathbf{dp}\}$ around the point $\{\mathbf{r}, \mathbf{p}\}$ in the six-dimensional position-momentum space is

$$f(\mathbf{r}, \mathbf{p}) \frac{\mathbf{dr}\mathbf{dp}}{(2\pi\hbar)^3}. \tag{2.2}$$

Strictly speaking, the Heisenberg uncertainty principle states that we cannot simultaneously talk about the precise momentum and position of an electron, questioning the mere validity of the concept of the distribution function $f(\mathbf{r}, \mathbf{p})$. However, it can be defined in a coarse-grained manner, averaging the position and momentum values over some interval. This interval is related to the width of the electron wave packet. With such an approach we may thus describe only phenomena occurring at scales much larger than this width, roughly given by the electron Fermi wavelength λ_F. Examples of the effects thrown out with this approximation are the interference effects such as localization or resonant tunnelling. The resultant $f(\mathbf{r}, \mathbf{p})$ essentially describes the classical kinetics of the electrons under the applied fields.[5]

Assume that at time t the electrons are described by the distribution function $f(\mathbf{r}, \mathbf{p}, t)$. If there is no scattering, after time dt this must equal to the distribution $f(\mathbf{r} + \mathbf{v}dt, \mathbf{p} + \mathbf{F}dt, t + dt)$, where $\mathbf{v} = \mathbf{p}/m$ is the electron velocity, and $\mathbf{F}$ is a force acting on them (Lorentz force due to electromagnetic fields). The difference between them must therefore result from 'collisions', i.e., scattering from impurities, other electrons or phonons. Let us denote this difference by the term $I_{\rm coll}[f]dt$. The term $I_{\rm coll}[f]$ is called a collision integral, and it is generally a functional of $f(\mathbf{r}, \mathbf{p})$. We thus have

$$f(\mathbf{r} + \mathbf{v}dt, \mathbf{p} + \mathbf{F}dt, t + dt) - f(\mathbf{r}, \mathbf{p}, t) = I_{\rm coll}[f]dt. \tag{2.3}$$

Expanding to the first order in dt and taking the limit $dt \to 0$ we obtain

[3]This is in fact nothing but the electron–electron scattering between two remote conductors. However, it may be separated from the direct electron–electron scattering because the two remote conductors can in principle be separately controlled and measured.

[4]In pure samples or at high energies (temperature/voltage), the scattering length due to phonons may become comparable or even smaller than $\ell_{\rm el}$ caused by impurities. As a result, current starts to depend on the temperature (see Exercise 2.4). However, for simplicity, below we mostly concentrate on the regime given by eqn (2.1).

[5]The semiclassical Boltzmann equation can be rigorously derived from the full quantum-mechanical description in the appropriate limit using the quasiclassical Keldysh Green's-function technique, see, for example, (Rammer, 2007).

the Boltzmann equation

$$(\partial_t + \mathbf{v} \cdot \partial_{\mathbf{r}} + \mathbf{F} \cdot \partial_{\mathbf{p}}) f(\mathbf{r}, \mathbf{p}, t) = I_{\text{coll}}[f]. \qquad (2.4)$$

The terms $\partial_{\mathbf{x}} \equiv \partial/\partial_{\mathbf{x}}$ should be understood as partial derivatives: they act only on the respective coordinates of f. In what follows, we neglect the magnetic fields and take only the electric field $\mathbf{E}$ into account. Then $\mathbf{F} = e\mathbf{E} = -\partial_{\mathbf{r}}\mu$, where $\mu(\mathbf{r})$ is the scalar potential. At the typical length scales discussed in this chapter, the scalar potential equals the chemical potential (see Appendix A.5), and these are taken to be the same in what follows.[6]

In many metals the dispersion relation $\varepsilon(\mathbf{p})$ between the kinetic energy and the momentum around the Fermi level $\varepsilon(\mathbf{p}) \approx E_F$ is almost independent of the direction of the momentum.[7] In this case we may describe the momentum by its direction $\hat{p}$ and by the kinetic energy $\varepsilon(\mathbf{p}) \approx \varepsilon(|\mathbf{p}|)$. The total energy $E = \varepsilon - \mu(\mathbf{r})$ is then independent of $\hat{p}$. We may thus write[8]

$$\mathbf{v} \cdot \partial_{\mathbf{r}} f(\mathbf{r}, \mathbf{p}) - e\mathbf{E} \cdot \partial_{\mathbf{p}} f(\mathbf{r}, \mathbf{p})$$
$$= \mathbf{v} \cdot \partial_{\mathbf{r}} f(\mathbf{r}, \hat{p}, E) + \underbrace{(\partial_{\mathbf{r}}\mu(\mathbf{r}))}_{\partial_{\mathbf{r}} E} \cdot \underbrace{(\partial_{\mathbf{p}} E)}_{\mathbf{v}} \partial_E f(\mathbf{r}, \hat{p}, E) \qquad (2.5)$$
$$= \mathbf{v} \cdot [\partial_{\mathbf{r}} + (\partial_{\mathbf{r}} E)\partial_E] f(\mathbf{r}, \hat{p}, E) = \mathbf{v} \cdot \nabla f(\mathbf{r}, \hat{p}, E).$$

Here ∇ is the total position derivative. Thus, the Boltzmann equation becomes

$$(\partial_t + \mathbf{v} \cdot \nabla) f(\mathbf{r}, \hat{p}, E) = I_{\text{coll}}[f]. \qquad (2.6)$$

This equation is solved in Example 2.1 in the ballistic limit $I_{\text{coll}}[f] = 0$. For the general case one should specify $I_{\text{coll}}[f]$, solve eqn (2.6), and calculate the observables from the solution.

Example 2.1 Boltzmann equation for a ballistic wire

Consider the quasi-one-dimensional wire in Fig. 2.3. Let us assume that it contains no impurities, it is perfectly rectangular and its edges are smooth, so that the momentum in the x-direction is conserved. In the static limit the distribution function is hence a solution to

$$\mathbf{v} \cdot \nabla f(\vec{r}, \hat{p}, E) = 0.$$

Using the boundary conditions specified in Fig. 2.3, the solution can be directly written as

$$f(\vec{r}, \hat{p}, E) = \begin{cases} f_L(E), & \hat{p}_x > 0, & \text{'right-movers'} \\ f_R(E), & \hat{p}_x < 0, & \text{'left-movers'}. \end{cases} \qquad (2.7)$$

The electron system inside the wire thus consist of 'left-movers' and 'right-movers', which do not mix in the absence of scattering and which are distributed in energy according to the distribution functions in the reservoirs from where they come.

[6]Note that this assumption does not always hold in semiconductors or graphene, where charge screening is weaker than in metals. There the equations may have to be appended by the Poisson equation relating the scalar and chemical potentials.

[7]This property is typically denoted as a 'spherical Fermi surface'.

[8]Note that here we also neglect the term $e\mathbf{E} \cdot \partial_{\hat{p}} f(\mathbf{r}, \mathbf{p})$ which changes the trajectories of the electrons. However, this effect turns out to be generally quite weak, of the order of eV/E_F. See also the corresponding discussion in (Sukhorukov and Loss, 1999).

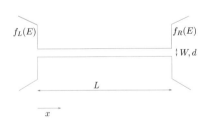

$f_L(E)$ $f_R(E)$ $\updownarrow W, d$ L x

Fig. 2.3 Quasi-one-dimensional wire discussed in Examples 2.1, 2.2 and 2.3.

2.2   Observables

As the number of electrons in the volume element $\mathbf{dp}$ is $n_{\mathbf{p}}\mathbf{dp} = 2f(\mathbf{p})\mathbf{dp}/(2\pi\hbar)^3$,[9] the charge current density at position $\mathbf{r}$ is

[9]The number 2 comes from spin.

$$\mathbf{j}_C(\mathbf{r}) = -e \int \frac{d^3p}{4\pi^3\hbar^3}\mathbf{v}(\mathbf{p})f(\mathbf{r},\mathbf{p}) = -e \int dE N(E) \int \frac{d\hat{p}}{4\pi}\mathbf{v}(\hat{p})f(\mathbf{r},\hat{p},E).$$
$$(2.8)$$

Here $N(E) = d|\mathbf{p}|/dE$ is the density of states. Note that in the absence of scattering ($I_{\text{coll}}[f] = 0$), eqn (2.6) can be considered as the continuity equation for a 'spectral current' $\mathbf{j_p} = N(E)\mathbf{v}f(\mathbf{r},\hat{p},E)/(4\pi^3\hbar^3)$:

$$\partial_t n_{\mathbf{p}} + \nabla \cdot \mathbf{j_p} = 0,$$

where $n_{\mathbf{p}} = N(E)f(\mathbf{r},\hat{p},E)/(4\pi^3\hbar^3)$ is the number of particles on the momentum/energy state $\hat{p}, E$. This continuity equation is a consequence of the fact that in the absence of scattering, elastic or inelastic, momentum and energy are conserved.

From $f(\mathbf{r},\hat{p},E)$ we may also calculate the heat current $\mathbf{j}_Q(\mathbf{r})$. Using the thermodynamic identity $dQ = dU - \mu dN$ between the change dQ in heat, dU in the internal energy, and dN in the number of particles, we obtain

$$\mathbf{j}_Q(\mathbf{r}) = \int dE N(E)(E - \mu) \int \frac{d\hat{p}}{4\pi}\mathbf{v}(\hat{p})f(\mathbf{r},\hat{p},E). \qquad (2.9)$$

This quantity is described in Sec. 2.8 in the context of thermoelectric effects.

2.3   Relaxation time approximation

For small deviations δf from equilibrium, the collision integrals can often be described using the relaxation time approximation. This is obtained by expanding the collision integral as

$$I_{\text{coll}}[f^0 + \delta f] \approx \left(\frac{\partial I_{\text{coll}}}{\partial f}\right)_{f=f^0} \delta f \equiv -\frac{1}{\tau}\delta f. \qquad (2.10)$$

The equilibrium state f^0 is chosen such that the term $I_{\text{coll}}[f^0] = 0$. The functional derivative $\partial I_{\text{coll}}/\partial f$ defines a *relaxation time* τ, which in general is energy dependent. The relaxation time, when it is well defined, is a useful quantity in characterizing the strength of the scattering effect. An example of such a relaxation time approach is given in Sec. 2.4 in the context of elastic scattering.

From the relaxation time we get a relaxation length by multiplying it by the speed of the electrons, $\ell = v\tau$, for the strongest type of scattering. In the diffusive limit where elastic scattering dominates (see Sec. 2.4), the relaxation length for other types of scattering process is obtained via $\ell = \sqrt{D\tau}$, where D is the diffusion constant.

In the following, we add different types of scattering process one by one, and for each type of scattering discuss how the description of the distribution function can be simplified in the limit of strong scattering. This idea is illustrated in Table 2.1.

Limit	*Length*	*Conserved quantity*
Ballistic	$L \ll \ell_{\mathrm{el}}, \ell_{\mathrm{e-e}}, \ell_{\mathrm{e-ph}}$	current for each momentum state
↓ elastic scattering: randomizes momentum direction		
Diffusive, non-equilibrium	$\ell_{\mathrm{el}} \ll L \ll \ell_{\mathrm{e-e}}, \ell_{\mathrm{e-ph}}$	current for each energy
↓ electron–electron scattering: mixes different energies		
Quasi-equilibrium	$\ell_{\mathrm{el}}, \ell_{\mathrm{e-e}} \ll L \ll \ell_{\mathrm{e-ph}}$	charge and energy currents
↓ electron–phonon scattering: energy exchange with environment		
Local equilibrium	$\ell_{\mathrm{el}}, \ell_{\mathrm{e-e}}, \ell_{\mathrm{e-ph}} \ll L$	charge current

Table 2.1 Different limits for the distribution function, defined by comparing the size L of the wire to the scattering lengths.

2.4   Elastic scattering and diffusive limit

Elastic scattering from impurities can be described by the collision integral of the form

$$I_{\mathrm{el}}[f] = \sum_{\mathbf{p}'}(J_{\mathbf{p}'\mathbf{p}} - J_{\mathbf{p}\mathbf{p}'}), \tag{2.11}$$

where

$$J_{\mathbf{p}\mathbf{p}'} = W_{\mathbf{p}\mathbf{p}'}f(\mathbf{r}, \mathbf{p}, t)(1 - f(\mathbf{r}, \mathbf{p}', t)) \tag{2.12}$$

describes scattering from the momentum state $\mathbf{p}$ to state $\mathbf{p}'$. For a description of elastic scattering in a system with a spherical Fermi surface, we have to have $|\mathbf{p}| = |\mathbf{p}'|$, and thus it is enough to describe the scattering between directions $\hat{p}$ and $\hat{p}'$. Such a description is based on the idea that the scatterer is 'fixed' (i.e., has an infinite mass), and the scattering can be pictorially represented as in Fig. 2.4.

Fig. 2.4 Elastic scattering diagram.

Function $W_{\hat{p}\hat{p}'}$ describes the strength and position dependence of the scattering potential. For a given configuration of impurities, described by the impurity potential $U_{\mathrm{imp}}(\mathbf{r})$, the scattering rate can be calculated using the Fermi golden rule (see Appendix A.3),[10]

[10]Here the momentum states $|\mathbf{p}\rangle$ are assumed to be plane wave states with a given momentum $\mathbf{p}$.

$$W_{\mathbf{p}\mathbf{p}'} = \frac{2\pi}{\hbar}|\langle\mathbf{p}|U|\mathbf{p}'\rangle|^2\delta[\epsilon(\mathbf{p}) - \epsilon(\mathbf{p}')] \equiv \frac{2\pi}{\hbar}|U_{\mathrm{imp}}(|p|(\hat{p} - \hat{p}'))|^2, \tag{2.13}$$

where the last equality defines the Fourier transform $U_{\mathrm{imp}}(\mathbf{p})$ and assumes a spherical Fermi surface.

The general solution of the Boltzmann equation with elastic scattering would require the knowledge of the position and type of each scatterer, i.e., of the function $U_{\rm imp}(\mathbf{p})$. However, the problem simplifies considerably if we assume that there are many scatterers and that it is enough to know $f(\mathbf{r}, \mathbf{p})$ only

- for a region much larger than the typical separation of scatterers, $\ell_{\rm el}$ (that is, we again use a coarse-grained description) and
- ensemble averaged over all positions of the scatterers.

In fact, these two requirements are analogous, as the classical $f(\mathbf{r}, \mathbf{p})$ is a self-averaging quantity (the limit of a large system is equivalent to an ensemble average). Due to the mixing of the momentum directions, in the presence of strong scattering all directions are almost equiprobable (see Fig. 2.5), and the distribution function becomes almost independent of the momentum direction $\hat{p}$.

We may then expand the dependence on the direction $\hat{p}$ in spherical harmonics,

$$f(\mathbf{r}, \hat{p}, E) \approx f_0(\mathbf{r}, E) + \delta\mathbf{f}(\mathbf{r}, E) \cdot \hat{p}. \tag{2.14}$$

These are the s- and p-wave terms of the spherical harmonics in the dependence on $\hat{p}$. As this is an expansion, we assume that the second term is much smaller than the first, and that higher-order terms can be disregarded. This assumption must of course be justified at the end of the calculation. For simplicity, in the following we assume that the scattering potential $U_{\rm imp}(\mathbf{p})$ is independent of the direction $\hat{p} = \mathbf{p}/|\mathbf{p}|$, and therefore so is W.[11]

As $f_0(\mathbf{r}, E)$ is independent of direction, it does not contribute to $I_{\rm el}[f]$: we would obtain $J_{\hat{p}\hat{p}'} = J_{\hat{p}'\hat{p}}$. We may hence write the collision integral in the form

$$I_{\rm el}[f] = W \sum_{\hat{p}'} [f(\vec{r}, \vec{p}', t) - f(\vec{r}, \vec{p}, t)] \approx W\delta\mathbf{f} \cdot \sum_{\hat{p}'} (\hat{p}' - \hat{p}). \tag{2.15}$$

The first term is an average of a vector around a circle, and must therefore vanish. We hence obtain

$$I_{\rm el}[f] = -\frac{1}{\tau}\delta\mathbf{f} \cdot \hat{p}, \quad \frac{1}{\tau} = W = \frac{2\pi|U_{\rm imp}|^2}{\hbar}. \tag{2.16}$$

The quantity τ is the elastic scattering time.

The Boltzmann equation can now be written as

$$(\partial_t + \mathbf{v} \cdot \nabla)(f_0(\mathbf{r}, E, t) + \delta\mathbf{f}(\mathbf{r}, E, t) \cdot \hat{p}) = -\frac{1}{\tau}\delta\mathbf{f}(\mathbf{r}, E, t) \cdot \hat{p} + I_{\rm inel}[f_0]. \tag{2.17}$$

Here $I_{\rm inel}[f]$ describes inelastic scattering and we assume it to be much weaker than elastic scattering, such that it essentially depends only on the lowest-order term f_0. Assuming a spherical Fermi surface, we can use $\mathbf{v} = \mathbf{p}/m = |\mathbf{p}|\hat{p}/m \approx v(p)\hat{p}$. Let us now integrate eqn (2.17) over the directions $\hat{p}$. This corresponds to integrating over the unit sphere, in spherical coordinates described by the angles θ and φ, as in Fig. 2.6. Let us choose the reference direction to $\mathbf{u}_z$. Then $\hat{p} \cdot \mathbf{u}_z = \cos(\theta)$ and

$$A_n \equiv \int d\hat{p}\,\hat{p}^n = \frac{1}{4\pi}\int_0^{2\pi} d\varphi \int_0^\pi d\theta \sin(\theta)\cos^n(\theta). \tag{2.18}$$

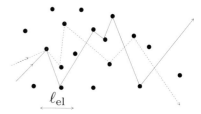

Fig. 2.5 Elastic scattering makes the particle lose the information about its initial direction. This takes place within the elastic mean free path $\ell_{\rm el}$.

[11] If this assumption is not made, the ensuing definition of the transport scattering rate should be modified to $\tau^{-1} = \int d\theta W(\theta)(1 - \cos(\theta))$, where θ is the angle between $\hat{p}$ and $\hat{p}'$. This can be shown to result from the diffuson contribution in the diagrammatic perturbation theory, see, for example, Ch. 1 in (Dittrich *et al.*, 1998).

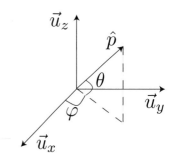

Fig. 2.6 Definition of the angles θ and φ.

Direct calculation shows that $A_0 = 1$, $A_1 = 0$, $A_2 = 1/3$, $A_3 = 0$, etc. Thus we obtain from $\int d\hat{p}(2.17)$

$$\left(\partial_t f_0 + \frac{1}{3}v\nabla \cdot \delta\mathbf{f}\right) = I_{\text{inel}}[f]. \tag{2.19}$$

We can also multiply by $\hat{p}$ and then integrate, i.e., $\int d\hat{p}\hat{p}(2.17)$, which yields

$$\frac{1}{3}\left(v\nabla f_0 + \partial_t \delta\mathbf{f}\right) = -\frac{1}{3\tau}\delta\mathbf{f} \Leftrightarrow \delta\mathbf{f} = -v\tau\nabla f_0 - \tau\partial_t\delta\mathbf{f}. \tag{2.20}$$

Let us make one more assumption: suppose the temporal changes in the distribution functions are much slower than the time τ describing elastic scattering. In this case we may throw out the last term from eqn (2.20). For what follows, let us define the three-dimensional *diffusion constant* characterizing the electron dynamics in a disordered solid,[12]

$$D = v^2\tau/3. \tag{2.21}$$

[12] The factor 3 comes from averaging in three dimensions. For a two-dimensional system we would obtain the same equation, but with $D = v^2\tau/2$, and in a one-dimensional system $D = v^2\tau$, see Exercise 2.1.

Namely, combining eqns (2.19) and (2.20) we obtain

$$(\partial_t - D\nabla^2)f_0(\mathbf{r}, E, t) = I_{\text{inel}}[f_0]. \tag{2.22}$$

Note that for $I_{\text{inel}}[f] = 0$, eqn (2.22) has the form of a diffusion equation, hence the term 'diffusive limit'.

The nature of the diffusion constant becomes clearer when analyzing the left-hand side of eqn (2.22). Let us assume that the distribution function changes by some factor over a length ℓ^*, i.e., we can replace ∇ by $1/\ell^*$. As eqn (2.22) has a unit of inverse time (rate), this length corresponds to a time $\tau_D = D/(\ell^*)^2$. This is the average time it takes for the electron to span a region with diameter ℓ^*. Conversely, the average radius of the region an electron spans within time τ is $\ell_D = \sqrt{D\tau}$.

Let us now check our expansion (2.14). Assume that f_0 changes on a scale ℓ^*, i.e., $\nabla f_0 \sim f_0/\ell^*$. From eqn (2.20) we then obtain

$$|\delta\mathbf{f}| \sim \frac{v\tau}{\ell^*}f_0 = \frac{\ell_{\text{el}}}{\ell^*}f_0 \ll f_0, \tag{2.23}$$

provided that $\ell^* \gg \ell_{\text{el}}$, i.e., all the changes in f_0 take place in scales much larger than ℓ_{el}. This is the validity regime of the diffusive approximation.

Example 2.2 shows a calculation of the non-equilibrium distribution function in a wire without inelastic scattering. For a description of the measurement of this function, see Example 2.4.

2.4.1   Currents in the diffusive limit

With the same prescription as in the derivation of eqn (2.22), we may now integrate the current densities, eqns (2.8) and (2.9), over $\hat{p}$. Then we obtain

$$\mathbf{j}_C(\mathbf{r}, t) = eN_F \int dE D(E)\nabla f_0(\mathbf{r}, E, t) \tag{2.24}$$

and

$$\mathbf{j}_Q(\mathbf{r}, t) = -N_F \int dE (E - \mu) D(E) \nabla f_0(\mathbf{r}, E, t). \qquad (2.25)$$

The diffusion coefficient $D(E)$ can be energy dependent due to the energy dependence of v, or the energy dependence of the scattering time τ. Typically such dependence is weak, but it is relevant for the thermoelectric effects that are described in Sec. 2.8.

Analogously to the ballistic case, eqn (2.22) can be viewed as the continuity equation for the 'spectral current' $D\nabla f_0(\mathbf{r}, E, t)/(4\pi^3\hbar^3)$.

Below we concentrate on the diffusive limit where it is enough to follow the position dependence of the s-wave term $f_0(\mathbf{r}, E)$, and therefore we drop the subscript 0.

Example 2.2 Distribution function in the diffusive limit
Let us now consider the wire of Fig. 2.3 in the diffusive limit. In the absence of inelastic scattering the static Boltzmann equation reads

$$D\partial_x^2 f(x, E) = 0,$$

where the wire is assumed quasi-one-dimensional, i.e., no changes take place in y and z directions. The solution to this and the boundary conditions is (see Fig. 2.7)

$$f(x, E) = f_L(E) \frac{L - x}{L} + f_R(E) \frac{x}{L}. \qquad (2.26)$$

This can be directly inserted into eqn (2.24) for the charge current density,

$$j_C = eN_F D \int_{-\infty}^{\infty} dE \partial_x f(x, E) = eN_F D \int_{-\infty}^{\infty} [f_R(E) - f_L(E)]/L$$

$$= \frac{eN_F D}{L}(\mu_R - \mu_L) = \frac{e^2 N_F D}{L} V = \sigma_D V/L. \qquad (2.27)$$

The integral is evaluated in eqn (A.64a). We hence obtain the Drude conductivity $\sigma_D = e^2 N_F D$ discussed in Sec. 1.2.1.

For $\mu_L = \mu_R$ but $T_L \neq T_R$ the heat current density is

$$j_Q = -DN_0 \int_{-\infty}^{\infty} dE (E - \mu) \nabla f(x, E)$$

$$= \frac{DN_0}{L} \int_{-\infty}^{\infty} dE (E - \mu)[f_L(E) - f_R(E)] = \frac{DN_0}{L} \frac{\pi^2 k_B^2}{6}[T_L^2 - T_R^2]$$

$$\overset{T_L - T_R \ll T_R, T_L}{\approx} \frac{\pi^2 T k_B^2 \sigma}{3e^2 L}(T_L - T_R) = \kappa_{\text{th}}(T_L - T_R)/L, \qquad (2.28)$$

which uses the integral from eqn (A.64b). This way we obtain the *Wiedemann–Franz law* for the heat conductivity in an electron system, $\kappa_{\text{th}} = \mathcal{L}T$, where the *Lorenz number* $\mathcal{L} = \pi^2 k_B^2/(3e^2) \approx 2.45 \times 10^{-8}$ WΩK^{-2} depends only on constants of nature.

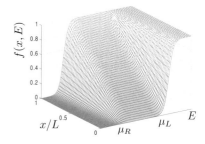

Fig. 2.7 Non-Equilibrium distribution function, eqn (2.26), interpolates at every energy between the Fermi functions in the reservoirs. Here $k_B T = (\mu_R - \mu_L)/20$.

2.5 Inelastic scattering

The main sources of inelastic scattering are those that scatter two electrons from states with energy E and E' to those with energy $E + \hbar\omega$ and

[13]At very low temperatures, a further inelastic scattering effect becomes relevant: that between the electrons and the photons of the electromagnetic environment; see (Meschke *et al.*, 2006). This effect is empirically well-known and often it can be efficiently reduced by filtering the wires that connect the sample to the measuring equipment. Therefore, the lower one aims in the temperature, the better the filtering has to be, and typically in sub-micron-size samples this effect can be clearly seen at temperatures below some 100 mK.

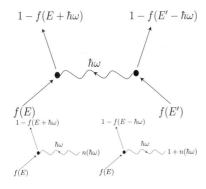

Fig. 2.8 Top: electron–electron scattering. Bottom: electron–phonon scattering (left: absorption, right: emission).

[14]For a detailed summary of electron–electron scattering in diffusive wires, see (Pierre, 2000) or (Rammer, 2007).

[15]In some references, this is also called the 'hot electron limit'.

[16]Here we assume that the collisions conserve the number of particles, and therefore the direct energy integral over the collision integral vanishes.

$E' - \hbar\omega$ (such that the total energy is conserved), and those that scatter an electron from state E to state $E \pm \hbar\omega$ and absorb/emit a phonon with energy $\hbar\omega$.[13] These can be pictorially represented as in Fig. 2.8.

The corresponding collision integrals are of the form (for brevity, the arguments $\mathbf{r}$ and t are dropped from the distribution function and the energy arguments are indicated as subscripts)

$$I_{e-e}[f] = -\int d\omega dE' K_{ee}(E, E', \omega)[f_E f_{E'}(1 - f_{E+\hbar\omega})(1 - f_{E'-\hbar\omega})$$
$$- (1 - f_E)(1 - f_{E'})f_{E+\hbar\omega} f_{E'-\hbar\omega}] \tag{2.29}$$

$$I_{e-ph}[f] = -\int_0^\infty d\omega K_{eph}(E, \omega)\{[f_E(1 - f_{E-\hbar\omega}) - f_{E+\hbar\omega}(1 - f_E)]$$
$$\times [n_{\hbar\omega} + 1] - [f_{E-\hbar\omega}(1 - f_E) - f_E(1 - f_{E+\hbar\omega})] n_{\hbar\omega}\}. \tag{2.30}$$

In the latter collision integral $n_E = n(E)$ denotes the distribution function of the phonons. Both collision integrals are written in the form 'scattering out' minus 'scattering in' from/to the state with energy E. The factor $-n(-\omega) = n(\omega) + 1$ in the electron–phonon 'scattering out' term describes phonon emission and contains both the stimulated and spontaneous emission terms. The minus sign in front of the collision integral is due to the fact that scattering out decreases the number of electrons at state E.

The scattering kernels $K_{ee}(E, E', \omega)$ and $K_{eph}(E, \omega)$ have to be computed separately for each type of scattering. Such a calculation is usually based on the Fermi golden rule, but its details are outside the scope of this book and only some of the results are presented in this section.

2.5.1   Electron–electron scattering

There are several sources that contribute to the electron–electron scattering collision integral.[14] In metals the most relevant are the direct (screened) Coulomb interaction between the electrons and the second-order effects from collisions between electrons and magnetic impurities.

If the energy dependence of the density of states can be disregarded, the kernel $K_{ee}(E, E', \omega)$ depends only on ω (see Complements 2.1 and 2.2). In this case it is straightforward to show that for a Fermi function $f(\mathbf{r}, E, t) = f^0(E, \mu(\vec{r}, t), T(\vec{r}, t))$ with any potential $\mu(\vec{r}, t)$ and temperature $T(\vec{r}, t)$, $I_{e-e}[f]$ vanishes (see Exercise 2.5). Thus, if the electron–electron scattering is sufficiently strong, i.e., the prefactor to the collision integral is large, the solution to the Boltzmann equation (2.22) tends towards a Fermi function with a position-dependent electron temperature $T(\vec{r})$. This is the *quasi-equilibrium* limit.[15]

The equations for the potential $\mu(x)$ and temperature $T_e(x)$ in the quasi-equilibrium limit can be found from eqn (2.22) by forcing the solution to have a Fermi function form. Let us assume this and integrate both sides of eqn (2.22) over energy. This leads to[16]

$$(\partial_t - D\nabla^2)\mu(\mathbf{r}) = 0. \tag{2.31}$$

An equation satisfied by the temperature is obtained by multiplying eqn (2.22) by $N_F E$ and then integrating over energy. This yields

$$C_e(\mathbf{r},t)\partial_t T_e - \nabla \cdot [\kappa(\mathbf{r},t)\nabla T_e] - \sigma(\nabla\mu/e)^2 = P_{\text{coll}}. \tag{2.32}$$

Here $C_e = \pi^2 N_F k_B^2 T_e/3$ is the electron specific heat, $\sigma = e^2 N_F D$ is the Drude conductivity, $\kappa = \sigma \mathcal{L} T_e$ is the electron heat conductivity and $\mathcal{L} = \pi^2 k_B^2/(3e^2)$ is the Lorenz number. Moreover,

$$P_{\text{coll}} = N_F \int dE E I_{\text{coll}}$$

is the power density (power/volume) flowing into other degrees of freedom due to the remaining inelastic scattering mechanisms — predominantly electron–phonon scattering.

In the quasi-equilibrium limit the charge/heat current densities are

$$j_C = \sigma \nabla \mu \tag{2.33a}$$
$$j_Q = \kappa \nabla T, \tag{2.33b}$$

respectively. Equation (2.31) is thus a continuity equation for charge current, whereas eqn (2.32) is the equivalent for the heat current, but in the presence of heat sources and sinks.

Example 2.3 One-dimensional wire in the quasi-equilibrium limit
Let us again consider the one-dimensional wire of Fig. 2.3, but now assume the quasi-equilibrium limit, i.e., the distribution function is a Fermi function with position-dependent potential and temperature. Let us assume that there is only internal electron–electron scattering, so that the only heat that flows away from the system is through the contacts to the reservoirs. Now we need to solve eqns (2.31) and (2.32) with the boundary conditions $\mu = \mu_{L/R}$, $T = T_{L/R}$ inside the reservoirs. We obtain

$$\mu(x) = \frac{L-x}{L}\mu_L + \frac{x}{L}\mu_R \tag{2.34a}$$
$$T(x) = \frac{1}{e^2\mathcal{L}}\frac{x}{L}\frac{L-x}{L}(\mu_L - \mu_R)^2 + \frac{L-x}{L}T_L^2 + \frac{x}{L}T_R^2. \tag{2.34b}$$

The potential is thus a linear function of the position. The first term in the equation for $T(x)$ represents Joule heating from the applied voltage. The currents obtained from eqns (2.34) are the same as in the case without electron–electron scattering (Example 2.2), and hence independent of the precise form of the distribution function. This is due to the fact that normal metals with constant density of states are linear systems: in them the charge current is a linear function of voltage. For non-linear systems, such as a tunnel junction between a normal metal and a superconductor (see Example 2.4 and Sec. 9.2), the currents do depend on the detailed shape of the distribution functions. Moreover, the non-equilibrium current noise (Ch. 6) depends on the form of the distribution function even in a normal metal.

Let us summarize the different forms of the Boltzmann equation in the different limits:

Limit	Equation
Ballistic	$(\partial_t + \mathbf{v} \cdot \nabla) \, f(\mathbf{r}, \hat{p}, E) = I_{\text{coll}}[f]$ (2.6)
Diffusive	
non-equilibrium	$(\partial_t - D\nabla^2) f(\mathbf{r}, E, t) = I_{\text{inel}}[f]$ (2.22)
Quasi-Equilibrium	$(\partial_t - D\nabla^2)\mu(\mathbf{r}) = 0$
(eqns (2.31,2.32))	$C_e \partial_t T_e - \nabla \cdot [\kappa \nabla T_e] - \sigma(\nabla \mu/e)^2 \quad = P_{\text{coll}}$

Once the distribution function has been found, the charge and heat currents can be calculated with eqns (2.8) and (2.9) in the ballistic, eqns (2.24) and (2.25) in the diffusive and eqns (2.33) in the quasi-equilibrium limit.

Complement 2.1 Kernel of electron–electron collision integral

The theory for electron–electron interactions in diffusive wires is due to Altshuler and Aronov (1985). This theory describes the remaining electron–electron collisions in the Landau Fermi liquid theory. Later it was found that there are also mechanisms that indirectly lead to collisions between electrons. One such mechanism is connected to magnetic impurities, and it can sometimes dominate over the direct interaction.

In the case of a direct electron–electron interaction in a diffusive wire, the kernel of the collision integral is of the form

$$K_{ee}(E, E', \omega) = \frac{2}{\pi N_F \Omega} \sum_{\mathbf{q}} \frac{1}{\omega^2 + (D\mathbf{q}^2)^2}, \tag{2.35}$$

where Ω is the volume of the wire and the sum goes over all electron momenta. The summand is a response function of the diffusion equation. If one, two or three of the wire dimensions are long compared to the scale $\xi_D \equiv \sqrt{\hbar D/\max(eV, k_B T)}$, where the frequency scale is given either by the voltage V or the temperature T, we may transform the sum into an integral by the prescriptions[17]

$$\sum_{\mathbf{q}} \mapsto \frac{L}{\pi} \int_0^\infty dq, \quad \text{one-dimensional wire}$$

$$\sum_{\mathbf{q}} \mapsto \frac{A}{2\pi} \int_0^\infty q\,dq, \quad \text{two-dimensional wire}$$

$$\sum_{\mathbf{q}} \mapsto \frac{\Omega}{2\pi^2} \int_0^\infty q^2\,dq, \quad \text{three-dimensional wire.}$$

Here L is the length and A is the cross section of the wire. As a result, we obtain a kernel of the form

$$K_{ee}(E, E', \omega) = \kappa_d \omega^{-\alpha}, \tag{2.36}$$

where $\alpha = 2 - d/2$ for a d-dimensional wire. The prefactor κ_d is

$$\kappa_1 = \frac{1}{\pi\sqrt{2D}\hbar^2 N_F A}, \quad \kappa_2 = \frac{1}{4\pi D N_F d}, \quad \text{and} \quad \kappa_3 = \frac{1}{8\pi^2\sqrt{2}\hbar^2 N_F D^{3/2}}.$$

[17] For the one-dimensional case, this transformation is explained in Sec. 3.1.1, for the two- and three-dimensional wires, we first go to polar or spherical coordinates and then use a similar scheme.

Here $A = \Omega/L$ is the cross-section of the wire in the one-dimensional case and $d = \omega/A$ is the thickness of the film in the two-dimensional case.

For the case of magnetic impurities, (Kaminski and Glazman, 2001) found

$$K_{ee}(E, E', \omega) = \kappa_m \omega^{-2}, \tag{2.37}$$

with $\kappa_m = \frac{\pi}{2} \frac{c_m}{\hbar N_F} S(S+1) \left[\ln\left(\frac{eV}{k_B T_K}\right)\right]^{-4}$. Here c_m is the concentration, S is the spin and T_K is the Kondo[18] temperature of the impurities.

[18]See Sec. 8.4.

Relaxation time and length

For $d = 3$, the bare electron–electron collision integral can be approximated by $-\delta f/\tau_{e-e}$, where $\tau_{e-e}^{-1} = \zeta(3/2)(k_B T)^{3/2}/(k_F \ell_{el} \sqrt{\hbar \tau} E_F)$ is the relaxation rate. For $d < 3$, a relaxation-time approach does not work directly, as the functional integral $\partial I_{\text{coll}}/\partial f$ has an infrared divergence. In this case often the full collision integral is needed.

Electron–electron scattering in the one-dimensional case has been actively studied in the recent years. Qualitatively, the findings are in line with the above theory, but a quantitative agreement is to some extent still lacking (Huard *et al.*, 2004).

2.5.2   Electron–phonon scattering

In most materials there are two main types of phonon: acoustic phonons whose spectrum is linear and described by the speed of sound v_S in the solid, i.e, $\epsilon_p = v_S|\mathbf{p}|$, and optical phonons which reside at the Debye energy ϵ_D. Typical Debye temperatures $T_D = \epsilon_D/k_B$ are a few hundreds of K, and at temperatures much lower than T_D and voltages lower than ϵ_D/e the electron scattering from optical phonons can be disregarded.

If the wire under consideration lies on a substrate, the phonons in the wire are typically fairly strongly connected to those in the lattice, and electron–phonon coupling does not strongly perturb their state. In this case the state of the phonons can be described via the Bose–Einstein distribution function $n(E) = [\exp(E/(k_B T_b)) - 1]^{-1}$, where T_b is the temperature of the substrate. The electron–phonon collision integral vanishes when $f(\mathbf{r}, E, t) = f^0(E; \mu(\mathbf{r}), T_b)$, i.e., when the electron temperature equals that of the phonons (see Exercise 2.5). Thus, the effect of the coupling to the phonons is to thermalize the electrons to the lattice temperature.

When the sample is driven with a large voltage, the electrons heat the phonons in the film more than the substrate can absorb. In this case, the temperature of the film phonons deviates from that of the substrate (or the phonon distribution function may become a strong non-equilibrium function). The final temperature of the electron gas has in this case to be calculated from the thermal balance of the heat flows between the sub-systems (electrons, film phonons and the substrate). An example of such a heat balance calculation is given in Exercise 2.2.

Due to the Bose factor in the collision integral, electron–phonon scattering is strongly dependent on the lattice temperature; see eqns (2.39) and (2.40). At low temperatures the corresponding scattering length becomes very long and its effect can often be disregarded altogether.

Complement 2.2 Relaxation time and dissipated power in electron–phonon scattering

The kernel of the electron–phonon scattering in a three-dimensional wire, obtained from the Fermi golden rule,[19] is related to an 'Eliashberg function' $\alpha^2 F(\omega)$ by $K_{\mathrm{eph}} = 2\pi\alpha^2 F(\omega)$. Here

$$\alpha^2 F(\omega) = \frac{|M_0|^2}{4\pi^2\hbar^2 v_S^3 N_F}\omega^2, \tag{2.38}$$

v_S is the speed of sound and $|M_0|^2$ is the prefactor of the matrix element for electron–phonon scattering. The latter is dependent on the microscopic details of the lattice, such as the form of the unit cell, and therefore it is often measured by relating it to certain observable quantities, such as the relaxation time or the power dissipated into the phonons at a certain temperature.

The power dissipated into the phonon system (i.e., the heat current times the volume Ω of the sample) equals in the quasi-equilibrium limit ($f(\mathbf{r}, E, t) = f^0(E; \mu(\mathbf{r}), T_e(\mathbf{r}))$)

$$\dot{Q}_{\mathrm{eph}} = N_F\Omega\int dE\, E I_{\mathrm{e-ph}} = \Sigma\Omega(T_e^5 - T_b^5), \tag{2.39}$$

where $\Sigma = 12\zeta(5)k_B^5|M_0|^2/(\pi\hbar^5 v_S^3)$ and $\zeta(x)$ is a Zeta function, $\zeta(5) \approx 1.0369$. Typical values for Σ in metals are around 10^9 Wm^{-3}K^{-5}.

From eqns (2.10) and (2.38) we determine that at $E = E_F$

$$\frac{1}{\tau_{e-ph}} = 4\pi\int_0^\infty d\omega\,\frac{\alpha^2 F(\omega)}{\sinh\left(\frac{\hbar\omega}{k_B T}\right)} = \alpha T^3, \tag{2.40}$$

where $\alpha = 7\zeta(3)|M_0|^2 k_B^3/(2N_F) = 7\zeta(3)\Sigma/(24\zeta(5)k_B^2 N_F)$ and $\zeta(3) \approx 1.2021$. With typical values at Cu, $\Sigma = 2\cdot 10^9$ WK^{-5}m^{-3}, $N_F = 1.6\cdot 10^{47}$ J^{-1}m^{-3}, $\tau_{e-ph} = 40$ ns at $T_e = 1$ K. The corresponding scattering length is obtained from $\ell_{e-ph} = \sqrt{D\tau_{e-ph}}$. With $D = 0.01$ m^2/s it yields $\ell_{e-ph} = 20$ μm at $T = 1$K and $\ell_{e-ph} = 700$ μm at $T = 100$ mK. This shows why electron–phonon scattering can be disregarded in nanoscale devices at low temperatures.

Recently, similar calculations of the electron–phonon power dissipation have been presented, for example, for lower-dimensional wires (Hekking *et al.*, 2008) and graphene (Bistritzer and MacDonald, 2009; Viljas and Heikkilä, 2010).

---

[19]See Sec. 10.8 in (Rammer, 1998).

2.6   Junctions

Tunnel junctions or other types of contact between materials with different properties (different mean free paths, etc.) are typically very small-scale objects, their thickness being of the order of the Fermi wavelength. These cannot therefore be directly treated with the Boltzmann equation. However, they can be described through an effective boundary condition on the distribution function (Nazarov, 1999*b*). In the diffusive limit the boundary conditions can be formed from (i) a statement of current conservation at the two sides of the junction for each energy and

(ii) from equating this current to the one flowing through the junction, in the case of a non-vanishing junction resistance R_I (see eqn (1.12)). In other words, the boundary condition is

$$\sigma_L A_L \nabla f(x = x_b^-) \cdot \hat{n} = \sigma_R A_R \nabla f(x = x_b^+) \cdot \hat{n} = \frac{f(x = x_b^+) - f(x = x_b^-)}{R_I},$$
(2.41)

where $\sigma_{L/R} = e^2 D_{L/R} N_F^{L/R}$, $D_{L/R}$, $N_F^{L/R}$, $A_{L/R}$ are conductivities, diffusion constants, densities of states and cross-sections of the wires at the left/right side of the junction, and $\hat{n}$ is the unit normal to the interface. In the case of a small R_I the latter part of eqn (2.41) leads to a statement of continuity of $f(x)$ across the interface.

In general, solving for the distribution function in the diffusive limit amounts to solving the second-order differential equation (2.22) in a given geometry with the mixed Dirichlet–Neumann boundary conditions described in eqn (2.41).

At quasi-equilibrium the boundary conditions may be integrated over energy to yield the Kirchhoff laws for charge and heat current,

$$\sigma_L A_L \nabla \mu_L \cdot \hat{n} = \sigma_R A_R \nabla \mu_R \cdot \hat{n} = \frac{\mu_L - \mu_R}{R_I}$$
(2.42a)

$$A_L \left[\kappa_L e^2 \nabla T_L + \sigma_L \mu_L \nabla \mu_L \right] \cdot \hat{n} = A_R \left[\kappa_R e^2 \nabla T_R + \sigma_R \mu_R \nabla \mu_R \right] \cdot \hat{n}$$
$$= \frac{1}{R_I} \left[\frac{\pi^2 k_B^2}{6} (T_R^2 - T_L^2) + \frac{1}{2} (\mu_R^2 - \mu_L^2) \right],$$
(2.42b)

where $\mu_{L/R} = \mu(x = x_b^\pm)$ and $T_{L/R} = T(x = x_b^\pm)$ are the potential and the electron temperature on both sides of the junction, and $\kappa_{L/R} = \mathcal{L} \sigma_{L/R} T$ are the heat conductivities. The different terms in eqn (2.42b) describe the heat flow due to a temperature gradient and the Joule power due to a potential gradient over a region with conductivity σ.

Example 2.4 Measurement of the distribution function with super-conducting contacts

As discussed in Example 1.1, the current through a tunnel barrier connecting normal electrodes is independent of temperature and, therefore, of the precise form of the distribution function. However, the situation changes when one or two of the electrodes become superconducting. In this case the current is given by eqn (1.12), with the superconducting density of states of the form

$$N_S(E) = N_F \frac{|\epsilon|}{\sqrt{\epsilon^2 - |\Delta|^2}} \theta \left(|\epsilon| - |\Delta| \right).$$
(2.43)

Here Δ is the superconducting energy gap, and $\theta(x)$ is the Heaviside step function, equal to unity for $x > 0$ and vanishing at $x < 0$. This formula is derived in Ch. 5 and the properties of the resultant tunnel currents are discussed in further detail in Ch. 9.

The energy-dependent density of states makes the tunnel current dependent on the precise form of the distribution function. This fact was first exploited by (Pothier *et al.*, 1997) to measure the non-equilibrium distribution function by placing a superconducting tunnel contact into a small normal-metal wire

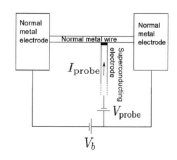

Fig. 2.9 Setup for measuring the distribution function in a wire driven out of equilibrium.

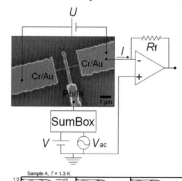

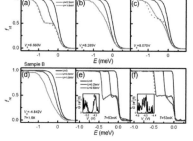

Fig. 2.10 Measurement of the non-equilibrium electron distribution function $f_{\rm nt}$ in a carbon nanotube, following the pioneering experiment by Pothier *et al.* (1997) made on metal wires. Top: Scanning electron micrograph of the setup, containing a single-walled nanotube placed between Cr/Au contacts, and a superconducting (Pb) tunnel probe, along with the measurement setup. Bottom: $f_{\rm nt}$, obtained for different charge densities and at different temperatures of the electrodes. The two-step functions in (a), (e) and (f) imply weak electron–electron relaxation. (Y.F. Chen *et al.*, *Phys. Rev. Lett.* **102**, 036804 (2009), Figs. 1 and 4.) © 2009 by the American Physical Society.

[20]The experimentalists needed to also take into account the environmental Coulomb blockade, discussed in Ch. 7. These yield a further nonlinearity into the tunnel current.

driven by a voltage (see Fig. 2.9).[20] This type of a measurement can be used to obtain information about the strength of energy relaxation in small disordered normal-metal wires. Besides the normal-metal wires, a similar scheme has been used, for example, to measure the effects of proximity-induced supercurrent on the distribution function (Crosser *et al.*, 2006) and the non-equilibrium distribution function in carbon nanotubes (Chen *et al.*, 2009) (see an example in Fig. 2.10).

2.7   Magnetic heterostructures

Boltzmann equations are a useful tool for analyzing the electronic transport properties of magnetic multilayers, i.e., structures containing several ferromagnets, with possibly different magnetization orientations, and non-magnetic conductors.[21] In these systems a defining feature of the metallic ferromagnets is their spin polarization: the density of states at the Fermi level and the Fermi velocity are different for majority (say, ↑) than for minority (say, ↓) spin electrons. This feature can be included from the outset in eqn (2.6) by introducing spin-dependent Fermi velocities $v_{F,s}$ and distribution functions $f_s(\vec{r}, \hat{p}, E)$, where $s \in \{\uparrow, \downarrow\}$.[22]

[21]See for example (Valet and Fert, 1993), who also describe the case where τ_{sf} is of the order of τ.

There are two types of scattering: those that preserve the spin and those that flip it. These can be described with the scattering times τ and τ_{sf}, respectively. Typically $\tau \ll \tau_{sf}$, so that in a sample much larger than $\max(v_{F\uparrow}\tau, v_{F\downarrow}\tau)$ the momentum direction randomizes due to elastic scattering before the different spin ensembles are mixed. In this case we may write the diffusive-limit Boltzmann equation, the spin-dependent analogue of eqn (2.22) in the form[23]

[22]For simplicity of presentation, in this discussion I consider only the case of collinear magnetization of the ferromagnets. In the case of non-collinear magnetization, the distribution function would become a 2×2 matrix in spin space as discussed in Sec. 3.6.3.

[23]Here $f_s(E)$ refers to the momentum direction-averaged, i.e., the *s*-wave component of the distribution function.

$$(\partial_t - D_s \nabla^2)f_s(\vec{r}, E, t) = \frac{f_{\bar{s}} - f_s}{2\tau_{sf}^s} + I_{inel}^s[f_s, f_{\bar{s}}]. \qquad (2.44)$$

Here $D_s = v_{Fs}^2 \tau/3$ and $\bar{s}$ denotes the spin opposite to s, and the spin-flip scattering is treated within the relaxation time approach. In what follows we disregard the possible energy dependence of τ_{sf}^s and note that in order to preserve the total current (see Example 2.5), the spin-flip scattering time inside a ferromagnet has to depend on spin.

For the description of the charge current, it is enough to derive the equation for the spin-dependent potential μ_s by integrating eqn (2.44). This yields[24]

[24]The corresponding equation for a spin-dependent electron temperature may also be written and in linear response it also yields reasonable results, but outside linear response its validity range is questionable, as the inelastic scattering responsible for equilibration also mixes the spin-dependent temperatures; see (Heikkilä *et al.*, 2010).

$$(\partial_t - D_s \nabla^2)\mu_s(\vec{r}, t) = \frac{\mu_{\bar{s}} - \mu_s}{2\tau_{sf}^s}. \qquad (2.45)$$

In the case of junctions this equation should be combined with the spin-dependent boundary conditions of the type in eqn (2.42a),

$$\sigma_{Ls} A_L \nabla \mu_{Ls} = \sigma_{Rs} A_R \nabla \mu_{Rs} = \frac{\mu_{Rs} - \mu_{Ls}}{R_{Is}}, \qquad (2.46)$$

[25]Note that elsewhere in the text the density of states is summed over spin, so that the total conductivity has the same form as here.

assuming that the junction does not directly mix the spins. Here $\sigma_{L/Rs} = e^2 N_{L/Rs} D_{L/Rs}$ are the spin-dependent conductivities on the left/right side of the junction.[25] With the solution $\mu_s(\vec{r}, t)$ to eqns (2.45),(2.46) we then obtain the charge current density

$$j = \sum_s \sigma_s \nabla \mu_s / e \qquad (2.47)$$

and its (in general non-conserved) spin polarization

$$P_j = (\sigma_\uparrow \nabla \mu_\uparrow - \sigma_\downarrow \nabla \mu_\downarrow)/j. \qquad (2.48)$$

Note that the charge current given by eqn (2.47) should be conserved even in the presence of spin-flip scattering. Because of this, it is straightforward to show that the spin-flip scattering times are linearly proportional to the spin-dependent densities of states, $\tau_{\mathrm{sf}}^s \propto N_s$.

Below, we solve eqns (2.45),(2.46) in an example system consisting of a normal metal and one or two ferromagnets.

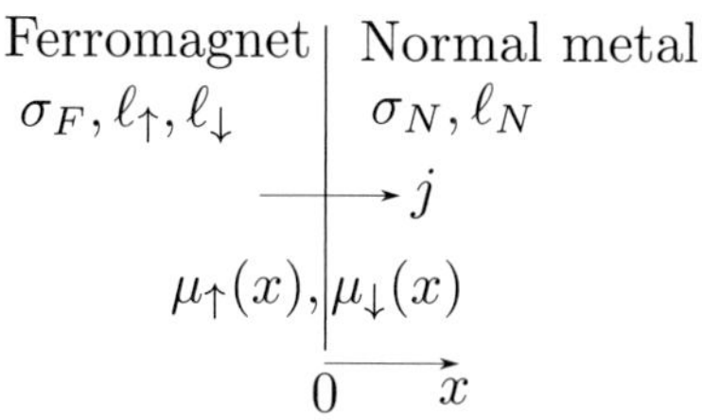

Example 2.5 Spin accumulation at a FN interface

Let us consider a junction between a ferromagnet (F) and a normal metal (N), driven with a constant current density j (see Fig. 2.11).[26] In the stationary case the spin-dependent potentials obey the second-order equations

$$\partial_x \mu_s = \frac{\mu_s - \mu_{\bar{s}}}{2\ell_s^2}, \qquad (2.49)$$

where $\ell_s = \sqrt{D_s \tau_{\mathrm{sf}}^s}$ are the spin-dependent spin-flip lengths. Including the fact that $\tau_{\mathrm{sf}}^s = \alpha N_s$ with some prefactor α, the current carried for spin component s is given by

$$j_s = e N_s D_s \partial_x \mu_s = e \alpha \ell_s^2 \partial_x \mu_s. \qquad (2.50)$$

A general solution of eqn (2.49) is of the form

$$\mu_{\uparrow/\downarrow} = \frac{\ell_\downarrow^2(C_1 + C_2 x) + \ell_\uparrow^2(C_3 + C_4 x)}{\ell_{\mathrm{tot}}^2}$$
$$\pm \frac{\ell_{\downarrow/\uparrow}^2}{\ell_{\mathrm{tot}}^2} \left[(C_3 - C_1)\cosh(x/\ell) + \ell(C_4 - C_2)\sinh(x/\ell) \right], \qquad (2.51)$$

where $\ell_{\mathrm{tot}}^2 = (\ell_\uparrow^2 + \ell_\downarrow^2)$, $\ell = \sqrt{2}\ell_\uparrow \ell_\downarrow / \ell_{\mathrm{tot}}$, and C_i are constants to be determined from boundary conditions.

For an FN junction with a vanishing boundary resistance, the boundary conditions correspond to the conservation of the spin-dependent current j_s at the boundary and the continuity of the spin-dependent potentials. Then, denoting the total current density carried through the junction by j and setting the zero level of the potential to $\mu_\uparrow(x=0) + \mu_\downarrow(x=0) = 0$, the potentials are

$$\mu_{\uparrow/\downarrow}(x > 0) = \frac{ejx}{2\sigma_N} \pm \frac{\delta\mu}{2} e^{-x/\ell_N}$$
$$\mu_{\uparrow/\downarrow}(x < 0) = \frac{ejx}{2\sigma_F} - \delta\mu \left(1 - e^{x/\ell}\right) \frac{\ell_\downarrow^2 - \ell_\uparrow^2}{2\ell_{\mathrm{tot}}^2} \pm \frac{\delta\mu}{2} e^{x/\ell}, \qquad (2.52)$$

where $\sigma_N = e^2 N_N D_N$ and $\sigma_F = e^2(N_\uparrow D_\uparrow + N_\downarrow D_\downarrow)/2$ are the conductivities of the normal metal and the ferromagnet, respectively. These functions are plotted in Fig. 2.12. The solutions for the two spin components differ by the *spin accumulation*, $\mu_\uparrow - \mu_\downarrow$, which at the interface is given by

$$\delta\mu = \frac{2ej\ell_N(\ell_\downarrow^2 - \ell_\uparrow^2)}{2\ell_N \ell \sigma_F + \ell_{\mathrm{tot}}^2 \sigma_N}. \qquad (2.53)$$

Ferromagnet | Normal metal

$\sigma_F, \ell_\uparrow, \ell_\downarrow$ | σ_N, ℓ_N

$\mu_\uparrow(x), \mu_\downarrow(x)$

$0 \quad x$

Fig. 2.11 Junction between a ferromagnet and a normal metal.

[26] This system was analyzed, for example, in (van Son *et al.*, 1987).

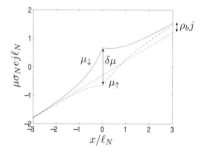

Fig. 2.12 Example spin-dependent potential profile of an FN junction. Solid line shows $\mu_\downarrow(x)$ and dashed line $\mu_\uparrow(x)$. Here $\sigma_F = \sigma_N$ and hence the slopes of the potentials are equal far from the interface. The dotted line is an extrapolation of the average voltage profile, showing the additional potential step related to the boundary resistivity.

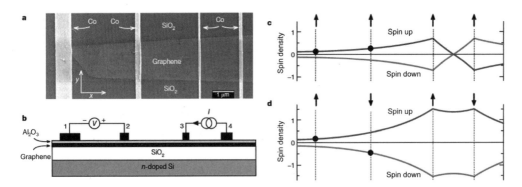

Fig. 2.13 Spin injection into graphene, analogous to the pioneering experiments by Jedema *et al.* (2001) made in metals: a) Micrograph of the device showing a single-layer graphene flake in contact with four ferromagnetic electrodes of different widths and therefore different coercive fields, allowing the experimentalists to change their magnetization orientation one by one. b) System schematics and measurement setup showing the Al$_2$O$_3$ layer, so that the connections to the electrodes are of the tunnelling type. The spin accumulation is measured as induced voltages between the different electrodes (say, 1 and 2), when current is driven between other electrodes (3 and 4). The resultant spin accumulation along the graphene flake is shown in c) and d). (Tombros *et al.*, *Nature* **448**, 571 (2007), Fig. 1.) Reprinted by permission from Macmillan Publishers Ltd., © 2007.

Antiparallel configuration

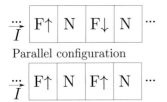

Parallel configuration

Fig. 2.14 Magnetic multilayer in the CPP configuration: Fσ is a ferromagnetic region with the majority spin σ.

[27] The discussion here concerns the case where the current flows perpendicular to the plane of the junctions (denoted as CPP). Most commercial products use the current in plane (CIP) configuration, which requires structure size smaller than the elastic scattering length, and therefore its theory is somewhat more involved; see, for example, (Camley and Barnaś, 1989).

The spin accumulation decays into the bulk ferromagnet and the normal metal over length scales ℓ and ℓ_N, respectively. The creation of spin accumulation by driving the current j is called spin injection, and it can be directly observed by placing a ferromagnetic tunnel contact close to the FN interface; see for example (Jedema *et al.*, 2001; Tombros *et al.*, 2007) and Fig. 2.13.

The presence of spin accumulation leads to an additional boundary resistance for the FN interface, indicated by a shift in the spin-averaged potential on the ferromagnetic side of the interface, the second term in $\mu(x < 0)$ in eqn (2.52). This resistance (per unit area) equals

$$\rho_b = \frac{\ell_N (\ell_\downarrow^2 - \ell_\uparrow^2)^2}{\ell_N \ell \ell_{\rm tot}^2 \sigma_F + \ell_{\rm tot}^4 \sigma_N} \tag{2.54}$$

and it is naturally independent of the sign of spin accumulation.

Example 2.6 Giant magnetoresistance in magnetic multilayers
The above example shows that the change of the relative magnitude of current carried in different spin channels at an FN interface leads to an additional boundary resistance in the system. The resistance related to this current conversion effect can be made tunable in magnetic multilayers with many ferromagnets (Binasch *et al.*, 1989; Baibich *et al.*, 1988). Typically such structures contain non-ferromagnetic (N) spacer layers to allow for individual manipulations of the magnetic configuration of the multilayer. Varying the thickness d_N of the spacer layer leads to an alternating ferromagnetic and antiferromagnetic coupling of the adjacent ferromagnets. Choosing d_N to correspond to the antiferromagnetic coupling (Baibich *et al.*, 1988), the magnetization of neighbouring ferromagnets becomes antiparallel in the absence of a magnetic field. It can then be made parallel by applying a large enough magnetic field.

Let us analyze a magnetic multilayer FNFNFNFN... of the type described in Fig. 2.14.[27] For simplicity, let us assume that $\ell_{\rm sf}$ is much larger than the size of the system, so that we can totally disregard spin-flip scattering. In this case eqns (2.45) and (2.46) reduce to current conservation equations separately for each spin. This allows us to depict the conductance of the multilayer with a circuit that includes separately contributions from each spin, as in Fig. 2.15. Let us also assume that each ferromagnetic section has a resistance $R_M = 2R_f/(1 + P)$ for the majority spin and $R_m = 2R_f/(1 - P)$ for the minority spin, where R_f is the total resistance of each ferromagnet.[28] The quantity P describes the *spin polarization*, the spin dependence of both the density of states and the diffusion constant. Moreover, each normal-metal section has the resistance $2R_N$ (per spin). Now, assume we can have either an antiferromagnetic configuration of alternating majority spins (this is the *antiparallel* configuration) or a ferromagnetic configuration, where the majority and minority spins point in the same direction in each ferromagnet (*parallel* configuration). It is straightforward to compute the resistance of a segment shown in Figs. 2.14 and 2.15. In the parallel configuration we obtain

$$R_P = 2\frac{R_f^2 + R_f R_N + R_N(1 - P^2)}{R_f + R_N(1 - P^2)} \xrightarrow{R_f \gg R_N} 2R_f, \qquad (2.55)$$

whereas in the antiparallel case the resistance is

$$R_{AP} = 2R_N + \frac{2R_f}{1 - P^2} \xrightarrow{R_f \gg R_N} \frac{2R_f}{1 - P^2}. \qquad (2.56)$$

As mentioned above, if the ground-state configuration at a low magnetic field is antiparallel, it can be switched to the parallel configuration by applying a magnetic field. This reduces the resistance of the multilayer from R_{AP} to R_P, and the change can typically be several tens of percent. This *Giant Magnetoresistance* (GMR) was discovered by two groups in the 1980s (Baibich *et al.*, 1988; Binasch *et al.*, 1989), and is used in magnetic read-heads of most commercial hard disk drives. In 2007 this discovery was rewarded with the Nobel prize in physics for Peter Grünberg and Albert Fert.

[28]Note that the contributions from the spin channels are in parallel, so that the conductances should be summed over.

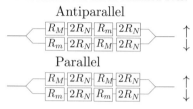

Fig. 2.15 Circuit description for the resistance of magnetic multilayers in different configurations. The upper path is for the 'up' spin and the lower path for the 'down' spin, chosen parallel to the spin quantization direction of the ferromagnets.

Complement 2.3 Spin relaxation mechanisms

All the spintronics experiments are hampered by spin relaxation. Therefore there has been a large amount of studies of different spin relaxation mechanisms in metals and semiconductors. As a result, four relevant mechanisms have been found: Elliot–Yafet mechanism arises from the spin-orbit interactions combined with momentum scattering, D'yakonov–Perel' mechanism is due to the coupling of spin-orbit interaction with the fact that spin up and spin down states for a given momentum are not exactly degenerate in systems lacking inversion symmetry, Bir–Aronov–Pikus mechanism is present in p-doped semiconductors, and the hyperfine interaction mechanism is due to the magnetic interaction between the electron magnetic moments with those of the nuclei. These processes are explained in detail by Žutić *et al.* (2004). The relative importance between these mechanisms depends on the temperature and the material in question. Quite generally, the resultant spin-flip scattering times are shorter at higher temperatures. For metals, the temperature dependence is rather weak and typical scattering times are of the order of tens or hundreds of ps (Jedema *et al.*, 2001), resulting in spin-flip lengths $\ell_{\rm sf} = \sqrt{D\tau_{\rm sf}}$ of a few hundred nanometres. In graphene, for example, the spin-orbit scattering is weak, and therefore the spin-flip times should there be much longer than in metals. Initial experiments (Tombros *et al.*, 2007) yielded $\ell_{\rm sf} \sim$ a few μm at room temperature, but also much longer spin-flip lengths are expected.

2.8   Thermoelectric effects

One of the direct applications of the Boltzmann-equation approach is the full treatment of thermoelectric effects, i.e., finding a relation between the electric and heat current to the applied voltage and temperature gradient. This relation can be described by the matrix equation

$$\begin{pmatrix} I \\ \dot{Q} \end{pmatrix} = \begin{pmatrix} G & -G\alpha \\ -G\Pi & G_{\text{th}} \end{pmatrix} \begin{pmatrix} L\nabla\mu/e \\ -L\nabla T \end{pmatrix}. \tag{2.57}$$

Here the diagonal terms G and G_{th} are the electrical and thermal conductances of the electron system. The length L of the wire is included in order to get the correct dimension for the coefficients. The off-diagonal components are generally called thermoelectric coefficients.

The coefficient α is called a *Seebeck* coefficient, and it describes a thermoelectric power. Assume we have a two-terminal system and we apply a temperature gradient across it. This temperature gradient gives rise to a current $I = -G\alpha L\nabla T$. Apply now an opposite voltage V that exactly cancels this current. From eqn (2.57) we see that this voltage has to be $V = \alpha L\nabla T$.

The *Peltier* coefficient Π describes the heat current $\dot{Q}$ for a given voltage V. It indicates the amount of heat transferred through the system as a response to a small voltage. As this effect is linear in voltage, the sign of the heat current follows that of the voltage, and as a result one can cool one part of the system and heat another one. At larger voltages, however, Joule heating (quadratic in V) starts to heat the sample throughout.

Boltzmann theory allows us to relate the various thermoelectric coefficients to more microscopic quantities. Let us concentrate on the linear regime, i.e., assuming only a small voltage or a small temperature gradient such that $eV/k_B, \Delta T \ll T$. Then the deviation from equilibrium is small and therefore necessarily $f(\mathbf{r}, E) \approx f^0(E; \mu(\mathbf{r}), T(\mathbf{r}))$. In this case, $\nabla f(\mathbf{r}, E) = \partial_\mu f^0 \nabla\mu + \partial_T f^0 \nabla T$. From eqns (2.24) and (2.25) multiplied by the cross-section A of the wire we get

$$G = \frac{A}{L} e^2 \int dE N(E) D(E) \partial_\mu f^0(E, \mu, T) \tag{2.58a}$$

$$G\alpha = \frac{A}{L} e \int dE N(E) D(E) \partial_T f^0(E, \mu, T) \tag{2.58b}$$

$$G\Pi = \frac{A}{L} e \int dE (E - \mu) N(E) D(E) \partial_\mu f^0(E, \mu, T) \tag{2.58c}$$

$$G_{\text{th}} = \frac{A}{L} \int dE (E - \mu) N(E) D(E) \partial_T f^0(E, \mu, T). \tag{2.58d}$$

In a normal metal the density of states and the diffusion constant are typically almost energy independent. We may express them in the form $N(E) \approx N_F + c_N(E - \mu)$ and $D(E) \approx D_0 + c_D(E - \mu)$ and retain only the lowest relevant order in the coefficients c_N and c_D.

For the diagonal coefficients we may neglect the small coefficients c_N

and c_D and obtain for the electrical conductance

$$G = e^2 N_F D_0 A/L, \qquad (2.59)$$

i.e., Drude conductance of a disordered wire, discussed in Sec. 1.2.1. The thermal conductance is given by the Wiedemann–Franz relation

$$G_{\text{th}} = \frac{A}{L} \frac{\pi^2}{3} k_B^2 N_F D_0 T = \mathcal{L}GT, \qquad (2.60)$$

as already discussed in Example 2.2.

The off-diagonal components rely on the small coefficients c_D and c_N. We obtain

$$\alpha = e\mathcal{L}G'T/G, \qquad (2.61)$$

where $G' \equiv e^2(c_D N_F + D_0 c_N)A/L$.[29] This is the *Mott law*. Finally, the Peltier coefficient satisfies the *Kelvin–Onsager relation* $\Pi = \alpha T$. The appearance of the coefficients c_D and c_N breaks the particle-hole symmetry between the energies above and below the Fermi level, and therefore the thermoelectric effects are sometimes used to justify (or unjustify) theoretical predictions for the electronic band structure in metals.[30] From these equations one can see that the thermoelectric coefficients are proportional to the factor $k_B T/E_F$ (as shown in Exercise 2.8), and are therefore quite small even at the room temperature.

A novel direction of research is to combine thermoelectrics with spintronics, i.e., the topics of Secs. 2.7 and 2.8. The recent experiments have, for example, shown a large *spin Seebeck effect* (Uchida *et al.*, 2008): spin accumulation created by a temperature gradient; and a *spin-dependent Peltier effect* (Flipse *et al.*, 2012), where a voltage across a ferromagnet–normal metal–ferromagnet system leads to cooling of the ferromagnet in a way that depends on the magnetic configuration of the system.

[29] As G' probes the energy dependence of the density of states and diffusion constant close to the Fermi sea, the quantity $G'/G = (\ln G)'$ is often called the logarithmic energy derivative of G. However, it should not be thought of as some kind of voltage derivative of the conductance.

[30] However, there are also other factors contributing to thermoelectric effects. The most important of them in normal conductors is phonon drag.

Further reading

Useful references on the semiclassical approach for electron transport are Chapters 13–16 in (Ashcroft and Mermin, 1976), Sec. 10.4 in (Plischke and Bergersen, 1994), (Nagaev, 1995) and (Sukhorukov and Loss, 1999). Thermal effects in nanoelectronics are reviewed in (Giazotto *et al.*, 2006). The fundamentals of spintronics have been reviewed by Žutić *et al.* (2004).

Exercises

(2.1) Show that the diffusion constant (see eqn (2.21) for the three-dimensional case) in two- and one-dimensional cases equals $D = v^2\tau/2$ and $D = v^2\tau$, respectively.

(2.2) Phonons in thin films may not always be strongly coupled to the phonons in the substrate. The thermal link between these two phonon systems is described by the Kapitza thermal resistance, which between two bulky phonon systems (dimensions $\gg$ phonon wavelength) is proportional to T^{-3}, i.e., $R_K = r_k T^{-3}$ with a system-dependent coefficient r_k. Assume an electron system in the mesoscopic film is heated with a small constant power P, resulting in an electron temperature $T_e = T_b + \Delta T_e$ and film phonon temperature $T_f = T_b + \Delta T_f$, $\Delta T_e, \Delta T_f \ll T_b$ slightly higher than the temperature T_b of the substrate. Linearize eqn (2.39) and find ΔT_e and ΔT_f.

(2.3) Assume a conductor with an unknown type of scattering described via the relaxation time τ. Starting from eqns (2.4), (2.8) and (2.10), show that in linear response the conductance is still given by the Drude form. Hint: Write the distribution function as $f(\mathbf{r}, \mathbf{p}) = f_0(\mathbf{p}) + g(\mathbf{p})$, where $f_0(\mathbf{p})$ is the Fermi function and to first approximation the correction $g(\mathbf{p})$ is independent of position. Note that around the Fermi level the momentum and the velocity are connected via the usual relation $\mathbf{p} \approx m^*\mathbf{v}$. Compute $g(\mathbf{p})$ from (2.4), and insert it in eqn (2.10) to obtain a formula for the current.

(2.4) Using the Drude formula with a general scattering time, show how the conductance depends on temperature when $\tau_{e-ph}(T)$ given by eqn (2.40) becomes shorter than the elastic scattering time τ_{el}. Often the relaxation rates $1/\tau$ for different types of scattering process add up. This is called the *Matthiessen's rule*.

(2.5) Assuming that the kernels $K_{ee}(E, E', \omega)$ and $K_{eph}(E, \omega)$ are functions of only ω, show that the electron–electron collision integral, eqn (2.29) vanishes if $f_E = f^0(E, T)$, a Fermi function, and that the electron–phonon collision integral, eqn (2.30) vanishes if $f_E = f^0(E, T_{ph})$, a Fermi function at the phonon temperature T_{ph}. For the latter case, assume that the phonon distribution n_E is the Bose function.

(2.6) Compute numerically the current–voltage curve of a normal-metal–superconductor tunnel junction,

when the normal metal is biased to the regime of strong non-equilibrium (Example 2.2). This corresponds to the results of the measurements in Example 2.4.

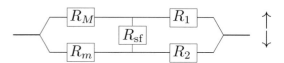

Fig. 2.16 Circuit describing giant magnetoresistance with some spin-flip scattering.

(2.7) **Spin-flip scattering effect on GMR**. The discussion of giant magnetoresistance in Example 2.6 disregards spin-flip scattering. The latter can be included in the circuit description by coupling the two spin channels by an additional resistor, as in Fig. 2.16, which also disregards the resistance from the normal-metal spacers. For an accurate description, however, one should split each resistor into many parts, each describing the resistance gathered within a single spin-flip scattering length. For a qualitative description, let us analyze only the case in Fig. 2.16. Calculate the resistance R_P in the parallel ($R_1 = R_M$, $R_2 = R_m$) and R_{AP} in the antiparallel ($R_1 = R_m$ and $R_2 = R_M$) case, and show how the GMR ratio $(R_{AP} - R_P)/R_P$ behaves as R_{sf} is varied. Note that with the same model you can describe the spin accumulation and its decay. How? What should be chosen for R_{sf} to describe a given spin relaxation time τ_{sf}?

(2.8) Calculate the thermopower from the Mott law by assuming the energy-dependent density of states obtained from a simple quadratic dispersion relation $\epsilon_p = p^2/(2m)$. How does it depend on the characteristics of the sample? Compare your result to the fundamental scale $\sim k_B/e = 86$ $\mu eV/K$ of the thermopower.

(2.9) **Question on a scientific paper**. Try to understand what was measured in Jedema *et al.*, *Nature* **410**, 345 (2001) by looking at its figures. The sample and measurement geometry are illustrated in Fig. 1. What causes the magnetic field dependence of the signal in Fig. 2? How is it related with spin accumulation? What is the length λ_N indicated in Fig. 3?

Scattering approach to quantum transport

The whole theory of mesoscopic electron transport started from the observation that the current through low-dimensional objects can be directly related to their scattering properties. The original idea is due to Rolf Landauer and dates back to his paper of 1957. This concept was refined by Markus Büttiker (1986) who provided the framework of the scattering theory explained in this chapter. This theory is often called by their originators the Landauer–Büttiker theory. The viewpoint in the scattering approach is completely different from the more traditional semiclassical approach outlined in the previous chapter. Whereas in the semiclassical approach one is typically interested in the local *conductivity* of some bulky material, the scattering approach directly calculates the *conductance* of a given conductor. This then allows the consideration of the properties of ballistic samples where conductivity is not well defined.

The essential steps in the scattering theory is to identify the scattering states inside the reservoirs, define the scattering matrix connecting the amplitudes of these scattering states via a scattering region under interest, and connect the properties of these scattering states to the current flowing through the scattering region. Below we go through these steps, although the rigorous definition of the scattering matrix is done only after the connection of the transmission probabilities to the current is made. Whereas the main idea usually is obviously to learn about the properties of the scattering region, during this process we also find out that for a given voltage there is a maximum current that a given scattering state can support. This maximum current corresponds to the maximum conductance for a single state, quantum of the conductance, $G_K = e^2/h$, or 38.74 microsiemens, or as it is usually specified, $1/G_K = 25812.8$ ohms.

The scattering theory is a versatile tool that can as such be used to study any kind of non-interacting conductors.[1] Inelastic scattering effects discussed in the previous chapter are outside this theory, but Sec. 3.5 shows how these can be phenomenologically included. However, contrary to the semiclassical Boltzmann equation approach, the scattering approach can deal with any kind of interference effects. In this chapter we discuss an example of this connected to resonant tunnelling (Sec. 3.4), and more examples are given in Ch. 4.

As shown in Ch. 6, besides the average current, the scattering theory

---

[1]Interactions can be included in a level of self-consistent mean field theory (see Secs. 3.6.2 and 8.2), where changes in the charge distribution alter the scattering potential.

can also be used to study the current fluctuations. Moreover, it serves as a starting point for many treatments of interacting systems.

3.1 Scattering region, leads and reservoirs

The scattering theory describes a structure with some number M reservoirs,[2] connected to each other via the *scattering region* (see Fig. 3.1). Between the reservoirs and the scattering region, the theory assumes semi-infinite *leads*. These leads are in a way a mathematical construction: they allow for defining the scattering states, and they mimic the physical properties of the reservoirs. As in the previous chapter, the quasiparticle reservoirs are characterized by their chemical potentials μ_i and temperatures T_i. We assume that the quasiparticles undergo inelastic scatterings inside the reservoirs such that the outgoing particles are uncorrelated and their energy distribution can be described by the Fermi function $f(E; \mu, T)$. Furthermore, we assume that the number of particles in the reservoirs is infinite, such that they can act as constant sources of particles, and their chemical potential is not affected by the removal or addition of particles. The leads are assumed semi-infinite and clean, to describe the fact that an electron entering the lead cannot scatter from them back into the scattering region.

3.1.1 Transverse modes in semi-infinite leads

Let us consider one of the semi-infinite leads. For simplicity, we model it as a waveguide, a perfect rectangular 3d wire with a finite width and thickness (say, x and y directions), but with an infinite length (in the z-direction), as in Fig. 3.2. The Schrödinger equation for the electrons in this wire in the free-electron approach is

$$\left[-\frac{\hbar^2}{2m} \nabla^2 + U(x,y,z) \right] \psi(x,y,z) = E\psi(x,y,z). \tag{3.1}$$

Here $U(x,y,z)$ describes the lead edges. To avoid backscattering within the leads, we assume that $U(x,y,z)$ is independent of the coordinate z. In this case, the equation separates into the transverse and longitudinal directions and the solution is of the form $\psi(x,y,z) = \chi(x,y)e^{ik_z z}$. The transverse functions obey the equation

$$\left[-\frac{\hbar^2}{2m}(\partial_x^2 + \partial_y^2) + U(x,y) \right] \chi_n(x,y) = \epsilon_n \chi_n(x,y). \tag{3.2}$$

After solving this with appropriate boundary conditions describing the edges of the leads, the wave number in the longitudinal direction is obtained from

$$k_z^n = \pm\sqrt{\frac{2m}{\hbar^2}(E - \epsilon_n)}. \tag{3.3}$$

The wave numbers k_z with finite imaginary parts correspond to evanescent waves which do not carry a current, and the real k_z describe currents

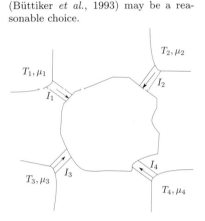

Fig. 3.1 Scattering theory. Quasiparticle reservoirs are connected to the scattering region via the ballistic (disorder-free) leads.

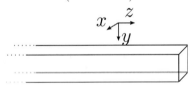

Fig. 3.2 Semi-infinite lead.

carried in the two directions in the wire, depending on the sign of k_z. This can be seen by considering the group velocity of state n in the direction of transport:

$$v_n = \frac{1}{\hbar}\frac{dE}{dk_z} = \frac{\hbar k_z}{m}, \qquad (3.4)$$

which has the same sign as k_z. Below, we refer to the states with $k_z > 0$ as 'incoming' states and to those with $k_z < 0$ as 'outgoing' states.

For the transverse modes one could envisage many different types of potential describing a wire with finite width and thickness. Often, which model is considered does not have a strong effect on the resultant physics. The simplest model is a wire with hard walls, i.e., $U(x,y) = 0$ inside the wire and infinity elsewhere. This yields the boundary conditions

$$\chi(0,y) = \chi(L_x, y) = \chi(x,0) = \chi(x, L_y) = 0, \qquad (3.5)$$

where L_x and L_y are the dimensions of the wire in the two directions. The resultant transverse mode wave functions are (see Fig. 3.3)

$$\chi_n(x,y) = A_n \sin(k_{n_x} x)\sin(k_{n_y} y), \qquad (3.6)$$

where $k_{n_{x/y}} = n_{x/y}\pi/L_{x/y}$. The corresponding energies are

$$\epsilon_n = \frac{\hbar^2}{2m}(k_{n_x}^2 + k_{n_y}^2). \qquad (3.7)$$

Conventionally the normalization of the states is chosen to be $A_n = \sqrt{v_n}$, such that each state carries a unit probability flux. Below, we refer to the transverse modes mostly only by their index n, referring to one particular combination of $\{n_x, n_y\}$.

Due to the assumption of an infinite length of the wire, in the z-direction the different modes have different dispersion relations (see Fig. 3.4)

$$\epsilon_{k_z^n} = \epsilon_n + \frac{\hbar^2 k_z^2}{2m}. \qquad (3.8)$$

A particular transverse state n can contribute to transport only if the transverse energy ϵ_n is below the Fermi level of the reservoir that feeds the particles into the states. We can thus define a function $M(E)$ that counts the number of modes with ϵ_n below some energy E:

$$M(E) \equiv \sum_n \theta(E - \epsilon_n), \qquad (3.9)$$

where $\theta(x)$ is the Heaviside step function.[3] Moreover, each mode can have a different probability T_n of transmission through a given scattering region. We can take this into account by defining

$$\bar{T}(E) \equiv \sum_n T_n \theta(E - \epsilon_n), \qquad (3.10)$$

i.e., the total transmission for modes with $\epsilon_n < E$. Likewise, the total reflection probability for the current-carrying modes is $\bar{R}(E) = M(E) - \bar{T}(E)$.

Fig. 3.3 Wave functions for the lowest-energy transverse modes in a two-dimensional wire (i.e., $L_y \to 0$) with hard-wall potentials.

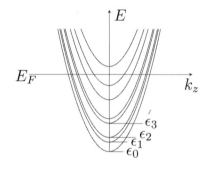

Fig. 3.4 Dispersion relation for the longitudinal k-vector in the different transverse modes. Only the modes with a cutoff energy $\epsilon_n < E_F$ can contribute to electric current. Here we thus have $M(E_F) = 7$.

[3] $\theta(x > 0) = 1$, $\theta(x < 0) = 0$.

Quasi-continuous k_z-values

In what follows, we aim to sum the contributions to the current from all the occupied k_z-states. In the case of an infinite wire, the spectrum is continuous and we need to find a way to convert the k_z-sums into integrals over k_z. Therefore, consider a finite one-dimensional wire of length L, and assume periodic boundary conditions, $\psi_z(z) = \psi_z(z + L)$. The Schrödinger equation is now solved with $\psi_z(z) = e^{ik_z z}$, $k_z = 2n\pi/L$. Therefore, the wave number spacing is $2\pi/L$, and hence any sum over the k_z-states (and spin σ) below can be converted with the prescription

$$\sum_{k,\sigma} g(k) \rightarrow 2 \text{ (for spin)} \times \frac{L}{2\pi} \int dk g(k), \qquad (3.11)$$

where $g(k)$ is an arbitrary function of k. This approximation works provided the function $g(k)$ is smooth on the scale of the wave number spacing and independent of spin.

3.1.2   Current carried by a transverse mode

Consider now a single incoming transverse mode m, occupied with some probability $f_L(E_k)$, the energy distribution function inside the (say, left) reservoir feeding the lead. A system with n electrons per unit length[4] carries a charge current env, where v is the velocity of the electrons. The electron density associated with a single electron in one $k = k_z^m$-state in a conductor of length L is $1/L$, and thus the incoming current in the wire (towards the scattering region) is

$$I_{\text{in}}^{L,m} = \frac{e}{L} \sum_k v_k f_L(E_k) = \frac{2e}{2\pi} \int dk \frac{1}{\hbar} \underbrace{\frac{dE}{dk}}_{v_k} f_L(E_k). \qquad (3.12)$$

Here we used eqn (3.11) for converting the sum over k into an integral over it. It is convenient to convert this integral into an integral over energy, defining the density of states for the one-dimensional case, $N(E) \equiv \frac{dk}{dE} = 1/(dE/dk)$. We find that the velocity and density of states cancel, yielding

$$I_{\text{in}}^{L,m} = \frac{2e}{h} \int_{\epsilon_m}^{\infty} dE f_L(E). \qquad (3.13)$$

The lower limit for the integral is the mode cutoff ϵ_m, below which this mode does not carry a current. Now summing over the modes yields the total incoming current,

$$I_{\text{in}}^{L} = \frac{2e}{h} \int_{-\infty}^{\infty} dE f_L(E) M(E), \qquad (3.14)$$

where $M(E)$ is the function counting the number of propagating modes, defined in eqn (3.9).

[4]Note that in such plane-wave states the electrons are completely delocalized over the lead length L.

3.1.3 Wire between two reservoirs

Fig. 3.5 The sum of the 'in' and 'out' currents amounts to the average current flowing in each lead.

Calculating the incoming current I_{in} alone would lead to a diverging integral unless $M(E \to -\infty) \to 0$ fast enough. However, it turns out that we do not need to know the behaviour of $M(E)$ at energies much below the Fermi level, as for the total current we need to include also the current carried in the opposite direction, due to the particles transmitted from the other lead and those reflected back into the same lead. Thus, the outgoing current is a sum of backscattered incoming particles and the current from the other reservoirs, transmitted with the probability $\bar{T}(E)$. In the case of two reservoirs, we hence get

$$I_{\text{out}}^L = \frac{2e}{h} \int_{-\infty}^{\infty} dE[\bar{R}(E)f_L + \bar{T}(E)f_R(E)], \qquad (3.15)$$

which assumes that the two leads have the same function $\bar{T}(E) = M - \bar{R}(E)$.[5] The total current is thus (see Fig. 3.5)

$$I = I_{\text{in}}^L - I_{\text{out}}^L = \frac{2e}{h} \int dE\bar{T}(E)(f_L(E) - f_R(E)) \overset{k_B T = 0}{=} \frac{2e}{h} \int_{\mu_R}^{\mu_L} dE\bar{T}(E). \qquad (3.16)$$

The last equality is obtained in the limit of a vanishing temperature. We hence see that even if all electrons in the conduction band contribute to the current, its average value is dictated by those with energy within the 'transport window', $E \in [\mu_R, \mu_L]$. Consider now the case where $\bar{T}(E)$ is constant for energies between μ_R and $\mu_L = \mu_R + eV$, or within the window of a few $k_B T$ around $\mu_{L/R}$. Then the resultant current is[6]

$$I = \frac{2e^2}{h}\bar{T}V. \qquad (3.17)$$

This is the *Landauer formula* relating transmission to the conductance $G = I/V$.

In of a ballistic wire $T = M$. In this case each spin-degenerate mode adds a contribution $G_0 = 2e^2/h = (12.9 \text{ k}\Omega)^{-1}$ to the total conductance of this ballistic wire. But this does not depend on the actual length of the wire! This is due to the lack of backscattering inside it. Such a property indeed leads to an infinite conductivity (conductance per unit length), but not to an infinite conductance! The finite resistance is due to the contacts of the larger reservoirs to the ballistic wires: also the dissipation takes place at these contacts. This resistance is hence typically called the *contact resistance*.

[5] As shown in Exercise 3.2, this is is in general valid and results from current conservation.

[6] This result is independent of temperature, as long as the above assumption is satisfied. This can be seen by using integral (A.64a).

Example 3.1 Number of modes for different types of mesoscopic conductor

Let us estimate the number of modes for some typical structures. This in principle depends on the assumption of the potential $U(x, y)$ describing the transverse modes, but we get a fair estimate already from the analysis with infinite walls. At a vanishing temperature, all the states up to the Fermi energy in principle contribute to the current. Thus, the requirement is $\epsilon_n < E_F$ or, for a three-dimensional wire,

$$M^{(3D)} = \frac{V}{(2\pi)^3} \int_0^{k_F} k^2 dk \int \sin\phi d\theta d\phi = \frac{4\pi V}{3\lambda_F^3} = \frac{4\pi A}{3\lambda_F^2}, \quad (3.18)$$

where k_F is the Fermi wave vector and $\lambda_F = 2\pi/k_F$ is the Fermi wavelength. This is the number of momentum states inside a spherical Fermi surface of radius p_F. The volume V should be chosen as the size of a wave packet $\sim \lambda_F$ in the z-direction times the cross-section A of the wire. In a metal, the typical λ_F is of the order of a few Ångström, and thus a 100 nm × 100 nm contact has some hundred thousand modes, indicating a contact resistance of the order of 100 mΩ, much smaller than that related to impurity scattering inside metal wires. In semiconductors and especially the two-dimensional electron gases (2DEGs) formed on their interfaces, or carbon nanostructures, however, λ_F can be a few tens of nanometres or longer, and the number of propagating modes can be reduced to only a few or in some cases just one. Moreover, with gates it is typically possible to control the local potential profile of the 2DEGs. With such control, the width W and therefore the number of modes can be controlled.

It should be mentioned that although the description of the scattering matrix is based on the modes in the leads, the number of the relevant modes (i.e., those with a non-zero transmission) to be included in it is fixed by the size of the scattering region (Scheer *et al.*, 1998). This is exemplified next.

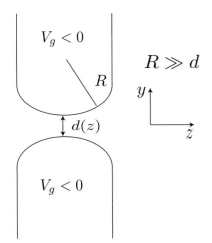

Fig. 3.6 Potential profile of a quantum point contact can be controlled via metallic gates. Because of these gates, the confining potential depends on the z-coordinate.

[7]In an adiabatic constriction the potential changes so smoothly that the mixing between the different transverse modes is weak.

[8]This type of an adiabatic ansatz is commonly referred to as the Wentzel–Kramers—Brillouin or WKB approximation.

3.1.4   Quantum point contacts

The conductance quantization can be clearly observed using quantum point contacts (QPC), where one uses metallic gates patterned on top of a two-dimensional electron gas to control the local electrostatic potential seen by the electrons (see Fig. 1.6 for a recent example).

Let us consider the adiabatic constriction depicted in Fig. 3.6.[7] This can be described by the Schrödinger equation of the form

$$\left[-\frac{\hbar^2}{2m} \left(\frac{\partial^2}{\partial y^2} + \frac{\partial^2}{\partial z^2} \right) + U(y, z) \right] \psi_E(y, z) = E\psi_E(y, z). \quad (3.19)$$

Here $U(y, z)$ describing the confining potential is assumed to be smooth in the z-direction, i.e., the changes in this potential are small within the length scale related to the width d of the constriction. The idea essentially is that the gate electrodes define a z-dependent width $d(z)$ of the constriction. Following the idea of these adiabatic changes, let us try an ansatz[8]

$$\psi_E(y, z) = \sum_n \chi_{nz}(y)\psi_{nE}(z), \quad (3.20)$$

where $\chi_{nz}(y)$ depends on the coordinate z parametrically. Inserting this into eqn (3.19) yields

$$\left[-\frac{\hbar^2}{2m} \left(\frac{\partial^2}{\partial y^2} \right) + U(y, z) \right] \chi_{nz}(y) = \epsilon_n(z)\chi_{nz}(y), \quad (3.21)$$

for the transverse mode wave functions. The treatment is hence very similar to that in Sec. 3.1.1, with the only difference that the transverse mode energies ϵ_n depend on z. Assuming hard-wall potential of the form

$$
U(y, z) = \begin{cases} 0, & \text{for } |y| < \frac{d(z)}{2} \\ \infty, & \text{otherwise} \end{cases} \tag{3.22}
$$

we find transverse modes of the form

$$
\chi_{nz}(y) = \sin\left(\frac{n\pi(y - \frac{d(z)}{2})}{d(z)} \right) \tag{3.23}
$$

with energies

$$
\epsilon_n(z) = \frac{\hbar^2}{2m} \left(\frac{\pi n}{d(z)} \right)^2 . \tag{3.24}
$$

Inserting eqn (3.20) into eqn (3.19) with the above form of $\chi_{nz}(y)$, multiplying the resultant equation by $\chi_{mz}(y)$ and integrating over y yields

$$
\left[-\frac{\hbar^2}{2m} \frac{\partial^2}{\partial z^2} + \epsilon_n(z) - E \right] \psi_{nE}(z) = \sum_m \Lambda_{nm} \psi_{mE}(z). \tag{3.25}
$$

The term

$$
\Lambda_{nm} = \frac{\hbar^2}{2m} \int dy \chi_{mz}(y) \left[2 \frac{\partial}{\partial z} \chi_{nz}(y) \frac{\partial}{\partial z} + \frac{\partial^2}{\partial z^2} \chi_{nz}(y) \right] \tag{3.26}
$$

mixes the different longitudinal functions ψ_{nE} and is a result of the fact that the potential depends on the z-coordinate.

But now we can employ the adiabatic approximation and assume that the changes in the transverse eigenfunctions in the z-direction are small, $\partial_z \chi_{nz}(y) \approx 0$.[9] In this limit $\Lambda_{nm} \approx 0$ and the WKB approximated solution for the longitudinal wave function is

$$
\psi_{nE}(z) \approx \psi_{nE}^{\text{WKB}}(z) = \frac{1}{[2m(E - \epsilon_n(z))]^{1/4}} e^{\left(i \int_{-\infty}^{\infty} dz' \sqrt{2m(E - \epsilon_m(z'))} \right)}. \tag{3.27}
$$

From this solution we can see that the transmission probability for modes with $E > \max_z[\epsilon_m(z)]$ is unity, whereas for other modes we get an exponentially decaying factor and the transmission probability is close to zero. Therefore, we can use the energies from eqn (3.24) to estimate the number M of propagating modes similarly to what is done in Sec. 3.1.1. The number is fixed in the position where the constriction is thinnest, $d_0 = \min_z d(z)$, which is where the eigenenergies $\epsilon_n(z)$ are largest. If $E > \epsilon_n(d_0)$, the mode is propagating and can carry a current. In the opposite case $E < \epsilon_n(d_0)$ we get an evanescent wave which has a negligible transmission probability. Controlling $d(z)$ through the gate voltage V_g in the electrodes thus allows for varying the number of modes, and hence the conductance of the QPC in a quantized fashion.

At the position where $E \approx \epsilon_n(d_0)$, i.e., where the number of modes is changing, we may expect some smearing of the conductance. At this

[9] These changes should be compared to the wavelength λ_m which is of the order of $1/\sqrt{2m(E - \epsilon_m)}$, i.e., $\lambda \partial_z \chi_{nz}(y) \ll 1$. This approximation fails at $E \approx \epsilon_m$, i.e., at the onset of a new mode, and hence there the transmission probability for the new mode is between 0 and 1.

point there indeed is some backscattering, which depends on the exact details of the potential profile. For the infinite-wall model the transmission probability is (Glazman *et al.*, 1988)

$$T_n = \frac{1}{1 + \exp\left[-z\pi^2\sqrt{2R/d_0}\right]}, \tag{3.28}$$

where $z = d_0\sqrt{2m(E-\epsilon_n)/\hbar}/\pi$ describes the deviation from the mode energy and $R = 2/(d^2M(z)/dz)$ the curvature of the constriction.

Another useful model for describing this effect is to expand the voltage profile in y- and z-directions (Büttiker, 1990; Kawabata, 1989),

$$U(y,z) = U_0 + \frac{1}{2}m\omega_y y^2 - \frac{1}{2}m\omega_z z^2. \tag{3.29}$$

The limit of an adiabatic constriction is realized for $\omega_y \gg \omega_z$, and the transverse eigenenergies are given by $\epsilon_n = U_0 + \hbar\omega_y(n+1)$. In this case the transmission function has a Fermi-function form

$$T_n(E) = \frac{1}{\exp[-2\pi(E-\epsilon_n)/(\hbar\omega_z)]+1}. \tag{3.30}$$

Hence the 'width' of the step in this case is roughly $\hbar\omega_z/(2\pi)$.

The conductance quantization was first observed in 1988 (van Wees *et al.*, 1988; Wharam *et al.*, 1988) in a 2DEG formed between GaAs and AlGaAs and patterned with gates (see Fig. 3.7).[10] Recently QPCs have been used as sensitive electrometers, which utilize the strong dependence of the QPC conductance on the changes in the (gate) potential in the environment. For example, with the setup in Fig. 1.6 one can measure the (time-dependent) charge state of the quantum dot by monitoring the current through the QPC (see also (Vandersypen *et al.*, 2004) and the paper studied in Exercise 8.8, where this technique is explained in detail). Moreover, QPCs can be used as electronic beam splitters (Liu *et al.*, 1998), which split the motion of an electron beam into the transmitted and the reflected parts.

[10] The quantization of the heat conductance G_{th} in similar type of systems was measured in the year 2006, see (Chiatti *et al.*, 2006). Where the conductance followed the simple quantized behaviour, the heat conductance was obtained by the Wiedemann-Franz law.

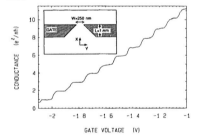

Fig. 3.7 Quantized conductance of a quantum point contact, as a function of the gate voltage that deforms the potential in the contact. (B. van Wees *et al.*, *Phys. Rev. Lett.*, **60**, 848 (1988), Fig. 2.) © 1988 by the American Physical Society.

3.2 Scattering matrix

To describe the transmission probability for the different propagating modes, it is useful to define a *scattering matrix* $s_{nm}^{\alpha\beta}$ that connects the amplitude of an incoming transverse mode m in lead β to that of an outgoing mode n in lead α. For an N-probe system with M_α propagating modes in lead α, this scattering matrix is thus $M_T \times M_T \equiv \sum_{\alpha=1}^{N} M_\alpha \times \sum_{\alpha=1}^{N} M_\alpha$-dimensional.

To be specific, consider a two-probe system with $M_1 = 2$ and $M_2 = 1$ (Fig. 3.8). Denote the amplitudes of the incoming modes by $a_{\alpha,n}$ and of the outgoing modes by $b_{\alpha,n}$. Then we have

Fig. 3.8 Scattering states of a two-probe system with $M_1 = 2$ and $M_2 = 1$.

$$\begin{pmatrix} b_{11} \\ b_{12} \\ b_{21} \end{pmatrix} = \begin{pmatrix} s_{11}^{11} & s_{12}^{11} & s_{11}^{12} \\ s_{21}^{11} & s_{22}^{11} & s_{21}^{12} \\ s_{11}^{21} & s_{12}^{21} & s_{11}^{22} \end{pmatrix} \begin{pmatrix} a_{11} \\ a_{12} \\ a_{21} \end{pmatrix}. \tag{3.31}$$

In many systems the dimension of the scattering matrix is huge, and it is better to refer to it by simply $s_{nm}^{\alpha\beta}$. The first index refers to the transverse mode n in lead α and the second to the transverse mode m in lead β. Below the lead indices are sometimes dropped for simplicity.

The scattering matrix may in principle be calculated if the potential profile $U(x, y, z)$ and the vector potential $\mathbf{A}$ within the total scattering region are known. Examples for such calculations are given in the exercises and in Complement 3.1. However, for now it is enough to know that the scattering matrix for a given system exists and can be rigorously defined.

3.2.1 Some properties of the scattering matrix

Transmission and reflection probabilities

From the scattering matrix it is easy to find the transmission and reflection probabilities T_{nm} and R_{nm} between modes n and m. They are given by

$$
\begin{aligned}
T_{nm}^{\alpha\beta} &\equiv |s_{nm}^{\alpha\beta}|^2, \quad (\alpha \neq \beta) \\
R_{nm}^{\alpha} &\equiv |s_{nm}^{\alpha\alpha}|^2.
\end{aligned}
\tag{3.32}
$$

As T_{nm} and R_{nm} are probabilities, they must lie between 0 and 1.

Unitarity

An important property of the scattering matrices can be shown from the requirement that in the stationary state the total amplitude for outgoing probability current is the same as that for the incoming probability current. For this, let us write the amplitudes of the incoming and outgoing states in the vector notation, $\{a\}$ and $\{b\} = S\{a\}$, where S is the total scattering matrix. The total amplitude for the probability currents hence is

$$
\sum_{\beta} \sum_{m \in \beta} |a_{\beta m}|^2 \equiv \{a\}^{\dagger}\{a\} = \{b\}^{\dagger}\{b\} \equiv \{a\}^{\dagger} S^{\dagger} S \{a\}.
\tag{3.33}
$$

Thus we obtain

$$
S^{\dagger} S = S S^{\dagger} = I,
\tag{3.34}
$$

where I denotes the unit matrix. Hence, any proper scattering matrix has to be unitary. Writing in terms of the scattering states, this implies

$$
\sum_{n,\alpha} |s_{nm}^{\alpha\beta}|^2 = 1 = \sum_{m,\beta} |s_{nm}^{\alpha\beta}|^2.
\tag{3.35}
$$

The first equality implies that the total probability for scattering from a given incoming state to one of the outgoing states is unity. The latter is the reverse: it tells that for any outgoing state, the total probability that it scattered from one of the incoming states is 1.

Time-reversal symmetry

Assume we have obtained the scattering matrix S from the solution of the Schrödinger equation[11]

[11]See Appendix A.4 on describing the magnetic field via the vector potential in quantum mechanics.

$$\left[\frac{(i\hbar\nabla + e\mathbf{A})^2}{2m} + U(\vec{r}) \right] \psi_B(\vec{r}) = E\psi_B(\vec{r}), \qquad (3.36)$$

describing our setup. From this we thus have $\{b\} = S\{a\}$. Now take the complex conjugate of eqn (3.36) and invert the magnetic field, $\mathbf{B} \to -\mathbf{B}$, i.e., $\mathbf{A} \to -\mathbf{A}$. We thus obtain

$$\left[\frac{(i\hbar\nabla + e\mathbf{A})^2}{2m} + U(\vec{r}) \right] \psi^*_{-B}(\vec{r}) = E\psi^*_{-B}(\vec{r}). \qquad (3.37)$$

As the differential operators for the two cases are the same, we find that $[\psi(\vec{r})^*]_{-B} = [\psi(\vec{r})]_B$. Taking the complex conjugate makes an outgoing wave an incoming one, and vice versa.[12] Thus, we have

[12]This is a natural consequence of the fact that $(e^{ik_z z})^* = e^{-ik_z z}$.

$$\{b\} = [S]_B\{a\}, \text{ and } \{a\} = [S^*]_{-B}\{b\}, \qquad (3.38)$$

or in other words,

$$[S^{-1}]_B = [S^*]_{-B}. \qquad (3.39)$$

Combining this with unitarity, $S^{-1} = S^\dagger$ yields

$$[S]_B = [S^T]_{-B}, \qquad (3.40)$$

where the superscript T indicates a transpose. For the transmission probabilities, this implies

$$[T_{nm}]_B = [T_{mn}]_{-B}. \qquad (3.41)$$

Thus, in the absence of a magnetic field, we have to have $T_{nm} = T_{mn}$. This symmetry is called *time-reversal symmetry*: for the system it is not relevant whether we calculate the probability for particles from modes m to travel to modes n in the forward or backward time direction.

This discussion assumes that the electrostatic potential $U(\vec{r})$ stays constant upon reversing the magnetic field. This assumption may break down locally in magnetic structures (in the presence of hysteresis), where the inversion of the field may lead to a different magnetization in the sample.

Two-probe scattering matrix

The scattering matrix connecting two leads has the form

$$S = \begin{pmatrix} r & t' \\ t & r' \end{pmatrix}. \qquad (3.42)$$

Here r and t are reflection and transmission matrices for the left lead, and r' and t' the same for the right lead. The reflection matrices are square matrices with dimensions $N_1 \times N_1$ and $N_2 \times N_2$, whereas the transmission matrices have the dimensions $N_1 \times N_2$ and $N_2 \times N_1$.

Example 3.2 Scattering matrix for a symmetric three-way junction
As an example, let us look for a scattering matrix coupling three incoming modes into the outgoing ones in the three-junction geometry depicted in Fig. 3.9. Such a scattering element is used, for example, in studies of the Aharonov–Bohm effect in ring-like structures (see Sec. 4.1 and Exercise 4.2). Let us assume that each lead has a single mode and the system is entirely symmetric upon the interchange of the lead indices. In the presence of time-reversal symmetry, the scattering matrix has thus the form

$$s_{3p} = \begin{pmatrix} r & t & t \\ t & r & t \\ t & t & r \end{pmatrix}, \tag{3.43}$$

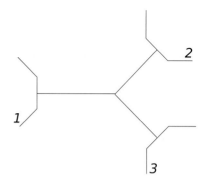

Fig. 3.9 Example three-junction geometry.

where r is the reflection and t the transmission amplitude. Unitarity of the scattering matrix gives us two conditions:

$$|r|^2 + 2|t|^2 = 1 \tag{3.44a}$$
$$tr^* + rt^* + |t|^2 = 0. \tag{3.44b}$$

The overall phase of the full scattering matrix has no observable consequences.[13] We can hence choose a real $r \in [0,1]$ and write $t = -t_0 e^{i\phi}$,[14] $t_0 \in [0,1]$. From eqn (3.44b) we then have

$$t_0[2r\cos(\phi) - t_0] = 0.$$

This has two solutions: $t_0 = 0$, in which case s_{3p} is an identity matrix, and $t_0 = 2r\cos(\phi)$. As the previous solution is included as a special case of the latter ($\phi = \pm\pi/2$), we can write down a parametrization for the reflection and transmission amplitudes in terms of the phase $\phi \in [-\pi/2, \pi/2]$:

$$t = -\frac{2\cos(\phi)}{\sqrt{1 + 8\cos(\phi)^2}}e^{i\phi}, \quad \text{and} \quad r = \frac{1}{\sqrt{1 + 8\cos(\phi)^2}}.$$

This means that for this type of the system there must always be backscattering: $r \in [1/3, 1]$.

[13] This is not the case when combining several scattering matrices, as shown below: in that case the phases of the individual scattering matrices matter!

[14] The minus sign is chosen for convenience.

3.2.2 Combining scattering matrices: Feynman paths

Sometimes it is convenient to separate the scattering problem into two or more parts containing scatterers with known scattering matrices. Then it is interesting to see how the total scattering matrix is related to the individual ones. Consider the example shown in Fig. 3.10. The left part of the system satisfies

$$\begin{pmatrix} b_L \\ b_{iL} \end{pmatrix} = \begin{pmatrix} r_L & t'_L \\ t_L & r'_L \end{pmatrix} \begin{pmatrix} a_L \\ a_{iL} \end{pmatrix} \tag{3.45}$$

and analogously for the right part after replacing the index L by R.

It is evident that the total scattering matrix is not simply a product of the two scattering matrices.[15] Assume the amplitude of the incoming

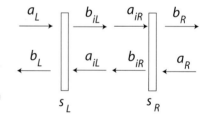

Fig. 3.10 System with scatterers in series.

[15] This property is characteristic for a *transfer matrix*, which connects the states from the left to those on the right side of a scatterer in contrast with the scattering matrix which connects incoming to outgoing modes.

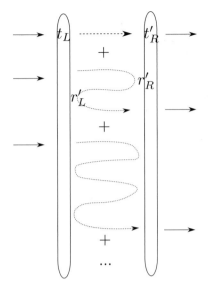

Fig. 3.11 Different 'Feynman' paths for an electron traversing between two scatterers.

state on the left part of the 'island' equals the amplitude of the outgoing state in the right part, i.e., we have $a_{iL} = b_{iR}$ and similarly for the incoming state of the right part, $a_{iR} = b_{iL}$. We may now eliminate a_{iR} and a_{iL} and find the total scattering matrix

$$\begin{pmatrix} b_L \\ b_R \end{pmatrix} = \begin{pmatrix} r & t' \\ t & r' \end{pmatrix} \begin{pmatrix} a_L \\ a_R \end{pmatrix}, \qquad (3.46)$$

where

$$\begin{aligned} t &= t'_R[I - r'_L r'_R]^{-1} t_L, \quad r = r_L + t'_L[I - r'_R r'_L]^{-1} r'_R t'_L \\ t' &= t'_L[I - r'_R r'_L]^{-1} t_R, \quad r' = r_R + t'_R[I - r'_L r'_R]^{-1} r'_L t_R. \end{aligned} \qquad (3.47)$$

Expanding the matrix inverses in geometric series shows that this result can be interpreted by the picture of the different 'Feynman paths' for the electron (see Fig. 3.11). For example, the transmission amplitude is (read the products in this expression from the right to the left)

$$t = t'_R t_L + t'_R (r'_L r'_R) t_L + t'_R (r'_L r'_R)(r'_L r'_R) t_L + \cdots . \qquad (3.48)$$

Hence, a transmission of a particle through this double-barrier structure can be seen as a sum of transmitting directly through both scatterers, transmitting from the left, reflecting from both scatterers and then transmitting to the right, and so on.

3.3   Conductance from scattering

For the calculation of the average current for a general scattering matrix, we could include the mode-dependent transmission in eqns (3.15,3.16) and obtain for the two-probe conductance in the case of energy-independent scattering

$$G = \frac{2e^2}{h} \sum_{n,m} T_{nm}. \qquad (3.49)$$

The generalization of this formula to a many-probe case is straightforward and is done below with the more general approach.

Chapter 6 uses the scattering theory to calculate fluctuations in mesoscopic conductors. To facilitate this discussion, I introduce here a somewhat more general approach connecting the current operators in the different leads to the scattering matrix by the use of the second quantized description of quantum mechanics (see Appendix A.1). Let us thus define the creation and annihilation operators $\hat{a}^\dagger_{\alpha n}$ and $\hat{a}_{\alpha n}$ for the incoming state n in lead α and similar operators $\hat{b}^\dagger_{\alpha n}$ and $\hat{b}_{\alpha n}$ for the outgoing state. These operators are related via the energy-dependent scattering matrix $s(E)$,[16]

$$\hat{b}_{\alpha n}(E) = \sum_\beta \sum_{m \in \beta} s^{\alpha\beta}_{nm}(E) \hat{a}_{\beta m}(E) \qquad (3.50a)$$

$$\hat{b}^\dagger_{\alpha n}(E) = \sum_\beta \sum_{m \in \beta} (s^{\alpha\beta}_{nm})^*(E) \hat{a}^\dagger_{\beta m}(E). \qquad (3.50b)$$

[16]In some cases it is necessary to describe the causality of scattering, i.e., the fact that the 'in' state precedes the 'out' state. This can be done by adding a small positive imaginary factor to the energy in the scattering matrix appearing in $\hat{b}_{\alpha n}$ and a corresponding negative imaginary factor in $\hat{b}^\dagger_{\alpha n}$. For the description of causality and its connection with the scattering matrices, see (Salo *et al.*, 2006).

In this relation, the operators $\hat{b}(E)$ and $\hat{a}(E)$ are Fourier transforms of the corresponding Heisenberg operators $\hat{b}(t)$ and $\hat{a}(t)$. The fact they are evaluated at the same energy is a consequence of the assumption of only elastic scattering inside the scattering region.

The current operator in lead α at time t is expressed via these operators as (see Appendix B for the derivation)

$$\hat{I}_\alpha = \frac{2e}{h} \sum_{n \in \alpha} \int dE dE' e^{i(E-E')t/\hbar} \left[\hat{a}^\dagger_{\alpha n}(E) \hat{a}_{\alpha n}(E') - \hat{b}^\dagger_{\alpha n}(E) \hat{b}_{\alpha n}(E') \right],$$
(3.51)

where the notation $n \in \alpha$ refers to the modes n in lead α. This equation can be written in terms of only operators $\hat{a}$ and $\hat{a}^\dagger$ by defining a matrix[17]

$$A^{\beta\gamma}_{mn}(\alpha, E, E') = \delta_{mn}\delta_{\beta\alpha}\delta_{\gamma\alpha} - \sum_j (s^{\alpha\beta}_{jm})^*(E) s^{\alpha\gamma}_{jn}(E').$$
(3.52)

This matrix satisfies (proven in Exercise 3.4)

$$\mathrm{Tr}[A^{\beta\beta}(\alpha, E, E)] \equiv \sum_n A^{\beta\beta}_{nn}(\alpha, E, E) = -\mathrm{Tr}[(s^{\alpha\beta}(E))^\dagger s^{\alpha\beta}(E)]$$
$$\equiv -\bar{T}_{\alpha\beta}(E), \quad \alpha \neq \beta \qquad (3.53a)$$
$$\mathrm{Tr}[A^{\alpha\alpha}(\alpha, E, E)] = M_\alpha(E) - \bar{R}_\alpha(E),$$

where $\bar{T}_{\alpha\beta}$ is the total transmission probability from β to α (summed over the modes), and $\bar{R}_\alpha = \mathrm{Tr}[(s^{\alpha\alpha})^\dagger s^{\alpha\alpha}]$ is the total reflection probability for electrons in lead α. The current operator can thus be expressed by

$$\hat{I}_\alpha(t) = \frac{2e}{h} \sum_{\beta\gamma} \sum_{m \in \beta, n \in \gamma} \int dE dE' e^{i(E-E')t/\hbar} \hat{a}^\dagger_{\beta m}(E) A^{\beta\gamma}_{mn}(\alpha, E, E') \hat{a}_{\gamma n}(E').$$
(3.54)

In this chapter we concentrate only on the average current $\langle \hat{I}_\alpha \rangle \equiv \mathrm{Tr}[\rho \hat{I}_\alpha]$, where ρ is the density matrix for the system. For this, we only need to know the statistical average of the number operator,

$$\langle \hat{a}^\dagger_{\alpha n}(E) \hat{a}_{\beta m}(E') \rangle = \delta_{mn}\delta_{\alpha\beta}\delta(E - E') f_\alpha(E).$$
(3.55)

This defines the distribution function $f_\alpha(E)$ in reservoir α. Thus, the average current in lead α is

$$\langle \hat{I}_\alpha \rangle = \frac{2e}{h} \sum_\beta \int dE \mathrm{Tr}[A^{\beta\beta}(\alpha, E, E)] f_\beta(E)$$

$$= \frac{2e}{h} \int dE [(M_\alpha - \bar{R}_\alpha) f_\alpha(E) - \sum_{\beta \neq \alpha} \bar{T}_{\alpha\beta} f_\beta(E)] \qquad (3.56)$$

$$\approx \frac{2e}{h} \left[(M_\alpha - \bar{R}_\alpha)\mu_\alpha - \sum_{\beta \neq \alpha} \bar{T}_{\alpha\beta}\mu_\beta \right],$$

where the second equality uses eqn (3.53). The latter approximate equality is valid in the linear response limit where the total scattering probabilities are independent of energy within the energy region of width

[17]Here we only consider the case $E = E'$ related with stationary average current. However, description of finite-frequency noise (eqn (6.16) in Ch. 6) or time-dependent transport (see Sec. 3.6.1) in general assumes $E' \neq E$.

$\max(k_B T, |\mu_\alpha - \mu_\beta| = e|V_{\alpha\beta}|)$ around the Fermi level. Equation (3.56) is the most general form for the average current in a system containing only elastic scattering, and can be used to construct the entire conductance matrix $G_{\alpha\beta} = edI_\alpha/d\mu_\beta$ once the scattering matrix is known.

For a two-probe system we have $M_1 - \bar{R}_1 = \bar{T}_{12}$ and thus

$$\langle \hat{I}_\alpha \rangle = \frac{2e}{h} \int dE \bar{T}_{12}(f_\alpha - f_\beta) \approx \frac{2e^2}{h} \bar{T}_{12} V, \qquad (3.57)$$

i.e., the same as obtained in eqn (3.17) following more heuristic arguments.

Example 3.3 Many-probe examples

Voltage probe

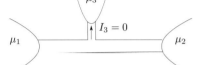

Fig. 3.12 Voltage probe is a probe into which the current vanishes.

A voltage probe is a terminal into which the average current vanishes. Consider a three-terminal system, where probe 3 is a voltage probe (see Fig. 3.12). In the linear response regime, the current I_3 is

$$I_3 = \frac{2e}{h}[(M_3 - \bar{R}_3)\mu_3 - \bar{T}_{31}\mu_1 - \bar{T}_{32}\mu_2]. \qquad (3.58)$$

Using unitarity of the total scattering matrix, we have $M_3 - \bar{R}_3 = \bar{T}_{31} + \bar{T}_{32}$. Now requiring that $I_3 = 0$ we get the potential in the probe,

$$\mu_3 = \frac{\bar{T}_{31}\mu_1 + \bar{T}_{32}\mu_2}{\bar{T}_{31} + \bar{T}_{32}}. \qquad (3.59)$$

This is a result that could have also been obtained from elementary circuit analysis with conductances proportional to $\bar{T}_{\alpha\beta}$, and it can be extended in a straightforward manner to the case of more than three terminals.[18] However, this result applies also to ballistic systems, not only classical resistors! Besides modelling real voltage probes, this type of extra probes are used to model dephasing, i.e., loss of quantum interference effects as discussed in Sec. 3.5.

[18]See Exercise 3.7.

Two-probe vs. four-probe setup

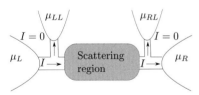

Fig. 3.13 Four-probe setup.

Originally, Landauer did not obtain the result of eqn (3.49) (Landauer, 1981). Rather, he calculated the four-probe resistance, which can be derived as follows. Consider the four-probe setup of Fig. 3.13. Assume we attach two voltage probes in the leads, in the region before the scattering takes place. The electron distribution function of mode n in the left lead is of the form[19]

$$f_{LL}(E_n, k_z > 0) = f_L(E_n),$$
$$f_{LL}(E_n, k_z < 0) = \sum_m |r_{nm}|^2 f_L(E_n) + \sum_m |t'_{nm}|^2 f_R(E_n),$$

[19]Note that as the lead is considered ballistic, Boltzmann equation (2.6) is linear and therefore the energy distribution function of a given mode with a given direction depends only on the distribution function in one of the reservoirs, as discussed in Example 2.1.

where $f_{L/R}$ is the distribution function of the left/right reservoir, and r_{nm} and t'_{nm} are the elements of the scattering matrix connecting other modes to mode n. Averaging this over the directions yields

$$f_{LL}(E) = \frac{1}{2M_L}[(M_L + \bar{R}_L)f_L(E) + \bar{T}f_R(E)]. \qquad (3.60)$$

Now, the chemical potential of the lead can be defined through

$$\mu = \int_{-\infty}^{\infty} dE(f(E) - f^0(E, \mu = 0)). \tag{3.61}$$

This yields for the left lead $2M_L\mu_{LL} = (M_L + \bar{R}_L)\mu_L + \bar{T}\mu_R$ and for the right lead $2M_R\mu_{RL} = (M_R + \bar{R}_R)\mu_R + T\mu_L$. Hence we obtain

$$\mu_{LL} - \mu_{RL} = \frac{M_L\bar{R}_R + M_R\bar{R}_L}{2M_L M_R}. \tag{3.62}$$

The current divided by the potential difference between the leads is thus given by

$$G_{4p} = \frac{2e^2}{h}\bar{T}\frac{\mu_L - \mu_R}{\mu_{LL} - \mu_{RL}} = \frac{2e^2}{h}\frac{2M_L M_R\bar{T}}{M_L\bar{R}_R + M_R\bar{R}_L} \stackrel{M_L=M_R,\bar{R}_L=\bar{R}_R}{=\!=\!=} \frac{2e^2}{h}\frac{M\bar{T}}{\bar{R}}. \tag{3.63}$$

This is the original Landauer formula. It could in principle be measured by attaching extra voltage probes in the leads as indicated in Fig. 3.13.[20] Therefore, this conductance is referred to as the four-probe conductance. It has the property that as the reflection probability $R \to 0$, the conductance becomes infinite. Now, the two-probe formula can be obtained by summing the resistances of the contacts and the four-probe resistance.[21]

[20] Note, however, that the measurement result depends on the exact coupling of these voltage probes to the different modes in the leads.

[21] In the asymmetric case $M_L \neq M_R$ the contact conductance is $G_{\text{contact}} = 2M_L M_R/(M_L + M_R) \times 2e^2/h$, i.e., as if there would be two half-contact resistances $h/(e^2 M_{L/R})$ in series.

3.3.1 Diffusive wire and Drude formula

The Drude conductivity (1.9) can be roughly argued from the Landauer–Büttiker formula (3.49) by assuming that the average transmission probability for an eigenmode of scattering is[22]

$$\langle T \rangle \sim \frac{\ell_{\text{el}}}{L + \ell_{\text{el}}} \stackrel{L \gg \ell_{\text{el}}}{\approx} \frac{\ell_{\text{el}}}{L}. \tag{3.64}$$

[22] A scattering eigenmode is an eigenvector of the scattering matrix with eigenvalue T_n. Expressed in the basis of its eigenmodes, the scattering matrix is obviously diagonal.

Combining this with the number of modes in the wire (eqn (3.18)), $M = 4\pi A/(3\lambda_F^2)$, in the limit $L \gg \ell_{\text{el}}$ we obtain the Drude conductivity:

$$\sigma = GL/A = \frac{4e^2\ell_{\text{el}}}{3\hbar\lambda_F^2} = \frac{e^2 m p_F v_F \ell_{\text{el}}}{3\pi^2\hbar^3} = e^2 N_F D, \tag{3.65}$$

which uses the density of states calculated in eqn (1.5) and the diffusion constant from eqn (2.21). Besides the average, the whole distribution of transmission eigenvalues can be calculated with the *random matrix theory* (Nazarov, 1994; Beenakker, 1997). Its idea is based on the fact that the random potential on the wire leading to scattering results into a random scattering matrix. Therefore, conductances of systems containing such randomness[23] can be linked to the properties of ensembles of random matrices satisfying certain symmetries, such as unitarity, or symmetry under transpose in the time-reversal case. For example, for a diffusive wire, the random matrix theory gives the probability density of transmission T,

[23] Besides diffusive wires, where randomness is due to the random potential, chaotic cavities where the shape of the scattering region leads to random electron trajectories are typically considered.

$$\rho(T) = \frac{\pi\hbar G}{e^2}\frac{1}{T\sqrt{1-T}}, \tag{3.66}$$

where G is the conductance of the wire. This distribution is useful, for example, in the description of current fluctuations; see Ch. 6.

[24]For a more thorough treatment, see (Datta, 1995).

Complement 3.1 Calculating the scattering matrix

Above and in the exercises, there are examples of calculating the scattering matrices of some model systems. But also the general problem for an arbitrary system can be calculated, at least numerically. The numerical methods typically rely on Green's-function methods on which we give a brief idea in the following.[24]

The (retarded) Green's function for the Schrödinger equation describing the whole scattering system obeys the equation

$$(E + i\eta - H(x))G^R(x, x') = \delta(x - x').$$ (3.67)

Here $H(x)$ is the Hamiltonian specified in the real-space coordinates. Now any numerical method would discretize the spatial coordinates, or start from the tight-binding description (see, e.g., (Ashcroft and Mermin, 1976) of the electrons in the conductor. In this case eqn (3.67) is a matrix equation with a formal solution

$$G^R = [(E + i\eta)I - H]^{-1},$$ (3.68)

where I is a unit matrix. The problem is that in the presence of the infinite leads H is an infinite-dimensional matrix, and therefore the problem is not yet computationally well defined. The solution to this is to divide the matrix G into parts,

$$G^R = \begin{pmatrix} G_l & G_{lS} \\ G_{Sl} & G_S \end{pmatrix} \equiv \begin{pmatrix} (E + i\eta)I - H_l & \tau_{lS} \\ \tau_{Sl} & (E + i\eta)I - H_S \end{pmatrix}^{-1}.$$ (3.69)

Here $(E + i\eta)I - H_l = g_l^{-1}$ represents the isolated leads with Green's functions g_l^{-1}, τ_{lS} and τ_{Sl} are the connections between the leads and the scattering region, and $(E + i\eta)I - H_S = g_S^{-1}$ describes an isolated scattering region. As a consequence the problem is reduced to a part with an infinite matrix, but a well-known system (clean, semi-infinite lead), and to a non-trivial but finite-dimensional scattering region. The previous can be calculated exactly, yielding (Datta, 1995)

$$G_l = \sum_\alpha G_l^\alpha, \quad G_l^\alpha(\vec{r}_\perp, z; \vec{r}_\perp', z) = -\sum_{n \in \alpha} \frac{2\sin(k_n^\alpha z)}{\hbar v_n^\alpha} |\chi_{n,\alpha}(\vec{r}_\perp)\rangle e^{ik_n^\alpha z} \langle\chi_{n,\alpha}(\vec{r}_\perp')|,$$ (3.70)

where v_n^α and k_n^α are the group velocity and the longitudinal wave number for mode n in lead α, respectively, z is the coordinate in the longitudinal and $\vec{r}_\perp = (x, y)$ in the transverse direction of the lead, and $|\chi_{n,\alpha}\rangle$ are the transverse mode wave functions. On the other hand, the now finite Green's function G_S of the scattering region can be calculated using numerical methods; for example, by inverting directly the finite-dimensional matrix of the type in eqn (3.68).[25] The coupling matrices G_{lS} and G_{Sl} to the leads are also finite-dimensional, as only the ends of the leads need to be coupled to the ends of the scattering region.

Once the total Green's function is obtained, the scattering matrix can be calculated from[26]

$$s_{nm}^{\alpha\beta} = -\delta_{nm}^{\alpha\beta} + i\hbar\sqrt{v_n^\alpha v_m^\beta}\langle\chi_{n,\alpha}|G^R|\chi_{m,\beta}\rangle.$$ (3.71)

This method allows calculating the transmission through any given scattering region. These types of method have been used for simulating many kinds of mesoscopic junctions, in particular molecular junctions and those based on

[25]Also faster 'recursive' methods exist, taking into account the typical nearest-neighbour coupling in the tight-binding model.

[26]For derivation, see (Datta, 1995) or (Fisher and Lee, 1981).

graphene.[27] Moreover, the same method can, for example, be used to study the properties of strongly disordered systems (Kramer and MacKinnon, 1993).

An alternative approach to evaluating properties of disordered systems using the scattering matrix approach is to use the *random matrix theory* as discussed in Sec. 3.3.1

[27]There are publicly available tools for this; see for example (Waintal, 2011) or (Datta, 2009).

3.4   Resonant tunnelling

Section 3.2.2 discusses how the scattering matrix of a two-barrier system can be cast into the form of the sum over the 'Feynman' paths connecting the left to the right. Let us add a (seemingly) small twist to the game, assuming that the particle obtains a dynamic phase $\phi = \sqrt{2mE}d/\hbar$ while traversing between the two scatterers with distance d between them. For simplicity, let us consider the single-mode case[28] and denote the transmission and reflection amplitudes as in Sec. 3.2.2. The Feynman-path picture gives us for the transmission amplitude from left to right:

[28]In the many-mode case the total transmission probability is a sum over the different modes, each contributing with in principle different t_1 and t_2, but with the same dynamic phase.

$$t_{12} = t_2' e^{i\phi} t_1 + t_2' e^{i\phi} r_1' e^{i\phi} r_2' e^{i\phi} t_1 + \cdots = t_2' e^{i\phi} \sum_{n=0}^{\infty} (e^{2i\phi} r_1' r_2')^n t_1$$

$$= \frac{t_2' t_1 e^{i\phi}}{1 - e^{2i\phi} r_1' r_2'}.$$

The same calculation is done in Exercise 3.5 by starting from the full scattering matrix. The total transmission probability $T_{12} = |t_{12}|^2$ is hence

$$T_{12} = \frac{T_1 T_2}{1 + R_1 R_2 - 2\sqrt{R_1 R_2} \cos(2\phi)}, \tag{3.72}$$

where $T_i = |t_i|^2 = 1 - R_i = 1 - |r_i|^2$ $(i = 1, 2)$ are the transmission probabilities through the individual scatterers.

If the transmission probabilities for individual scatterers are small, $T_i \ll 1$, for most phases ϕ the combined transmission probability through the two scatterers is even smaller, of the order of $T_1 T_2$. However, whenever $\phi \approx n\pi$, the denominator of the expression (3.72) becomes very small and T_{12} becomes large, of the order of unity (see Fig. 3.14 for the case with symmetric scatterers). This is an example of *resonance*, and this system is known as the electronic Fabry–Perot interferometer in analogy with the optical counterpart.[29]

Let us investigate the transmission probability close to the resonance, $\phi \approx n\pi$, i.e. $E \approx \epsilon_n = (\hbar n\pi)^2/(2md^2)$. There we can expand the $\cos(\phi)$ term by using

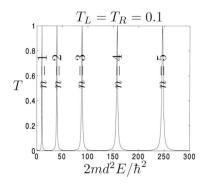

$T_L = T_R = 0.1$

Fig. 3.14 Transmission probability for resonant tunnelling as a function of electron energy.

[29]In fact, the same derivation applies for photons as well.

$$\cos(2\phi(E)) = \cos(2(\phi(E) - \phi(\epsilon_n))) \approx 1 - 2(\phi(E) - \phi(\epsilon_n))^2$$

$$\approx 1 - 2[\phi'(E)(E - \epsilon_n)]^2 = 1 - \frac{md^2}{\hbar^2 \epsilon_n}(E - \epsilon_n)^2. \tag{3.73}$$

Moreover, let us approximate

$$1 + R_1 R_2 - 2\sqrt{R_1 R_2} \approx (T_1 + T_2)^2/4,$$

valid for $T_i \ll 1$. Plugging these approximations into eqn (3.72) yields the transmission

$$T_{12} \approx \frac{T_1 T_2}{(T_1 + T_2)^2/4 + \frac{2md^2}{\hbar^2 \epsilon_n}(E - \epsilon_n)^2}.$$

We can proceed to define characteristic energy scales for the individual scatterers via $\Gamma_i = T_i \hbar v_n/(2d)$ and $v_n = \sqrt{2\epsilon_n/m}$. The factor $v_n/(2d)$ is an 'attempt frequency', i.e., the frequency with which the particle hits the scatterers once it is inside the island. Using these definitions we find a Breit–Wigner form for the transmission around the resonance:[30]

$$T_{12} = \frac{\Gamma_1 \Gamma_2}{(E - \epsilon_n)^2 + (\Gamma_1 + \Gamma_2)^2/4}. \tag{3.74}$$

Our model system is nothing else than a quantum dot with discrete energy levels and 'broadening' or inverse lifetime $\Gamma_1 + \Gamma_2$. I postpone further discussion about quantum dots to Ch. 8.

3.5   Models for inelastic scattering and dephasing

By construction, scattering theory assumes that the scattering region contains only elastic scattering, i.e., the incoming modes are connected to the outgoing modes with the same energy. Direct extensions to include a general scattering matrix $s(E, E')$ between different-energy states are complicated.[31] However, the inelastic effects can be modelled phenomenologically by coupling to the system extra fictitious probes (see Fig. 3.15) which mimic inelastic scattering. To be in any way physical, these probes have to conserve the average charge current—a constraint similar to that imposed on voltage probes. There may also be additional constraints, depending on the type of scattering one wishes to model. The three typically applied models are[32]

(1) 'Quasielastic' scattering, i.e., 'pure dephasing'. Adding voltage probes on the way of the current allows us to measure the potential at different points of the structure. Such measurements lead to dephasing. To model only 'pure' dephasing, these probes have to conserve the current for each energy. Then the entire distribution function in the fictitious probe(s) has to be determined from the requirement of current conservation for each energy (for an example, see Exercise 3.9).

(2) Electron–electron scattering can be modelled by assuming that the fictitious probe conserves both the charge and energy current, but not necessarily current for each energy. This constraint then fixes the potential and the temperature of the fictitious probe.

(3) Models where only the total net particle current into the fictitious probe is assumed to vanish describe effectively electron–phonon scattering or other types of inelastic scattering where the total

[30]See (Breit and Wigner, 1936). This work considered a completely different type of a resonance, occurring in neutron radiation.

[31]Except in the case of coherent driving; see Sec. 3.6.1.

[32]For further details, see (Blanter and Büttiker, 2000).

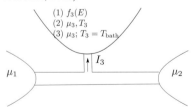

Fig. 3.15 Model for a fictitious dephasing probes (on top) and the different quantities fixed depending on which scattering process is described. To describe a situation where the length L between contacts 1 and 2 exceeds the scattering length ℓ described by the dephasing probe, L/ℓ such probes should be included in the setup.

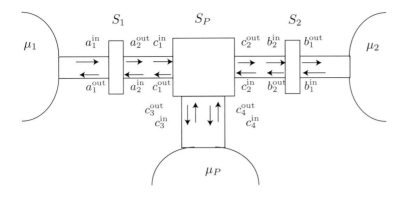

Fig. 3.16 Toy model for partial dephasing in a resonant tunnelling structure.

electron system loses or gains energy to/from some 'bath' with a fixed temperature. In this case, only the potential of the fictitious probe needs to be set from the current conservation.

An example of modelling the effect of dephasing with the extra probes is discussed in Exercise 3.9. It is based on the model described in Fig. 3.16. It has to be emphasized that such models are phenomenological. Using them one can describe some general features related to dephasing or relaxation. However, the quantitative details of these scattering effects have to be computed from the microscopic theory (see, for example, Complements 2.1, 2.2 and 6.1). Note that the dephasing probes can also mimic the effects of real probes in the setup (Tsuneta *et al.*, 2009).

The role of dephasing probes and probes describing inelastic scattering is especially relevant when considering higher-order fluctuations or correlations of the current (this is the topic of Ch. 6). For a further discussion, see the review (Blanter and Büttiker, 2000).

3.6 Further developments

The scattering approach is nowadays a well-established tool for modelling the (quantum) transport properties of various types of nanostructures. Whereas its basics were developed already by the 1980s, the research on its extensions continues even today. Below, I discuss briefly a few examples of the more recent developments. I introduce only a few specific ideas, and point an interested reader to the references.

3.6.1 Time-dependent transport

Consider a system where either the scattering potential oscillates in time or one of the leads is driven with a time-dependent periodic voltage. Denote the frequency of these oscillations by ω. In the case of stationary oscillations, all observables of the system must be periodic functions with the fundamental period $2\pi/\omega$. An electron with energy E entering the scattering region may in this situation gain or lose energy quanta in units of $\hbar\omega$. This thus connects incoming states at energy E with outgoing states at energies $E + n\hbar\omega$, where n is an arbitrary integer. Therefore, the generalization of the scattering approach to this case requires *Floquet scattering theory*, where the scattering matrices $S_{F,\alpha\beta}(E + n\hbar\omega, E)$ describe the probability amplitudes for electrons with energy E entering the scattering region through lead β having absorbed n energy quanta before exiting to lead α. The case of an ac potential in the leads is discussed in Exercise 3.8, the general formulation is presented in (Büttiker *et al.*, 1994), and the relation to quantum pumping in (Brouwer, 1998) and (Moskalets and Büttiker, 2002). A specific example of the calculation of the Floquet scattering matrix is presented, for example, by Moskalets and Büttiker (2008).

3.6.2 Non-linear transport

The Landauer–Büttiker conductance formula, eqn (3.17), typically implicitly assumes that the transmission probability does not depend on the voltage. This is true for the *linear response* regime, where the applied voltage has no effect on the distribution of charge in the device. An extension to the non-linear regime has been discussed, for example, by Christen and Büttiker (1996). In general, the response of a current in lead α, eqn (3.56), to a change in voltages $V_\beta = \mu_\beta/e$ in leads β is of the form

$$I_\alpha = \sum_\beta G_{\alpha\beta} V_\beta + \sum_{\beta\gamma} G_{\alpha\beta\gamma} V_{\beta\gamma} + \ldots, \tag{3.75}$$

where the neglected terms are of the third or higher order in the voltages. The non-linear conductances $G_{\alpha\beta\gamma}$ are related to energy or voltage dependencies of the scattering matrix via

$$G_{\alpha\beta\gamma} = \frac{e^2}{h} \int \frac{\partial f(E)}{\partial E} \left(2\partial_{V_\gamma} \text{Tr}[A^{\beta\beta}(\alpha, E, E)] + e\partial_E \text{Tr}[A^{\beta\beta}(\alpha, E, E)]\delta_{\beta\gamma} \right). \tag{3.76}$$

An energy-dependent scattering process (the second term) naturally leads to non-linear transport. A more interesting question is how to access the first term: a voltage-dependent scattering matrix. Namely, variations of the electrochemical potentials μ_β in the contacts induce varying charge densities inside the scatterer, which on the other hand change the potential configuration responsible for the scattering. For an example of such a model, see (Sánchez and Büttiker, 2004).

3.6.3 Application to magnetic systems

The scattering matrix approach can be used as a tool to describe transport through magnetic systems where the conductances become spin-dependent. Consider for example a structure of the type shown in Fig. 3.17, where the scattering region (assumed non-magnetic) connects two or more ferromagnets with non-collinear magnetization. Because of spin polarization in the ferromagnets, the scattering matrices describing this structure become non-trivial 2×2 matrices in the spin space. The

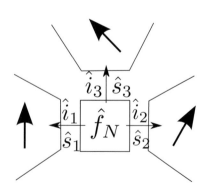

Fig. 3.17 Structure containing ferromagnets with non-collinear magnetization, i.e., non-parallel magnetization vectors, connected by a non-magnetic scattering region.

generalization of the current formula (e.g., eqn (3.56)) is straightforward, as spin is then included as another mode of transmission and the prefactor 2 is suppressed from the equations. However, in this case also the *spin current* becomes an interesting quantity. For a two-probe system the scattering matrix is of the form eqn (3.42), with spin-dependent reflection and transmission matrices.

Assume we can pick a node, a point in the scattering region through which the current flows. The non-equilibrium distribution function inside this node is in general a 2×2 matrix $\hat{f}_N$ in spin space, its particular form depending on the chosen reference for the spin quantization axis. Related to this distribution function, we can define a current matrix from the node into lead α as (Brataas *et al.*, 2000)

$$\hat{i}_\alpha = \frac{e^2}{h} \left\{ \sum_{nm} [\hat{r}_{nm} \hat{f}_\alpha (\hat{r}_{nm}) + \hat{t}_{nm} \hat{f}_N (\hat{t}_{nm})^*] - M_N \hat{f}_N \right\}. \tag{3.77}$$

Here $\hat{r}$ and $\hat{t}$ are reflection and transmission matrices connecting the node to lead α and M_N is the number of propagating modes in the node. In the absence of inelastic scattering, the non-equilibrium distribution function inside the node is determined by current conservation at each energy,

$$\sum_\alpha \hat{i}_\alpha = \hat{i}_{\mathrm{sf}}^N,$$

where $\hat{i}_{\mathrm{sf}}^N$ describes spin-flip scattering in the normal node, disregarded in the following. The total charge current is obtained as an energy integral and a trace over spin space,

$$I = \int d\epsilon \, \mathrm{Tr}[\hat{i}]. \tag{3.78}$$

Because the trace of a matrix is invariant under unitary transformations, this is hence independent of the choice of the spin quantization axis. On the other hand, the spin current describing the asymmetry of the currents carried between different spin channels,

$$\hat{I}_S = \int d\epsilon \, \mathrm{Tr}[\hat{\sigma}\hat{i}] \tag{3.79}$$

depends on the chosen spin basis. Here $\sigma = (\sigma_x, \sigma_y, \sigma_z)$ is the vector of Pauli matrices (see Appendix A.6) in this basis.

Let us transform into the spin basis of the ferromagnet α with magnetization pointing into the direction $\vec{m}$. This can be done by using the rotation matrices $\hat{u}_{\uparrow/\downarrow} = (\hat{1} \pm \hat{\sigma} \cdot \vec{m})/2$. In this basis, the reflection matrices are diagonal in spin space, i.e., $\hat{r}_{nm} = \sum_{s=\uparrow,\downarrow} \hat{u}_s r_{nm}^s$. The current matrix can now be written as

$$\hat{i} = \sum_{s=\uparrow,\downarrow} \left[G_s \hat{u}_s (\hat{f}_\alpha - \hat{f}_N) \hat{u}_s \right] - G_{\uparrow\downarrow} \hat{u}_\uparrow \hat{f}_N \hat{u}_\downarrow - (G_{\uparrow\downarrow})^* \hat{u}_\downarrow \hat{f}_N \hat{u}_\uparrow. \tag{3.80}$$

The spin-dependent conductances G_s

$$G_s = \frac{e^2}{h} \left[M_N - \sum_{nm} |r_{nm}^s|^2 \right] \tag{3.81}$$

[33]Note that in the case when magnetization axis is the same in all ferro-
magnets, the projections $\hat{u}_\uparrow \hat{f}_N \hat{u}_\downarrow$ and
$\hat{u}_\downarrow \hat{f}_N \hat{u}_\uparrow$ vanish.

can be used to describe, for example, the giant magnetoresistance effect
discussed in Example 2.6. In the case of non-collinear magnetization
between two or more ferromagnets inside the structure,[33] there is a third,
in general complex, spin mixing conductance

$$G_{\uparrow\downarrow} = \frac{e^2}{h} \left[M_N - \sum_{nm} r_{nm}^\uparrow (r_{nm}^\downarrow)^* \right]. \tag{3.82}$$

This mixing conductance is important because it describes the difference
in the spin currents entering the ferromagnet and those far inside it. The
latter is parallel with the magnetization inside the ferromagnet, and this
difference between the spin currents is absorbed at the interface. The
resultant angular momentum transfer to the ferromagnet gives rise to a
spin transfer torque on the ferromagnet (Waintal *et al.*, 2000; Xiao *et al.*,
2008), which can be used for switching the magnetization by current.

Further reading

- A classic treatment of the scattering approach on
  quantum transport is the book by Datta (1995),
  which also dwells on the Green's function approach
  for calculating the scattering matrices.

- For details of the random matrix theory, you can
  have a look at the book by Mehta (2004) or the
  review by Beenakker (1997) on its use in quantum
  transport problems.

Exercises

(3.1) Prove the sum rules for the total transmission and
reflection probabilities

$$\sum_{\alpha \neq \beta} \bar{T}_{\alpha\beta} + R_\beta = M_\beta \tag{3.83a}$$

$$\sum_{\beta \neq \alpha} \bar{T}_{\alpha\beta} + R_\alpha = M_\alpha. \tag{3.83b}$$

Using these, show that the total current into the
scattering region vanishes.

(3.2) Show that for a two-probe system $\bar{T}_{12} = \bar{T}_{21}$ even in
the absence of time-reversal symmetry.

(3.3) Consider a symmetric single-mode four-probe system, where the scattering matrix is symmetric upon
interchanging any of the lead indices. Write down
the scattering matrix in terms of the (complex)

transmission and reflection amplitudes t and r, find
a relation between them and look for the range of
possible values for them.

(3.4) Prove eqns (3.53).

(3.5) **Resonant tunnelling**. Assume we have two scatterers connected in series in a single-mode setup.
Denote the separation of the scatterers by d. Assume that between the scatterers, the wave amplitudes acquire a dynamical energy-dependent phase
$\phi = kd = \sqrt{2mE}d/\hbar$, i.e., $a_{iR} = e^{i\phi} b_{iL}$ and
$a_{iL} = e^{i\phi} b_{iR}$. For simplicity, assume that the two
scatterers have the same scattering matrix with
$t = t' = \sqrt{T}$ and $r = -r' = \sqrt{R}$, where $R = 1-T$ is
the reflection probability. Calculate the total transmission probability through this system. Plot this
result as a function of ϕ for $T = 0.01$.

(3.6) For a multi-probe system with energy-independent scattering, prove the Onsager reciprocity relation

$$R_{\alpha\beta,\gamma\delta}(B) = R_{\gamma\delta,\alpha\beta}(-B), \qquad (3.84)$$

where $R_{\alpha\beta,\gamma\delta} \equiv e^{-1}d(\mu_\gamma - \mu_\delta)/dI_\alpha$, and current flows only between electrodes α and β. Hint: write the resistances in terms of the conductances and use their symmetry.

(3.7) Show that in the general case of N probes, the voltage probe potential (3.59) becomes

$$\mu_p = \frac{\sum_{\alpha \neq p} T_{p\alpha}\mu_\alpha}{\sum_\alpha T_{p\alpha}},$$

where the sum goes over all the other probe indices α.

(3.8) **Photon assisted electron transport.** Assume one of the leads, say α, is driven with a time-dependent potential $V = V_0 \cos(\omega_0 t)$. The effect of this potential can be described by replacing the annihilation operator of mode n in this lead by (Pedersen and Büttiker, 1998; Tien and Gordon, 1963)

$$\hat{a}_{n\alpha}(E) = \sum_{l=-\infty}^{\infty} J_l(eV_0/\hbar\omega_0)\hat{a}'_{n\alpha}(E - l\hbar\omega_0), \quad (3.85)$$

where $\hat{a}'_{n\alpha}(E)$ is the annihilation operator for mode n in the corresponding time-independent system and $J_l(x)$ is the lth order Bessel function. Substituting this annihilation operator in eqn (3.54), find the expression for the average dc current in lead α.

(3.9) **Partial dephasing:** Consider the setup in Fig. 3.16, which models the effect of dephasing on resonant tunnelling. Resonant tunnelling is modelled by assuming $c_1^{\text{in}} = e^{i\varphi/2}a_2^{\text{out}}$, $a_2^{\text{in}} = e^{i\varphi/2}c_1^{\text{out}}$, $b_2^{\text{in}} = e^{i\varphi/2}c_2^{\text{out}}$, $c_2^{\text{in}} = e^{i\varphi/2}b_2^{\text{out}}$ and the dephasing can be described with the matrix S_P which connects the outgoing and incoming states c through

$$\begin{pmatrix} c_1^{\text{out}} \\ c_2^{\text{out}} \\ c_3^{\text{out}} \\ c_4^{\text{out}} \end{pmatrix} = S_P \begin{pmatrix} c_1^{\text{in}} \\ c_2^{\text{in}} \\ c_3^{\text{in}} \\ c_4^{\text{in}} \end{pmatrix},$$

$$S_P = \begin{pmatrix} 0 & \sqrt{1-\epsilon} & 0 & -\sqrt{\epsilon} \\ \sqrt{1-\epsilon} & 0 & -\sqrt{\epsilon} & 0 \\ \sqrt{\epsilon} & 0 & \sqrt{1-\epsilon} & 0 \\ 0 & \sqrt{\epsilon} & 0 & \sqrt{1-\epsilon} \end{pmatrix}.$$

Here $\epsilon \in [0,1]$ describes the strength of dephasing. Assume for simplicity that the scattering matrices of the left and right interfaces are the same, $S_1 = S_2$, and of the form (3.42) with $t = t' = \sqrt{T}$ and $r = -r' = \sqrt{R}$, as in Exercise 3.5. a) Find the formula for the current I_P flowing into the probe. Then require $I_P = 0$ and find the potential μ_P. b) With this μ_P, find the current flowing between the leads 1 and 2 as functions of the transmission t, phase φ and dephasing strength ϵ. Interpret the results in the limits $\epsilon = 0$ and $\epsilon = 1$.

(3.10) **Question on a scientific paper.** Take the paper Tombros *et al.*, *Nature Phys.* **7**, 697 (2011) presenting measurements of the quantized conductance of graphene nanoribbons (detailed more in Sec. 10.5). The broadened conductance steps are shown in Fig. 2. a) Find out what range of the control voltage had to be chosen for Figs. 2a and 2b. b) How can the width of the constriction be estimated from the data in Fig. 2?

Quantum interference effects

[1]Here, broadly speaking, I refer to interference effects when the effect is intrinsically related to the wave character of electrons. In some of the examples, like the quantized conductance and persistent currents, the interference of the waves is less visible than in others.

Interference effects found and often intensively searched in nanoelectronics are a fingerprint of the quantum-mechanical character of the charge carriers. Such effects cannot be described with the semiclassical approach presented in Ch. 2. The aim of the present chapter is to introduce the most relevant interference effects observed in mesoscopic conductors. As discussed in Ch. 2, the semiclassical theory of electron dynamics is valid when the length scales of the system far exceed the Fermi wavelength λ_F. Quantum effects should thus be proportional to some positive power of λ_F/ℓ^*, where ℓ^* is a relevant length scale—width, length, mean free path, etc. As discussed in Sec. 3.1.4, the *quantized conductance of a point contact* is observable if its minimum width d_0 is not much larger than λ_F. The previous section also discusses another quantum effect on transport, *resonant tunnelling* (Sec. 3.4), where the constructive interference of different electron paths leads to an enhanced transmission for certain energies.

Apart from resonant tunnelling and the quantized conductance, there are four main types of interference effects observed in nanoelectronic systems:[1] Aharonov–Bohm effect that arises due to a magnetic field in multiply connected structures, (Anderson) localization due to the enhanced backscattering in a disordered wire, universal fluctuations of conductance in disordered wires and persistent currents in small isolated metal rings. Most of these may be described by using the fact that the overall transmission probability is obtained as a squared sum of transmission amplitudes, rather than the sum of squares, for different paths an electron takes within a conductor, and hence they are strongly dependent on phase coherence. At non-zero temperature there are many effects that break this phase coherence. Typically such effects are due to the coupling of the individual electrons to other degrees of freedom, for example to other electrons or phonons. The loss of phase coherence is thus intimately connected to inelastic scattering. Phase breaking or *dephasing* is typically described with a dephasing time τ_φ or a dephasing length ℓ_φ, such that conductors much shorter than ℓ_φ exhibit the strongest interference effects, and conductors much longer than ℓ_φ behave in a 'classical' way, i.e., exhibiting no interference.

Most of the interference effects can be rigorously described only with microscopic calculations involving two-particle correlators (Green's functions, see Further Information 4.1 for a brief introduction) or applying random matrix theory. Both of these are outside the scope of this book. However, quite much can be said already by following simple arguments based on the scattering theory and the idea of different Feynman paths (see Sec. 3.2.2). This is the approach we follow in this chapter.

4.1 Aharonov–Bohm effect

Aharonov–Bohm effect (Aharonov and Bohm, 1959) is a variant of the Young's double-slit experiment which is typically used to describe the particle–wave duality of quantum mechanics. The idea is to control the constructive and destructive interference of two electron paths via an external field. Besides being important to show that an electron really is a wave, there is another interesting aspect to this experiment as it shows that rather than electric or magnetic fields, the fields that affect the electron motion are the scalar and vector potentials.

Consider the structure shown in Fig. 4.1. Experimentally the conductance through such a device has been found to oscillate as a function of the magnetic field B (or the flux Φ) through the ring with a period corresponding to a flux equal to 'flux quantum' $\Phi_0 \equiv h/e$ through the ring.[2] For an example, see Fig. 4.2.

In the simplest way this observation can be understood using a similar 'Feynman path' idea that we found when combining scattering matrices for example for the resonant tunnelling effect (see Sec. 3.2.2). The transmission amplitude t between given modes in the left and right leads is separated into two parts, $t = t_1 + t_2$, where t_1 describes transmission via the upper path and t_2 via the lower path (see Fig. 4.1). The total transmission probability for this mode is[3]

$$T_{AB} = |t_1 + t_2|^2 = T_1 + T_2 + 2\mathrm{Re}(t_1^* t_2) = T_1 + T_2 + 2\sqrt{T_1 T_2}\cos(\phi), \quad (4.1)$$

where $T_i = |t_i|^2$ and ϕ is the phase difference between t_1 and t_2. The oscillations in the conductance occur due to the last, 'interference' term in this equation. In the classical case, only the probabilities for the two paths would be added, i.e.,

$$T_{AB}^{\mathrm{cl}} = T_1 + T_2. \quad (4.2)$$

This would be the result from a classical circuit theory, i.e., the total conductance of two conductors in parallel is the sum conductance.

The effect of a magnetic field can be explained in terms of the vector potential $\mathbf{A}$. It can be included in the description via the minimal substitution[4]

$$\mathbf{p} \mapsto \mathbf{p} - e\mathbf{A}. \quad (4.3)$$

This has the effect of changing the dynamic phase $\mathbf{p} \cdot \mathbf{r}$ obtained by an electron wave function when traversing a distance $\mathbf{r}$. The phase obtained within the transmission through the upper arm is

$$\phi_1 = \frac{1}{\hbar} \int_{\mathrm{upper\ arm}} \vec{p} \cdot \vec{dl} \approx \frac{\sqrt{2mE}\pi r}{\hbar} - \frac{e}{\hbar} \int_{\mathrm{upper\ arm}} \mathbf{A} \cdot \vec{dl}. \quad (4.4)$$

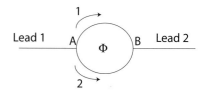

Fig. 4.1 Aharonov–Bohm ring.

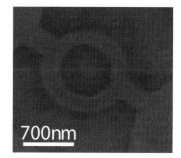

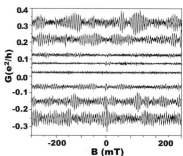

Fig. 4.2 Aharonov–Bohm oscillations in the conductance (lower figure) of a graphene ring (scanning electron micrograph of the device on top). The different curves, offset for clarity, are with different charge densities, the density being smallest in the middle. (S. Russo *et al.*, *Phys. Rev. B* **11**, 085413 (2008), Figs. 1 and 3.)

[2]As shown in Example 4.1, all observables related to such rings are periodic functions of the flux, with the periodicity given by Φ_0.

[3]See Exercise 4.2 for a more careful analysis of such a probability.

[4]See Appendix A.4.

Here r is the radius of the ring. Thus, the transmission amplitude for the upper arm is $t_1 = t_1^0 \exp(i\phi_1)$, where t_1^0 is the amplitude for $E = 0$, $\mathbf{A} = 0$. For the lower arm, we have the phase ϕ_2, which is otherwise the same as ϕ_1 but where the integration is carried out over the lower arm. This all assumes that the ring is very thin compared to its diameter, so that the areas enclosed by the outer and inner circumferences are approximately the same. The difference in the phases ϕ_1 and ϕ_2 is proportional to the magnetic flux enclosed by the ring:

$$\phi_2 - \phi_1 = \frac{e}{\hbar} \oint \mathbf{A} \cdot d\vec{l} = \frac{e}{\hbar} \int_S \mathbf{B} \cdot d\vec{S} = \frac{2\pi\Phi}{\Phi_0}. \tag{4.5}$$

The second equality uses the Stokes theorem and the fact that the magnetic field is given by $\mathbf{B} = \nabla \times \mathbf{A}$, and in the last equality the magnetic flux is defined as the surface integral of the magnetic field. Hence,

$$T_{12} = T_1 + T_2 + 2\sqrt{T_1 T_2} \cos\left(\frac{2\pi\Phi}{\Phi_0} + \varphi\right), \tag{4.6}$$

where φ is the phase difference in the absence of a field. The relative oscillations are greatest when $T_1 = T_2$, in which case the total transmission oscillates between 0 and $4T_1$.

Note that this calculation does not assume anything about the position dependence of the magnetic field. The Aharonov–Bohm effect can be realized also if the actual field within the wires vanishes, as long as it is finite inside the ring. The transmission phase only depends on the total magnetic flux penetrating the ring.

The above calculation takes into account only a subset of the electron paths. When there is some backscattering (finite reflection amplitude) within the system, there are also paths that traverse many times around the ring (similar to the higher-order paths in eqn (3.48)). Associated with these paths there are additional oscillations in the conductance vs. flux, with periods h/Ne, $N = 1, 2, \ldots$ being an integer.

The Aharonov–Bohm effect has been applied to measure phase shifts produced by other mesoscopic systems (see Fig. 4.3), such as quantum dots, placed in one of the ring arms. Besides the absolute value of the transmission probability, this type of a measurement allows measuring also the usually inaccessible phase of the transmission amplitude.

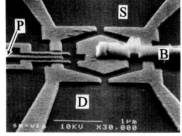

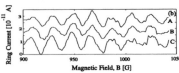

Fig. 4.3 Top: Scanning electron micrograph of a setup for measuring the transmission phase shift of the island residing on the left path. Bottom: Aharonov–Bohm oscillations in the current as a function of the magnetic field for three different gate voltages controlling the island. (A. Yacoby *et al.*, *Phys. Rev. Lett.* **74**, 4047 (1995), Figs. 1 and 3.) © 1995 by the American Physical Society.

4.2   Localization

Electron wave functions in ballistic, disorder-free wires can be well described by plane waves extending throughout the wires, and the conductance of such systems is independent of length. Such plane waves with well-defined momentum also allow for the ray optics-type semiclassical description of electron motion inside the conductors (see Fig. 4.4). Scattering from impurities, dislocations and other deviations from the perfect lattice mixes the different plane waves and the electron motion becomes diffusive, yet the idea of the electron motion in terms of semiclassical trajectories bouncing back and forth from the impurities remains a valid picture. For weak enough disorder, the electron wave functions are still extended over the entire conductor, and hence current may be carried via these states. In this limit conductance is inversely proportional to the length of the conductor. Increasing the strength of disorder, say, by increasing the density of impurities, leads to a qualitative change to this picture (Anderson, 1958). In the presence of strong disorder, the system is characterized by a rapidly varying potential, and electrons get trapped inside the potential wells. The trapping can no longer be depicted within the semiclassical model, but it results from the wave nature of the electrons: trapped electron states are standing waves. In the extreme limit all electrons are localized and the system has become an insulator, where the conductance decays exponentially as the length of the system is increased. This crossover may be analyzed with the scattering theory.

Let us concentrate first on the single-channel case and consider the transmission through two scatterers (see eqn (3.72))[5]

$$T = \frac{T_1 T_2}{1 + R_1 R_2 - 2\sqrt{R_1 R_2}\cos(2\phi)}. \tag{4.7}$$

A many-channel wire can be described with the same approach if T is viewed as the transmission of one of the non-mixing eigenmodes of the transmission matrix. In that case the total transmission probability is nothing but the sum of the contributions from the individual channels. The dynamical phase ϕ depends on the distance between successive scatterers. Therefore, we may assume that an ensemble of such scatterers averages the effects related with ϕ. Let us discuss this ensemble average for the quantity that closest resembles the classical resistance in the scattering theory: the four-probe resistance discussed in Example 3.3 (contact resistance should be included only once after the ensemble average), which scales linearly with the length in the absence of interference effects.[6] The ensemble average over the phase is thus[7]

$$\rho_{12} = \left\langle \frac{1-T}{T} \right\rangle = \int_0^{2\pi} \frac{d\phi}{2\pi} \frac{1 + R_1 R_2 - T_1 T_2 - 2\sqrt{R_1 R_2}\cos(2\phi)}{T_1 T_2}$$
$$= \frac{1 + R_1 R_2 - T_1 T_2}{T_1 T_2} = \frac{R_1 + R_2}{T_1 T_2} = \frac{R_1}{T_1} + \frac{R_2}{T_2} + \frac{2 R_1 R_2}{T_1 T_2}. \tag{4.8}$$

Here the averaging sign $\langle \cdot \rangle$ refers to the averaging over ϕ. Defining the individual resistances of the scatterers via

$$\rho_1 = \frac{R_1}{T_1}, \quad \rho_2 = \frac{R_2}{T_2}, \tag{4.9}$$

we get

$$\rho_{12} = \rho_1 + \rho_2 + 2\rho_1 \rho_2. \tag{4.10}$$

Compared to adding resistances in a classical manner, we hence get an extra term $2\rho_1\rho_2$.

Let us use this result to obtain a resistance as a function of length. Assume that $\rho_1 = \rho(L)$ describes a conductor of length L, and add to

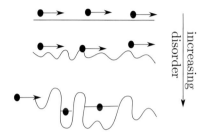

Fig. 4.4 Picture of Anderson localization: (Top) In a ballistic wire, electrons move in rectilinear paths that extend across the whole wire. Centre: Mild disorder leads to a diffusive motion, but still allows for the presence of extended states. Bottom: Strong disorder localizes electrons. Note that the effective strength of the disorder is determined by comparing the amplitude of potential variation to the (Fermi) energy of the electrons.

[5]Note that eqn (4.7) explicitly requires phase coherence, as otherwise the total transmission probability is $T_1 T_2/(T_1 + T_2)$, as obtained for a classical addition of resistances $\propto 1/T_1$ and $\propto 1/T_2$; see also Exercise 3.9.

[6]This viewpoint was first suggested by Landauer (1970). It was also discussed at length by Anderson *et al.* (1980), who show that, whereas the discussion given here is valid for the average resistance, the more proper scaling variable is $f(\rho_{12}) = \ln(1+\rho_{12})$, whose average is additive and whose distribution is close to a Gaussian.

[7]In this discussion we describe the dimensionless four-probe resistance $\rho_{12} = 2e^2 R_{12}/h$, as the prefactor is unessential for the phenomenon.

it a small piece of length ΔL and resistance $\rho_2(\lambda) = \Delta L/\lambda$. From eqn (4.10) we obtain

$$\rho(L + \Delta L) = \rho(L) + (1 + 2\rho(L))\frac{\Delta L}{\lambda}. \qquad (4.11)$$

In the limit $\Delta L \to 0$ we get a differential equation

$$\frac{d\rho}{dL} = \frac{\rho(L + \Delta L) - \rho(L)}{\Delta L} = \frac{1 + 2\rho}{\lambda}. \qquad (4.12)$$

The solution to this equation is

$$\rho(L) = \frac{1}{2}\left(e^{2L/\lambda} - 1\right). \qquad (4.13)$$

Thus, the resistance of a disordered wire grows exponentially with its length. This is the phenomenon of *localization* (Anderson, 1958). If the wire is much shorter than λ, we get an 'ohmic' result,

$$\rho(L) \approx \frac{L}{\lambda}, \quad L \ll \lambda. \qquad (4.14)$$

Comparing this scaling to the semiclassical conductivity, eqn (1.9), we find that λ should be of the order of the mean free path. This is correct for a single-mode conductor. The localization length for the many-mode case is discussed in Sec. 4.2.2.

Figure 4.5 shows a schematic picture of the behaviour of the conductance as the length of a phase coherent conductor is varied. It illustrates how the wire turns from the quasi-ballistic (where G is almost independent of length) via the diffusive regime (where $G \propto L^{-1}$) to a localized regime (where G decays exponentially with L). This picture is valid for any two-dimensional conductor, such as a 2DEG, at a vanishing temperature. For three-dimensional conductors, λ_{loc} may become infinite (see Complement 4.3), and at $k_B T > 0$ the dephasing length ℓ_φ may become smaller than λ_{loc}, in which case localization cannot be observed.

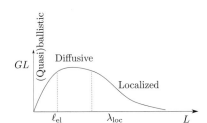

Fig. 4.5 Behaviour of the conductance times the length of a disordered conductor with a finite localization length λ_{loc} as the length is varied: in the (quasi-)ballistic regime $L < \ell_{\mathrm{el}}$, conductance is almost independent of the length, in the diffusive regime $\ell_{\mathrm{el}} \ll L \ll \lambda_{\mathrm{loc}}$, it behaves classically, and for $L \gg \lambda_{\mathrm{loc}}$ it decays exponentially. For a numerical simulation showing this behaviour, see for example (Heikkilä *et al.*, 1999).

4.2.1 Weak localization

Expanding the resistance of eqn (4.13) into second order yields

$$\rho(L) \approx \frac{L}{\lambda} + \left(\frac{L}{\lambda}\right)^2 = \rho_{\mathrm{cl}} + \Delta\rho. \qquad (4.15)$$

The quantum correction $\Delta\rho$ to the classical resistance is quadratic in the length L, provided the correction is weak. The corresponding *weak localization* correction to the classical conductance is independent of L:

$$\frac{G}{2e^2/h} \approx \frac{1}{\rho_{\mathrm{cl}} + \Delta\rho} \approx \frac{1}{\rho_{\mathrm{cl}}} - \frac{\Delta\rho}{\rho_{\mathrm{cl}}^2} = \frac{1}{\rho_{\mathrm{cl}}} - 1. \qquad (4.16)$$

The conductance of a disordered phase-coherent sample is thus reduced by $\Delta G \approx 2e^2/h$ compared to the classical conductance—provided the above expansion is realistic, i.e., the classical conductance far exceeds the correction. Although the derivation of this result above is somewhat crude, a similar finding can be made from the rigorous microscopic theory or by considering the possible Feynman paths an electron takes in a disordered wire.[8]

[8]See for example Ch. 1 in (Dittrich *et al.*, 1998).

4.2.2 Localization length

As discussed in Sec. 4.2.1, in the diffusive limit $\ell_{\mathrm{el}} \ll L \ll \lambda$, and thus $G \gg (2e^2/h)$, localization gives rise to a change in conductance which is of the order of one conductance quantum $2e^2/h$. In a many-mode conductor, the localization length λ_{loc} (the size of the localized wave within the scattering region) can be estimated from the condition that when the length L of the conductor is of the order of λ_{loc}, the classical dimensionless conductance $G_{\mathrm{cl}}/(2e^2/h)$ should be of the order of unity: in this case localization corrections dominate the conductance. From the Drude conductivity (see Sec. 3.3.1) we thus get for $\lambda_{\mathrm{loc}} \gg \ell_{\mathrm{el}}$ (Thouless, 1977)[9]

$$\frac{G_{\mathrm{cl}}(L = \lambda_{\mathrm{loc}})}{2e^2/h} = \frac{M\ell_{\mathrm{el}}}{\lambda_{\mathrm{loc}}} \approx 1 \qquad (4.17)$$

provided that

$$\lambda_{\mathrm{loc}} \approx M\ell_{\mathrm{el}} \sim \frac{2A\ell_{\mathrm{el}}}{3\pi^2\lambda_F^2}. \qquad (4.18)$$

Here A is the cross-section of the wire and M is the number of modes. Thus, in a 100 nm × 100 nm metal wire with 10^5 modes (see discussion below eqn (3.18)) and elastic mean free path some tens of nanometres, the localization length would be a few millimetres. This is much longer than the typical dephasing lengths in such wires, and thus Anderson localization in metal wires is difficult to observe. However, in low-dimensional semiconductor structures, graphene and carbon nanotubes the number of modes is much smaller, and hence also the localization length may be clearly below ℓ_φ.

4.2.3 Weak localization from enhanced backscattering

The weak localization effect can be understood by considering the different paths a particle takes in a disordered medium, similar to the Feynman paths discussed in Sec. 3.2.2.[10] Consider a process that scatters an incoming mode n (with wave vector $\vec{k}_n$) into an outgoing mode m (with wave vector $\vec{k}_m$) via multiple scattering events from impurities. The individual scattering events can be written as scattering between states with wave vectors $\vec{k}'_j$ and $\vec{k}'_k$. Thus, the total process—the 'path' between states n and m—is of the form (see Fig. 4.6)

$$\vec{k}_n \to \vec{k}'_1 \to \vec{k}'_2 \to \cdots \to \vec{k}'_N \to \vec{k}_m. \qquad (4.19)$$

Denote these paths between states n and m with index p and the probability amplitude for taking these paths with A_p^{mn}. Then the total scattering amplitude between states n and m is of the form

$$s_{mn} = \sum_p A_p^{mn}. \qquad (4.20)$$

[9]This estimate applies in the quasi-one-dimensional case where the width and thickness of the wire are much smaller than λ_{loc}. In the two-dimensional case $\lambda_{\mathrm{loc}} \sim \ell_{\mathrm{el}} \exp(k_F \ell_{\mathrm{el}})$ (Dittrich *et al.*, 1998).

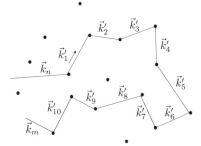

Fig. 4.6 Scattering path between states n and m.

[10]This semiclassical picture of weak localization is detailed, for example, in (Chakravarty and Schmid, 1986). It is also discussed in (Datta, 1995).

In a conductor much larger than the mean free path, the conductance is (almost) self-averaging:[11] the total conductance averages out the effects of individual scattering events, i.e., the forms or locations of the individual impurity potentials are not relevant, only their average effect. Then it is enough to consider the average transmission/reflection probability between modes n and m. This is

$$\langle |s_{mn}|^2 \rangle = \langle \sum_{pp'} A_p^{mn} (A_{p'}^{mn})^* \rangle = \langle \sum_{pp'} |A_p||A_{p'}|e^{i\varphi_{pp'}} \rangle. \tag{4.21}$$

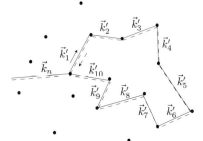

Fig. 4.7 Time-reversed paths with multiple scattering events.

Here $\varphi_{pp'}$ is the phase difference between the amplitudes A_p and $A_{p'}$. The disorder average of this phase factor $e^{i\varphi_{pp'}}$ vanishes except in two cases. First, if $p = p'$ are the same paths in which case $\varphi_{pp'} = 0$. These processes yield the classical conductance of the wire. The second case is realized when $p' = \bar{p}$ is the time-reversed path of p, i.e., it goes through the same wave vectors $\vec{k}_j'$ as the path p, but in reverse order (see Fig. 4.7). The latter corresponds to a closed path, and therefore contributes only when $m = n$. If the system is symmetric under time reversal, the amplitudes for the paths p and $\bar{p}$ have to equal.

Let us concentrate on the reflection probabilities between different states n, m in lead 1 of a two-probe system. For $n \neq m$, the time-reversed paths connect different states and therefore the localization correction does not enter. We hence obtain the classical case (denoted by the superscripts 'cl'),

$$\langle |s_{mn}|^2 \rangle = \sum_p \langle |A_p^{mn}|^2 \rangle \equiv R_{nm}^{cl}. \tag{4.22}$$

Summing up these reflection coefficients gives us the total classical reflection probability into, say, lead 1,

$$\bar{R}_1^{cl} = \sum_{nm} R_{nm}^{cl}. \tag{4.23}$$

For $n = m$ the enhanced backscattering is relevant, and we obtain

$$\langle |s_{nn}|^2 \rangle = \sum_p \langle |A_p^{nn}|^2 + A_p^{nn}(A_{\bar{p}}^{nn})^* \rangle = 2R_{nn}^{cl}. \tag{4.24}$$

Now the resultant two-probe conductance contains the classical terms and the correction from enhanced backscattering,

$$\begin{aligned} \frac{G}{2e^2/h} &= \bar{T}_{12} = M_1 - \bar{R}_1 = M_1 - \sum_{m,n} |s_{mn}|^2 \\ &= M_1 - \bar{R}_1^{cl} - \sum_n \left(|s_{nn}|^2 - R_{nn}^{cl} \right) \\ &= M_1 - \bar{R}_1^{cl} - \sum_n R_{nn}^{cl}. \end{aligned} \tag{4.25}$$

We hence get a correction of the form $\sum_n R_{nn}^{cl}$ to the total conductance. To get an estimate for its magnitude, let us make the crude assumption that the classical backscattering probability between different propagating channels is independent of the channel index. In a disordered wire, the classical conductance is obtained from the Drude conductivity (see the discussion in Sec. 3.3.1). This can be obtained by assuming[12]

$$R^{\text{cl}}_{nm} = \frac{1}{M_1}\left(1 - \frac{\ell_{\text{el}}}{L}\right) \quad \forall n, m \in \{1, 2, \ldots M_1\}. \tag{4.26}$$

Now we may substitute this in eqn (4.25) and get

$$\frac{G}{2e^2/h} = M_1\frac{\ell_{\text{el}}}{L} - 1 + \frac{\ell_{\text{el}}}{L} \approx M_1\frac{\ell_{\text{el}}}{L} - 1. \tag{4.27}$$

We thus again find the conductance correction $\Delta G = 2e^2/h$ in the diffusive limit $\ell_{\text{el}} \ll L$ and $M_1 \gg 1$.

Complement 4.1 Microscopic calculation with Green's-function techniques

The weak-localization correction to the conductance can be calculated more rigorously than above by using the equilibrium Green's function theory of condensed-matter physics. This also allows us to treat the effect of a magnetic field and of dephasing quantitatively. For linear response, the conductivity can be obtained from the equilibrium Green's function using the Kubo formula[13]

$$\langle \sigma \rangle = \frac{\hbar e^2}{\pi\Omega}\left(\frac{\hbar}{m}\right)^2 \sum_{kk'} k_z k'_z \langle\langle k|G^R|k'\rangle\langle k'|G^A|k\rangle\rangle. \tag{4.28}$$

Here Ω is the volume of the sample, and G^R and G^A are the retarded and advanced Green's functions of the electron system, respectively. The outer brackets refer to disorder averaging.

The semiclassical (Drude) conductivity can be found by approximating

$$\langle G^R G^A \rangle \approx \langle G^R \rangle \langle G^A \rangle. \tag{4.29}$$

These are strongly peaked functions of $k - k'$, allowing to concentrate only on the diagonal part $G^{R/A}_k = \langle k|G^{R/A}|k\rangle = (\epsilon_F - \epsilon_k \pm i\gamma)^{-1}$, where in the diffusive limit the broadening is described by elastic scattering, $\gamma = \hbar v_F/(2\ell_{\text{el}})$. Inserting this in eqn (4.28), converting the sum over k to an integral over energy, and approximating the density of states by that in the Fermi level then yields the Drude conductivity $\sigma = e^2 N_F v_F \ell_{\text{el}}/d$, where the dimensionality d comes from the angular part of the momentum integral.

It turns out that the major corrections to this result can be expressed with two types of Feynman diagrams: diffusons and cooperons.[14] The former only modifies the diffusion constant to take into account the possible average angular dependence of scattering (as discussed in Sec. 2.4), but in the spherically symmetric case its contribution vanishes. The latter contains the weak localization correction.

4.2.4 Dephasing

Weak localization is an interference effect. If the time-reversed paths are longer than the dephasing length ℓ_φ, their relative phase becomes

[12]Note that here we assume, instead of concentrating on scattering eigenmodes, that all the elements of the scattering matrix with different n, m are non-zero and of the same order of magnitude.

[13]See, for example, Ch. 1 in (Dittrich *et al.*, 1998) or Ch. 6 in (Bruus and Flensberg, 2004). The main point of the Kubo formula is to relate the equilibrium fluctuations to the linear conductance through the fluctuation–dissipation theorem explained in Ch. 6 and Appendix C.

[14]The calculation of these diagrams is beyond the scope of this book, see for example (Dittrich *et al.*, 1998) for the details.

randomized, and as a result their contribution to the enhanced backscattering is lost. If the width and thickness of such a wire are still smaller than ℓ_φ, it is straightforward to show that the correction to the conductance is (see Exercise 4.3)

$$\Delta G(L \gg \ell_\varphi) = \frac{2e^2}{h} \frac{\ell_\varphi}{L}. \tag{4.30}$$

In the two-dimensional case also the width of the sample exceeds the dephasing length. Then the correction to the conductance is (Chakravarty and Schmid, 1986)

$$\Delta G = \frac{2e^2}{\pi h} \ln\left(\frac{\ell_\varphi}{\ell_{\mathrm{el}}}\right). \tag{4.31}$$

Exploiting the fact that the weak localization correction also decays with an applied magnetic field through the system (see Sec. 4.2.5), measuring this correction is often used to measure the dephasing length.

Equations (4.30)–(4.31) are independent of the actual mechanisms that cause dephasing. Such mechanisms can be generally divided into two classes: reversible changes of the phase, for example due to a static magnetic field or due to spin-orbit interaction, and irreversible changes, due to the fluctuations in the positions of the scatterers or time-dependent changes in the electric or magnetic field. The irreversible changes are often also called *decoherence*, and are related to the random noise in quantities determining the electron motion. Different contributions to noise are discussed in Ch. 6. Typically the most important origin for random fluctuations is thermal noise, whose amplitude increases with an increasing temperature. Therefore, also the dephasing length ℓ_φ is generally temperature dependent, decreasing with an increasing temperature. Dephasing times and dephasing lengths are discussed more in Complement 6.1.

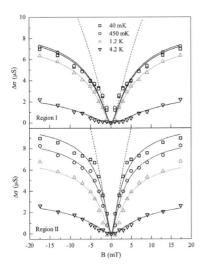

Fig. 4.8 Weak localization correction to the conductance as a function of magnetic field in bilayer graphene. The two regions correspond to different charge densities: region I to the Dirac point and region II far off from the Dirac point. Symbols are data at different temperatures, and the solid lines are the weak localization theory adjusted to take into account the graphene-specific forms of scattering; dashed lines are without these adjustments. (R.V. Gorbachev *et al.*, *Phys. Rev. Lett* **98**, 176805 (2007), Fig. 3.) © 2007 by the American Physical Society.

4.2.5   Magnetic field effect on weak localization

Assume we lift the time-reversal symmetry by applying a magnetic field perpendicular to the sample. Due to this magnetic field, the paths between states n and m obtain an additional phase,

$$A_p^{mn} \rightarrow A_p^{mn} \exp\left(i\frac{e}{\hbar}\int_p \mathbf{A} \cdot \vec{dl}\right). \tag{4.32}$$

This additional phase is irrelevant for the paths whose phase dependence averages out in disorder averaging. It also does not contribute to the terms $A_p^{nn}(A_p^{nn})^*$ in eqn (4.21) as the phase change induced by the vector potential cancels out in them. However, the contribution from a pair of a path p and its time reverse $\bar{p}$ changes to

$$A_p^{nn}(A_{\bar{p}}^{nn})^* = |A_p^{nn}|^2 \exp\left(2i\frac{e}{\hbar}\oint_p \mathbf{A} \cdot \vec{dl}\right) = |A_p^{nn}|^2 \exp\left(2i\frac{2\pi\Phi_P}{\Phi_0}\right). \tag{4.33}$$

Here Φ_P is the flux through the area enclosed by the path. As the different paths enclose different areas, the sum over the paths eventually washes out this additional backscattering term. The critical magnetic field required for this corresponds roughly to a flux Φ_P equal to one flux quantum through the path with the largest loop area. As only the paths whose length is below the dephasing length ℓ_φ contribute, we may estimate this area as roughly ℓ_φ^2. Thus, the critical field is of the order of[15]

$$B_c \sim \frac{h}{e\ell_\varphi^2}. \tag{4.34}$$

For $B > B_c$, the weak localization contribution decays on a length scale given by $\ell_m = \sqrt{\Phi_0/B}$, and this length scale should replace ℓ_φ in eqns (4.30),(4.31).[16]

Figure 4.8 shows an example of the measured weak localization correction in bilayer graphene.

Complement 4.2 Weak antilocalization

In materials with strong spin-orbit scattering, spin-dephased time-reversed paths interfere destructively, causing an enhanced conductivity contrary to the enhanced resistivity due to weak localization. As a magnetic field destroys the interference, such a setting results in a positive magnetoresistance. For further information on this effect, see for example (Bergmann, 1984). The measurement of this *weak antilocalization* effect yields information on spin-flip scattering mechanisms in metals.

Complement 4.3 Single-parameter scaling theory of strong localization

The discussion in the beginning of Sec. 4.2 shows that a one-dimensional conductor is never a conductor in the strict statistical limit, in which its length tends to infinity. This is because of the localization of electron wave functions within the localization length λ_{loc}. This rather surprising result has been generalized to two and three dimensions as well: for strong enough disorder, the localization length λ_{loc} is finite for a sample of any dimension. But λ_{loc} may be very long. For example, in the two-dimensional case it is (Dittrich *et al.*, 1998; Nazarov and Blanter, 2009)[17]

$$\lambda_{\text{loc,2d}} \approx \ell_{\text{el}} \exp\left(\frac{\pi}{2} k_F \ell_{\text{el}}\right). \tag{4.35}$$

For a weakly disordered sample, $\ell_{\text{el}} \gg \lambda_F$, this localization length would indeed become very large. However, it always stays finite,[18] whereas in three dimensions, there is a transition from a non-localized state ($\lambda_{\text{loc}} = \infty$) to a localized one (where λ_{loc} is finite) as the disorder strength is varied. This *Anderson transition* is a quantum phase transition, as it takes place at zero temperature.

The existence of the localization transition can be shown by using the single-parameter scaling theory (Abrahams *et al.*, 1979). The arguments are based on a single-parameter scaling hypothesis: assume that as the size of the system tends to infinity, its properties are described by a single relevant parameter, which in this case is the ensemble-averaged dimensionless conductance $g = \langle G \rangle/(2e^2/h)$.

[15]For $\ell_\varphi = 1\ \mu$m, this field equals 40 G or 4 mT.

[16]The exact form of the weak localization correction as a function of magnetic field is given, for example, in (Altshuler *et al.*, 1980).

[17]This is obtained similarly as eqn (4.18) in the one-dimensional case: equating the conductance correction from eqn (4.31) with the total conductance.

[18]This statement is valid for non-interacting electrons in the absence of spin-orbit scattering. Interactions or spin-orbit scattering may cause a metal–insulator transition (i.e., transition from an infinite λ_{loc} to a finite λ_{loc}) in a two-dimensional sample; see (Spivak *et al.*, 2010) for a recent review of such effects.

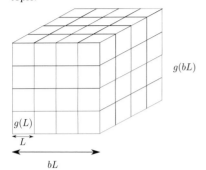

Fig. 4.9 Scaling of the conductance as the size of the system is scaled by a factor b.

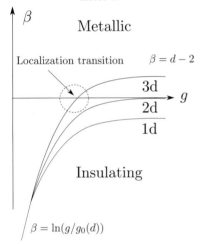

Fig. 4.10 Sketch of the behaviour of the β-function for different dimensionalities of the conductor.

The scaling arguments go as follows.[19] Consider the scaling of g in a d-dimensional system composed of b^d blocks of size L (as in Fig. 4.9). Combine these blocks into a single larger system with a size bL. According to the scaling hypothesis, there is a function f that satisfies

$$g(bL) = f(b, g(L)). \tag{4.36}$$

This hypothesis can also be written in the form of a logarithmic derivative,

$$\frac{d \ln g(L)}{d \ln L} = \beta(g(L)) \tag{4.37}$$

with some function β. In other words, the scaling of $g(L)$ at some length L depends only on the value of g for a conductor of that length.

At large and small g we can find the asymptotes of β from general physical arguments. For large g, the conductance follows Ohm's law,

$$G(L) = \sigma L^{d-2}. \tag{4.38}$$

Therefore, $\ln g = \ln(\sigma L^{d-2}/(2e^2/h)) \propto \text{const.} + (d-2) \ln L$. For the β-function, this means that

$$\beta(g) \equiv \frac{d \ln g}{d \ln L} \xrightarrow{g \to \infty} d - 2. \tag{4.39}$$

For $g \ll 1$, the system is in the regime of strong localization and g falls off exponentially:

$$g = g_0 e^{-L/\lambda_{\text{loc}}}. \tag{4.40}$$

Hence, in this limit

$$\beta = \frac{d(-e^{\ln L}/\lambda_{\text{loc}})}{d \ln L} = -L/\lambda = \ln(g/g_0(d)). \tag{4.41}$$

The rough behaviour of $\beta(g)$ for different dimensionalities $d=1,2,3$ is sketched in Fig. 4.10. For $\beta > 0$, the conductance grows with an increasing L, i.e., it behaves in a 'metallic' manner, whereas for $\beta < 0$ it decreases when increasing the scale L. Thus, the point $\beta = 0$ indicates a metal–insulator transition, only present in strictly three-dimensional systems.

Anderson localization is a general phenomenon that can also be found in systems other than mesoscopic conductors. For example, it has been observed for light waves (Albada and Lagendijk, 1985) and matter waves in a Bose condensate (Roati *et al.*, 2008).

4.3   Universal conductance fluctuations

The above text considers the behaviour of the conductance averaged over the different configurations of disorder. In a macroscopic sample (much larger than the dephasing length) the conductance G is self-averaging, i.e., it equals the disorder-averaged conductance, and changing the disorder configuration but keeping its strength (mean free path) constant leaves G unaltered.

This is different for mesoscopic conductors. The conductance depends significantly on the disorder configuration even in the diffusive limit $L \gg \ell_{\text{el}}$. This dependence shows up when varying the magnetic field or the Fermi velocity. The latter can be tuned by a gate voltage which tunes the total charge on the conductor and thereby affects its Fermi energy. Changing either the magnetic field or the gate voltage leads to changing phase shifts that the electrons acquire between the scattering events. This reproducible 'magnetofingerprint' is a typical property of mesoscopic samples whose conductance is not far from the conductance quantum $2e^2/h$. It turns out that the variance of the conductance for different disorder configurations is independent of G (in the regime $\ell_{\rm el} \ll L \ll \lambda_{\rm loc}$), and depends only on the global symmetry of the sample, such as the time-reversal symmetry. This is why these variations are called *universal conductance fluctuations.*

Let us estimate the variance using the scattering theory and the Feynman paths. As the 2-probe conductance can be written as $G = (2e^2/h)\bar{T} = (2e^2/h)(M - \bar{R})$ (dropping the lead index for simplicity), its variance is

$$\langle \delta G^2 \rangle = \langle G^2 \rangle - \langle G \rangle^2 =$$
$$= \frac{4e^4}{h^2} \left[\left(M^2 - 2M\langle \bar{R} \rangle + \langle \bar{R}^2 \rangle \right) - \left(M^2 - 2M\langle \bar{R} \rangle + \langle \bar{R} \rangle^2 \right) \right] \quad (4.42)$$
$$= \frac{4e^4}{h^2} \left[\langle \bar{R}^2 \rangle - \langle \bar{R} \rangle^2 \right] \equiv \frac{4e^4}{h^2} \langle \delta \bar{R}^2 \rangle.$$

Here $\bar{R} = \sum_{m,n} |r_{nm}|^2$ is the total reflection probability and $\langle \cdot \rangle$ denotes averaging over the different configurations of disorder. Let us calculate this variance under the assumption that the variations in the individual $|r_{nm}|$ are independent, and disregarding the small effect from enhanced backscattering (see Exercise 4.4). We thus find

$$\langle \delta \bar{R}^2 \rangle \approx M^2 \langle \delta R_{nm}^2 \rangle. \quad (4.43)$$

The variation of the reflection coefficients for individual channels can be expressed in terms of the different paths in the same manner as we treat weak localization in Sec. 4.2.3,

$$\langle R_{nm}^2 \rangle = \sum_{pp'p'p''} \langle A_p A_{p'} A_{p'}^* A_{p''}^* \rangle$$
$$= \sum_{pp'p'p''} |A_p|^2 |A_{p'}|^2 (\delta_{pp'}\delta_{p'p''} + \delta_{pp''}\delta_{p'p'}) = 2\langle R_{nm} \rangle^2. \quad (4.44)$$

The δ-functions arise because for other combinations of paths the phases between the different amplitudes average the terms to zero: only the terms containing pairs of the same paths survive. Therefore, we obtain with the help of eqn (4.26)

$$\langle \delta R_{nm}^2 \rangle \equiv \langle R_{nm}^2 \rangle - \langle R_{nm} \rangle^2 = \langle R_{nm} \rangle^2 \approx (L - \ell_{\rm el})^2/(M^2 L^2). \quad (4.45)$$

Substituting in eqns (4.43) and (4.42) we obtain

$$\sqrt{\langle \delta G^2 \rangle} = \frac{2e^2}{h} \frac{(L - \ell_{\rm el})^2}{L^2} \approx \frac{2e^2}{h}. \quad (4.46)$$

Thus, the average variation in the conductance is of the order of a single conductance quantum. The exact variance can be obtained with the

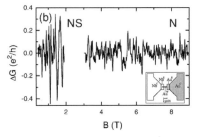

Fig. 4.11 Universal conductance fluctuations measured in a normal-superconducting contact sketched in the inset. Superconductivity enhances the magnitude of fluctuations, which is seen when the field is decreased below the critical field of the superconductor. (K. Hecker *et al.*, *Phys. Rev. Lett.* **79**, 1547 (1997), Fig. 1.) © 1997 by the American Physical Society.

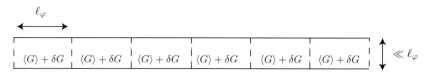

[20] Graphene is a special case, as
the conductance fluctuations are typically larger than in normal metals,
and they are sensitive to the relative
strengths of different types of scattering (Kharitonov and Efetov, 2008).

random matrix theory (Beenakker, 1997) or by numerical simulations.
The general result is $\sqrt{\langle \delta G^2 \rangle} = \sqrt{2/15}\beta^{-1}2e^2/h$, where the value of the
prefactor β^{-1} depends only on the symmetry class where the scattering
matrix belongs to (Altland and Zirnbauer, 1997), for example, $\beta = 1$ in
the presence of time-reversal symmetry and $\beta = 2$ in its absence.[20]

Figure 4.11 shows an example of measured conductance fluctuations
in a normal-superconducting contact and shows a transition between
two symmetry classes.

4.3.1 Effect of dephasing

Let us estimate the effect of dephasing on the magnitude of fluctuations.
Assume we have a conductor of length $L \gg \ell_\varphi$. Let us divide it into
$N = L/\ell_\varphi$ coherent sub-units, whose resistances we can add incoherently
(see Fig. 4.12). The variance of a resistance R_i in a single sub-unit is

$$\mathrm{var}(R_i) = \mathrm{var}(G_i)R_i^4. \tag{4.47}$$

The total resistance is $R = NR_i$ and the total resistance variation is
$\mathrm{var}R = N\mathrm{var}(R_i)$, as the variations in the different coherent sub-units
are independent of each other. The total conductance variation can be
obtained from

$$\mathrm{var}(G) = \frac{\mathrm{var}R}{R^4} = \frac{N\mathrm{var}(G_i)R_i^4}{N^4 R_i^4} = N^{-3}\mathrm{var}G_i = \left(\frac{\ell_\varphi}{L}\right)^3 c^2 \left(\frac{e^2}{h}\right)^2. \tag{4.48}$$

Thus, the fluctuations decay in a power-law fashion with the decreasing
dephasing length and can thus persist to quite high temperatures.

4.4   Persistent currents

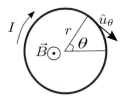

Fig. 4.13 A metallic ring carries a
tiny persistent current in the presence
of a magnetic field.

One manifestation of coherence in electron transport shows up in multi-
ply-connected conductors, where the periodic boundary conditions for
the electron wave function lead to the presence of non-vanishing currents
at equilibrium, i.e., without a bias. Similar *persistent currents* are found
in superconducting rings (see Ch. 9), but there the effect is much stronger
due to the superconducting coherence.

A non-vanishing average current with a definite sign is possible in the
presence of a magnetic field that breaks the symmetry between clockwise

and anticlockwise flow. To illustrate the effect, let us consider the case of a ballistic single-channel ring in the presence of a homogeneous magnetic field $\vec{B}$ applied perpendicular to the ring (see Fig. 4.13). This can be described by a vector potential

$$\vec{A} = \frac{Br}{2}\hat{u}_\theta \equiv A_0\hat{u}_\theta,$$

where r is the radius of the ring. In the single-channel case this vector potential is constant within the ring, and the electron wave function depends only on the polar angle θ. The Schrödinger equation in polar coordinates thus reads

$$\frac{(i\hbar\partial_\theta + eA_0r)^2}{2mr^2}\psi_n(\theta) = \epsilon_n\psi_n(\theta). \tag{4.49}$$

Note that the quantity $eA_0r = eBr^2 = \hbar\Phi/(2\Phi_0)$ is related to the flux Φ through the loop. Requiring single-valued wave functions, $\psi_n(\theta) = \psi_n(\theta + 2\pi)$, yields the eigenfunctions and -energies

$$\psi_n(\theta) = \frac{1}{\sqrt{2\pi}}e^{in[\theta - \Phi/(2\Phi_0)]}, \quad \epsilon_n = \frac{\hbar^2}{2mr^2}\left(n - \frac{\Phi}{2\Phi_0}\right)^2. \tag{4.50}$$

These energies are plotted in Fig. 4.14.

At a vanishing temperature, the energy levels are filled up to the Fermi energy and the total energy of the system is the sum

$$F = 2\sum_{n;\epsilon_n < E_F} \epsilon_n, \tag{4.51}$$

where the factor 2 is due to spin. As the single-electron energies depend on the flux Φ so does the total energy. The flux dependence gives rise to a current. This is straightforward to see by assuming the flux to vary slowly, such that F follows this time dependence adiabatically. In this case the rate of change of the energy (Joule power absorbed by the system) equals

$$\frac{dF}{dt} = \frac{dF}{d\Phi}\frac{d\Phi}{dt} = -\frac{dF}{d\Phi}V,$$

where $V = \dot{\Phi}$ is the voltage created by the time-dependent flux. Since the Joule power is $\frac{dF}{dt} = IV$, this yields the current

$$I = -\frac{\partial F}{\partial \Phi}. \tag{4.52}$$

For our example system, the current is

$$I = -2\sum_{n;\epsilon_n < E_F}\frac{dE_n}{d\Phi} = -\frac{\hbar e}{2\pi mr^2}\sum_n\left(n - \frac{\Phi}{2\Phi_0}\right). \tag{4.53}$$

This sum consists of contributions with alternating signs. This makes the total current rather small. As the energy is periodic with the flux Φ, so is the current. The order of magnitude of the maximum current

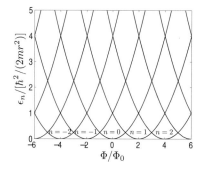

Fig. 4.14 Energies of the different eigenstates of a ballistic single-channel ring as a function of the magnetic flux threading the ring.

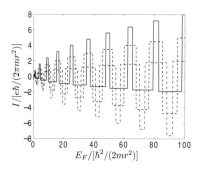

Fig. 4.15 Persistent current vs. Fermi energy in a ballistic single-channel ring of radius r. The current is plotted for three values of the flux Φ/Φ_0, 0.2 (dash-dotted line), 0.5 (dashed) and 0.8 (solid).

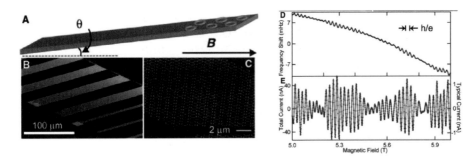

Fig. 4.16 Measurement of the persistent currents using cantilever torque magnetometry. In the presence of a magnetic field B, the circulating currents inside the rings produce a Lorentz force on the cantilever, and this force shifts the eigenfrequency of the cantilever vibrations. Measurements of such vibrations are described in Ch. 11. A shows schematics of the cantilever enclosing the rings, and B and C contain scanning electron micrographs of the cantilevers and the rings in them (C is magnified from B). D shows the measured frequency shift, and E the inferred oscillations of the persistent current. The magnitude of the average current (left axis) and the typical current of single rings (right axis) are in line with the non-interacting theory outlined here. For a more detailed discussion of this experiment, see Exercise 4.7. (A.C. Bleszynski-Jayich *et al.*, *Science* **326**, 272 (2009), Fig. 1.) Reprinted with permission from AAAS.

[21] For an exact result, see Exercise 4.5.

can be estimated by noting that I consists of an alternating series of terms whose magnitude increases linearly with the order of the terms (see Fig. 4.15).[21] The order of magnitude of the maximum current is thus that of the last term in the series, which is given by $n_{\max} \approx \sqrt{2mE_F r^2}/\hbar$. This yields for the magnitude of the current in the ballistic limit

$$|I_{\mathrm{pers,ball}}| \sim \frac{\hbar e}{2\pi m r^2} n_{\max} = \frac{e}{2\pi r}\sqrt{\frac{2E_F}{m}} = \frac{ev_F}{2\pi r}, \qquad (4.54)$$

where v_F is the Fermi speed. This result means that at maximum the average persistent current consists of one more electron traversing the ring, say, clockwise, than counterclockwise within the natural timescale given by $v_F/(2\pi r)$.[22]

[22] For example, for a typical value of $v_F = 10^6$ m/s, and a ballistic ring with a diameter of 500 nm, the current would be 50 nanoamperes.

One might think that in rings with more than one channel the persistent current is much larger than in the above example of a single-channel ring. However, this is not the case, as the contributions from the different channels tend to have varying signs, and the resultant persistent current increases only as a square root of the number of channels (Cheung *et al.*, 1989).

[23] For more discussion on avoided crossing, see Ch. 8 and especially Exercise 8.6.

In the presence of weak disorder, the picture remains otherwise the same, but as scattering mixes the eigenstates ψ_n, the crossings of the different eigenvalues $\epsilon_n(\Phi)$ become avoided crossings.[23] For stronger disorder the energies evolve into bands (Büttiker *et al.*, 1983), but retaining the periodicity in Φ (see Example 4.1 below). In the diffusive regime the order of magnitude of the typical persistent currents in non-interacting models is obtained by replacing the traversal time $v_F/(2\pi r)$ by the diffusion time $(2\pi r)^2/D$ along the ring (Cheung *et al.*, 1989), i.e.,[24]

[24] With a typical diffusion constant of a metal, $D = 0.01$ m^2/s, a diffusive ring with diameter of 500 nm would have a persistent current of 0.2 nA.

$$|I_{\mathrm{pers.,diff}}| \sim \frac{eD}{(2\pi r)^2}, \qquad (4.55)$$

independent of the number of channels.

Persistent currents require coherence of the electron wave functions across the entire perimeter of the ring, and therefore they decay when the systems become larger than the dephasing length. Moreover, they can only be measured in closed rings that are not connected to external electrodes, as the electrodes would generally lead to a vanishing coherence, and as the persistent currents are circulating. Therefore, the measurement of the persistent current generally needs to be done via the magnetization $M = -dF/dB = \pi r^2 I$.

There have been two kinds of measurements of the persistent currents: those of the single rings and those related with ensembles of rings. Since the sign of the currents within individual rings varies, their ensemble average—in the absence of interactions between the rings—is much smaller than the magnitude of a typical current in eqns (4.54,4.55). The typically measured root mean square current $\langle I^2 \rangle^{1/2}$ increases as a square root of the number of rings, rather than linearly.[25]

Recent experiments (Bleszynski-Jayich *et al.*, 2009; Bluhm *et al.*, 2009) (see Fig. 4.16) confirm these predictions of the non-interacting model, both for single rings and ensembles of rings in the diffusive limit. The same conclusion was already reached in the ballistic limit some two decades ago (Mailly *et al.*, 1993), but for the diffusive limit the sufficiently accurate measurement of the persistent currents required developing new technologies such as scanning magnetometer microscopes based on superconducting SQUID devices (see Ch. 9), and force measurements using mechanical resonators (see Ch. 11).[26]

[25]Thus, the mean current per ring, $\langle I \rangle = \langle I^2 \rangle^{1/2}/N$, tends to zero at large N.

[26]The first measurements of persistent currents in diffusive rings (Lévy *et al.*, 1990) found orders-of-magnitude larger currents than those predicted by theory. However, the new more accurate measurements quoted here are in disagreement with them, but agree with the theory.

Example 4.1 Flux periodicity of the observables

All observables of systems making up loops must be periodic functions of the magnetic flux through the area covered by the loop. This can be shown by describing the magnetic field in the Coulomb gauge where the vector potential satisfies $\nabla \cdot \vec{A} = 0$. The static Schrödinger equation for an electron in an arbitrary potential $U(\vec{r})$ reads

$$\left[\frac{(i\hbar\nabla + e\vec{A})^2}{2m} + U(\vec{r}) \right] \psi(\vec{r}) = \epsilon\psi(\vec{r}).$$

Using an ansatz of the type (Byers and Yang, 1961)

$$\psi(\vec{r}) = e^{i\phi(\vec{r})}\tilde{\psi}(\vec{r}), \tag{4.56}$$

with the phase chosen such that $\vec{A} = \hbar\nabla\phi/e$ leads to a Schrödinger equation

$$\left[\frac{\hbar^2}{2m}\nabla^2 + U(\vec{r}) \right] \tilde{\psi}(\vec{r}) = \epsilon\tilde{\psi}(\vec{r}), \tag{4.57}$$

which does not contain the vector potential. This is rather straightforward to show and is left as an exercise (see Exercise 4.6). The price to be paid from this transformation is that, unlike $\psi(\vec{r})$, which must be single-valued, the transformed function $\tilde{\psi}(\vec{r})$ obtains a phase factor upon traversal around the ring. In particular, written in polar coordinates,

$$\psi(r, \theta + 2\pi, z) = e^{i\phi(r,\theta+2\pi,z)}\tilde{\psi}(r, \theta + 2\pi, z) = \psi(r, \theta, z) = e^{i\phi(r,\theta,z)}\tilde{\psi}(r, \theta, z)$$

and therefore

$$\tilde{\psi}(r, \theta + 2\pi, z) = e^{i[\phi(r,\theta,z) - \phi(r,\theta+2\pi,z)]}\tilde{\psi}(r, \theta, z) = e^{-\frac{i\hbar}{e}\oint d\vec{l}\cdot\vec{A}}\tilde{\psi}(r, \theta, z)$$
$$= e^{-\frac{i\Phi}{2\pi\Phi_0}}\tilde{\psi}(r, \theta, z). \tag{4.58}$$

The relation above assumes that the enclosed flux Φ is independent of the position r inside the wire. This is satisfied if the magnetic flux inside the wire vanishes or if the wire is very thin compared to the diameter of the loop, so that the variation of the flux is negligible. Now, since adding an integer multiple of 2π to the phase of the wave function does nothing to the wave function, $\tilde{\psi}$ and therefore all observables of the system are periodic functions of the flux with the periodicity given by the flux quantum Φ_0.

Further reading

Many of the interference effects discussed in this chapter were thoroughly studied in the 1970s and 1980s. Therefore, there are many excellent treatments of some of the effects. Here I mention just a few:

- Localization and fluctuations are also discussed, for example, in (Datta, 1995), in Ch. 1 of (Dittrich *et al.*, 1998) and in (Imry, 2002).

- Persistent currents are discussed at length in (Imry, 2002).

- A useful elementary overview on the physics of disordered electron systems is given in (Al'tshuler and Lee, 1988)

- The semiclassical picture of weak localization is described in (Chakravarty and Schmid, 1986).

In this chapter I have tried to include examples of experiments from the more recent times. This is particularly the case in persistent currents, whose new measurements in disordered systems established the fact that the non-interacting theory predictions are very good descriptions of the effect. Moreover, especially the discovery of graphene (see more in Ch. 10) has reinitiated the studies of interference effects in it. Some of these developments are discussed in the recent review of the electronic properties of graphene (Castro Neto *et al.*, 2009).

Exercises

(4.1) Compare the transmission probabilities (rather than the amplitudes) of a two-scatterer system obtained by summing the transmission probabilities for individual 'Feynman paths' of the particle and the quantum-mechanical result obtained in Exercise 3.5. Show that the previous method results in a classical Ohm's law-type summing of the resistances.

Fig. 4.17 Incoming and outgoing wave amplitudes for an Aharonov–Bohm ring.

(4.2) Consider an Aharonov–Bohm ring, where both arms contain a single propagating mode. Assume that the scattering matrices for the three-way junction at both ends (positions A and B in Fig. 4.1) are described with the scattering matrices of the form (3.43), parametrized by the transmission and reflection amplitudes $t_{L/R}$ and $r_{L/R}$ for the left/right junctions, respectively. Use the fact that the incoming amplitudes within the wire can be expressed in terms of the outgoing amplitudes (Fig. 4.17) as

$$a_{UR} = e^{i(\phi_d+\phi)}b_{UL}, \quad a_{DR} = e^{i(\phi_d-\phi)}b_{DL}$$
$$a_{UL} = e^{i(\phi_d-\phi)}b_{UR}, \quad a_{DL} = e^{i(\phi_d+\phi)}b_{DR}.$$

Here ϕ_d is the dynamic phase and ϕ the phase induced by the vector potential. a) Calculate the total transmission amplitude t_{LR} from left to right by combining the scattering matrices (hint: you can use mathematical software for this). b) In the lowest order in transmission, the reflection amplitudes can be assumed to be close to unity, $r_{L/R} \approx 1$, and the transmission amplitudes $t_{L/R}$ can be assumed small. For $\phi_d \neq 0$, expand t_{LR} into the lowest order in $t_{L/R}$ and interpret your results with the Feynman paths. c) Another interesting special case is $r_L = t_L$ and $r_R = t_L$. Find the transmission coefficient in this case and interpret the result in terms of the Feynman paths.

(4.3) Calculate the weak localization correction ΔG to the conductance of a quasi-one-dimensional wire with length $L \gg \ell_\varphi$ and width $W \ll \ell_\varphi$. Hint: divide the wire into L/ℓ_φ coherent sections. Then sum up the resistances (with the weak localization correction) of these sections. Calculate also the relative correction $\Delta G/G$. This is then also the conductivity correction.

(4.4) Estimate the effect of the time-reversed paths on the conductance fluctuations.

(4.5) Calculate the total persistent current as a function of the Fermi energy by performing the sum in eqn (4.53).

(4.6) Show that the ansatz used in eqn (4.56) leads to a wave function $\tilde{\psi}$ that satisfies the Schrödinger equation without a vector potential, eqn (4.57).

(4.7) **Question on a scientific paper.** Take the paper A.C. Bleszynski-Jayich *et al.*, *Nature* **326**, 272 (2009) and try to answer the following questions. (i) The measured persistent currents (Fig. 1E) do not seem to be strictly periodic functions of the flux with the period $\Phi_0 = h/e$. Why is this? (ii) In the disordered case, the persistent currents are due to contributions from an energy interval within E_c around E_F. When temperature $T > E_c/k_B$, the currents are reduced by a factor $\exp(-\text{const.}(k_B T/E_c)^{1/2})$ as the relative contribution from that energy interval to the equilibrium persistent current becomes weaker. Try to estimate E_c from dimensional analysis, remembering that the persistent current is a quantum interference effect, and that the relevant time scale in the diffusive limit is $\tau_D = (2\pi r)^2/D$. Compare your estimate to Fig. 3.

5 Introduction to superconductivity

Superconductivity is a quantum-mechanical effect, but in contrast to the quantum phenomena discussed in the previous chapters, it does not disappear as the system size is enlarged. Therefore, it is not a mesoscopic effect. However, as many mesoscopic experiments are performed at low temperatures where some of the materials become superconducting, it strongly affects the mesoscopic systems and there are many mesoscopic phenomena only present in (partially) superconducting nanostructures.

A thorough review of superconductivity and superconducting effects is given in the excellent book by Tinkham (1996). In what follows, I mostly summarize the main phenomena and discuss some of the effects characteristic for normal-superconductor interfaces relevant for mesoscopic effects. The most relevant quantities or effects of interest related to the properties of these interfaces are the BCS density of states, eqn (5.18), and the Andreev reflection and Andreev bound states, discussed in Sec. 5.4. This chapter may be used for obtaining the necessary background information for the description of effects related to superconducting nanostructures, discussed in more detail in Ch. 9.

5.1 Cooper pairing

Superconductivity of metals was discovered as early as 1911 by H. Kamerlingh Onnes in Leiden, but it took until 1957 before a microscopic theory was laid out by Bardeen, Cooper and Schrieffer (1957) (BCS). The BCS theory is based on the possibility of some physical process to give rise to an effectively attractive interaction between the electrons. Due to this interaction, the ground state of the system changes such that it contains correlated pairs of electrons. The amount of these correlations is measured by a correlation function called a *pair amplitude* whose mathematical definition is based on the statistical average of a pair of second-quantized field operators (assumption of a local correlation is valid in the semiclassical limit described in Ch. 2, i.e., assuming that we are interested only in length scales far exceeding λ_F),

$$F_{\sigma\sigma'}(\vec{r}) = \langle \psi_\sigma(\vec{r})\psi_{\sigma'}(\vec{r}) \rangle. \tag{5.1}$$

For conventional superconductors (Al, Nb, Pb, Sn, Ti, etc.), only the spin singlet pair amplitude with $\sigma' = \bar{\sigma}$ is non-vanishing,[1] and $F_{\uparrow\downarrow}(\vec{r}) =$

[1] $\bar{\sigma}$ denotes the spin opposite to σ.

$-F_{\downarrow\uparrow}(\vec{r}) \equiv F(\vec{r})$. In what follows, we concentrate only on this case. Essentially, what eqn (5.1) means is that in the superconducting state with $|F(\vec{r})| > 0$, the positions of pairs of electrons are correlated.

The appearance of superconducting correlations is a phase transition, and is described by an order parameter $F(\vec{r})$. An alternative choice for the order parameter is the pair potential

$$\Delta(\vec{r}) = \lambda(\vec{r})F(\vec{r}), \qquad (5.2)$$

where $\lambda(\vec{r})$ characterizes the strength of the attractive interaction. The latter is a materials constant, but it may depend on position in heterostructures made of different materials. The pair potential and the pair amplitude are complex functions of $\vec{r}$, and often it is useful to present $\Delta(\vec{r})$ in the form

$$\Delta(\vec{r}) = |\Delta(\vec{r})|e^{i\varphi(\vec{r})}. \qquad (5.3)$$

In a bulk superconductor in the presence of a small magnetic field (in the absence of vortices), the absolute value $|\Delta|$ is constant, but the phase $\varphi(\vec{r})$ may be position dependent. This position dependence leads to supercurrent.

In conventional superconductors the effective attractive interaction is related to the electron–phonon coupling. This can be phenomenologically understood as follows. Consider an electron moving in a lattice. As it is negatively charged, its movement exerts a force on the positively charged ions, which then slightly move towards the electron. Now the other electrons in the system may sense this movement, and follow the movement of the ions, i.e., towards the first electron. Thus, an effectively attractive interaction is created, and with certain materials parameters and average distance between the electrons it can even beat the repulsive Coulomb interaction.

5.2 Main physical properties

5.2.1 Current without dissipation

The defining property of a superconductor is that it can carry current without dissipation, i.e., without inducing a voltage drop. In addition to the usual *quasiparticle current* (similar to the normal-state current), there is a possibility for a *supercurrent*. Whereas the previous is proportional to the potential gradient, the latter may be present due to a gradient in the phase $\varphi(\vec{r})$. The supercurrent density may be written phenomenologically as

$$j_S = 2en_s v_S, \qquad (5.4)$$

where n_s is the density of Cooper pairs, related to the superconducting order parameter Δ, and $v_S \propto \nabla\varphi$ is the superfluid velocity. The previous is generally temperature dependent, such that above a certain *critical temperature* T_c it vanishes.[2]

If a current is driven through a superconductor, it flows as a supercurrent, and as a result, no voltage builds up. This means that the

[2]There is also a critical current density j_c: currents exceeding this can break pairs in the superconductor, and thereby lead to a finite resistance.

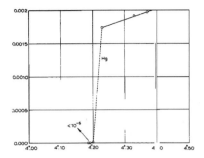

Fig. 5.1 Resistance of mercury around its critical temperature $T_c = 4.2$ K. From the Nobel lectures of H. Kamerlingh Onnes (1913). © The Nobel Foundation

[3]This is true for magnetic fields below a certain critical field H_c. Above this field, superconductivity is destroyed. In certain types of superconductors, above a given field value $H > H_{c1}$, there appears a topological excitation, a vortex, through which the magnetic field may penetrate the sample, and only above a much higher field H_{c2}, superconductivity is totally destroyed.

resistance of the superconductor vanishes as long as $T < T_c$. This behaviour is illustrated in Fig. 5.1, taken from the Nobel lectures of H. Kamerlingh Onnes.

5.2.2 Meissner effect

One of the hallmarks of superconductivity in bulk metals is the Meissner effect. A magnetic field gives rise to circulating supercurrents on the superconductor surface. These supercurrents on the other hand produce an opposing field, such that the total field decays into the bulk superconductor.[3] This is why a superconductor is a *perfect diamagnet*. The corresponding decay length of the magnetic field is called a *penetration depth* λ_p.

Often the penetration depth λ_p is of the same order or longer than the thickness of the films used for experiments in mesoscopic nanoelectronics. In these cases the magnetic field may be considered homogeneous. This is not true in the lateral direction: one of the interesting fields of mesoscopic superconductivity is the study of the vortex structure in small superconducting islands, resulting from the circulating supercurrents (Geim *et al.*, 1997). Moreover, the presence of circulating supercurrents due to a magnetic field is relevant, especially in superconducting loops or SQUIDs, which are discussed in Ch. 9.

5.2.3 BCS theory briefly

The BCS theory of superconductivity is based essentially on two assumptions: (i) attractive electron–electron interaction in a conductor leads to (Cooper) pairing, and (ii) this pairing can be described in a mean-field theory. A sketch of the resultant theory is given below, but for a more detailed treatment, see books specializing in superconductivity, e.g., (Tinkham, 1996).

In the language of second-quantized field operators, the attractive electron–electron interaction can be described with a Hamiltonian of the form

$$H = \sum_\sigma \int d\vec{r}\, \Psi_\sigma^\dagger(\vec{r}) H_0(\vec{r}) \Psi_\sigma(\vec{r})$$
$$+ \sum_{\sigma,\sigma'} \int d\vec{r}\, d\vec{r'}\, \lambda_{\sigma,\sigma'}(\vec{r},\vec{r'}) \Psi_\sigma^\dagger(\vec{r}) \Psi_{\sigma'}^\dagger(\vec{r'}) \Psi_{\sigma'}(\vec{r'}) \Psi_\sigma(\vec{r}),$$

(5.5)

where

$$H_0 = \frac{1}{2m}\left(\frac{\hbar}{i}\nabla - e\mathbf{A}\right)^2 + U(\vec{r}) - \mu$$

(5.6)

is the single-particle Hamiltonian of the electron gas and $\Psi_\sigma(\vec{r})$ annihilates an electron at position $\vec{r}$. For simplicity, let us assume a local spin singlet coupling, i.e.,

$$\lambda_{\sigma,\bar{\sigma}}(\vec{r},\vec{r'}) = \lambda(\vec{r})\delta(\vec{r}-\vec{r'}), \quad \lambda_{\sigma,\sigma}(\vec{r},\vec{r'}) = 0,$$

(5.7)

where $\bar{\sigma}$ is the opposite of spin σ. Now define the pair amplitude and pair potential according to eqns (5.1) and (5.2). The interaction term in eqn (5.5) may be written in terms of pairing amplitudes, the operators $\hat{\delta}(\vec{r})$ describing fluctuations around the mean field, and their complex conjugates,

$$
\begin{aligned}
\Psi_{\bar{\sigma}}(\vec{r})\Psi_\sigma(\vec{r}) &= F(\vec{r}) + \hat{\delta}_{\bar{\sigma}\sigma}(\vec{r}), \\
\Psi_\sigma^\dagger(\vec{r})\Psi_{\bar{\sigma}}^\dagger(\vec{r}) &= F^*(\vec{r}) + \hat{\delta}_{\bar{\sigma}\sigma}^\dagger(\vec{r}).
\end{aligned}
\tag{5.8}
$$

The fluctuation operators $\hat{\delta}_{\bar{\sigma}\sigma}(\vec{r}) = \Psi_{\bar{\sigma}}(\vec{r})\Psi_\sigma(\vec{r}) - F(\vec{r})$ by definition satisfy $\langle \hat{\delta}(\vec{r}) \rangle = 0$. Let us now assume that these fluctuations are small and expand eqn (5.5) into the first order in $\hat{\delta}$.[4] Assuming that the system is symmetric under spin rotation and plugging in the definition of $\hat{\delta}$, we then obtain

$$
\begin{aligned}
H \approx \sum_\sigma \int d\vec{r}\, \Psi_\sigma^\dagger(\vec{r})H_0(\vec{r})\Psi_\sigma(\vec{r}) \\
+ \int d\vec{r}\left[\Delta(\vec{r})^* \Psi_{\bar{\sigma}}(\vec{r})\Psi_\sigma(\vec{r}) + \Delta(\vec{r})\Psi_\sigma^\dagger(\vec{r})\Psi_{\bar{\sigma}}^\dagger(\vec{r}) \right] - E_0.
\end{aligned}
\tag{5.9}
$$

The last term $E_0 = \int d\vec{r}\Delta(\vec{r})F^*(\vec{r})$ describes the energy difference between the normal and superconducting states. To diagonalize this we introduce the *Bogoliubov transformation*

$$
\begin{aligned}
\Psi_\uparrow(\vec{r}) &= \sum_n \gamma_{n\uparrow}u_n(\vec{r}) - \gamma_{n\downarrow}^\dagger v_n^*(\vec{r}) \\
\Psi_\downarrow(\vec{r}) &= \sum_n \gamma_{n\downarrow}u_n(\vec{r}) + \gamma_{n\uparrow}^\dagger v_n^*(\vec{r}).
\end{aligned}
\tag{5.10}
$$

The fermion operators $\gamma_{n\sigma}$ are called *Bogoliubon* operators: they annihilate excitations from the superconducting state. The coefficients $u_n(\vec{r})$ and $v_n(\vec{r})$ are found by requiring that $\gamma_{n\sigma}$ diagonalize the Hamiltonian (5.9). To do this, they must satisfy the Bogoliubov–de Gennes equation,

$$
\begin{pmatrix} H_0(\vec{r}) & \Delta(\vec{r}) \\ \Delta^*(\vec{r}) & -H_0^\dagger(\vec{r}) \end{pmatrix}\begin{pmatrix} u_n(\vec{r}) \\ v_n(\vec{r}) \end{pmatrix} = E_n \begin{pmatrix} u_n(\vec{r}) \\ v_n(\vec{r}) \end{pmatrix}.
\tag{5.11}
$$

If $\Delta(\vec{r}) = 0$, the equations decouple into the form

$$
H_0 u_n = E u_n \tag{5.12a}
$$
$$
H_0^\dagger v_n = -E v_n. \tag{5.12b}
$$

The first is just the usual Schrödinger equation for the electrons (with the only difference that the energies are measured with respect to the Fermi energy). The latter equation describes time-reversed excitations, i.e., holes. In this way, the Bogoliubov–de Gennes equation can be understood as the Schrödinger equation for the superconducting state, where the pair potential couples electron- and hole-like excitations.[5]

Note that an important part of eqn (5.11), the pair potential Δ, is defined with respect to the average pair amplitude of the system described

[4] This expansion is possible in the weak-coupling limit $\lambda N_0 \ll 1$, where N_0 is the density of states at the Fermi level. Most conventional superconductors satisfy this inequality at least approximately, and the deviations from the BCS theory show up mostly in the fact that the critical temperature does not follow the zero-temperature pair potential as in the BCS relation, eqn (5.19).

[5] The electron–hole state space spanned by the vectors of the form $(u_n \;\; v_n)^T$ is typically called the Nambu space.

by the same equation (eqn (5.2)). Hence, the Bogoliubov–de Gennes equation has to be appended by the self-consistency equation for the pair potential. In terms of the eigenfunctions $u_n(\vec{r})$, $v_n(\vec{r})$ of eqn (5.11) with the corresponding eigenenergies E_n, it is written as

$$\Delta(\vec{r}) = \lambda(\vec{r})F(\vec{r}) = \lambda(\vec{r})\sum_n v_n^*(\vec{r})u_n(\vec{r})\tanh\left(\frac{E_n}{2k_BT}\right). \tag{5.13}$$

Equations (5.11) and (5.13) in the general case have to be solved iteratively: first assume an ansatz Δ, which is used in eqn (5.11) to calculate the eigenvalues u_n, v_n. Then calculate a new guess for Δ from eqn (5.13). This iteration continues until Δ has converged. In practice, either of these steps can often be disregarded. In bulk superconductors an analytical solution for u_n, v_n with respect to Δ can be found, and hence the self-consistency equation reduces to a non-algebraic equation which can be solved numerically. At interfaces between normal and superconducting material, the self-consistency changes the results only a little and therefore is often neglected.

Example 5.1 BCS eigensolutions

Let us look for the eigenfunctions of the Bogoliubov–de Gennes equation in a bulk superconductor ($\Delta(\vec{r}) = \Delta_0 e^{i\varphi} = $ const. with $\Delta_0 \in \mathbb{R}$). In a bulk system the solutions can typically be written in terms of plain waves,

$$u(\vec{r}) = u_{\vec{k}}e^{-i\vec{k}\cdot\vec{r}}, \quad v(\vec{r}) = v_{\vec{k}}e^{-i\vec{k}\cdot\vec{r}}.$$

In the absence of a magnetic field these solve the equation

$$\begin{pmatrix} \xi_{\vec{k}} - E_{\vec{k}} & \Delta_0 e^{i\varphi} \\ \Delta_0 e^{-i\varphi} & -\xi_{\vec{k}} - E_{\vec{k}} \end{pmatrix}\begin{pmatrix} u_{\vec{k}} \\ v_{\vec{k}} \end{pmatrix} = 0, \tag{5.14}$$

where $\xi_{\vec{k}} = \hbar^2 k^2/(2m) - \mu$. Moreover, to preserve the normalization of the field operators in the Bogoliubov transformation, eqn (5.10), we must require the normalization $|u_{\vec{k}}|^2 + |v_{\vec{k}}|^2 = 1$.

A non-trivial solution to eqn (5.14) exists if the determinant of the coefficient matrix vanishes. This is satisfied for eigenenergies (see Fig. 5.2)

$$E_{\vec{k}} = \pm\sqrt{\xi_k^2 + \Delta_0^2}. \tag{5.15}$$

The corresponding wave numbers are

$$\begin{aligned} k^+ &= \pm\sqrt{k_F^2 + \frac{2m\sqrt{E_k^2 - \Delta^2}}{\hbar^2}} \approx \pm k_F + \frac{\sqrt{E_k^2 - \Delta^2}}{\hbar v_F} \\ k^- &= \pm\sqrt{k_F^2 - \frac{2m\sqrt{E_k^2 - \Delta^2}}{\hbar^2}} \approx \pm k_F - \frac{\sqrt{E_k^2 - \Delta^2}}{\hbar v_F}, \end{aligned} \tag{5.16}$$

where the latter form is obtained in the usual limit $\sqrt{E_k^2 - \Delta^2} \ll \mu$. In the normal state ($\Delta = 0$), k^+ corresponds to an electron-like solution and k^- to a hole-like solution. Note that the group velocity of the hole-like solution points in the direction opposite to that of the momentum.[6] The corresponding normalized eigensolutions are

$$u_{\vec{k}} = \frac{1}{\sqrt{2}}\sqrt{1 + \frac{\xi_{\vec{k}}}{E_{\vec{k}}}}e^{i\phi}, \quad v_{\vec{k}} = \frac{1}{\sqrt{2}}\sqrt{1 - \frac{\xi_{\vec{k}}}{E_{\vec{k}}}}, \tag{5.17}$$

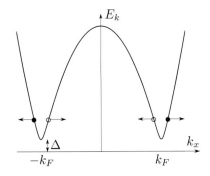

Fig. 5.2 BCS energy spectrum in one direction (x) of momenta; in a bulk system with rotation symmetry spectrum is the same in all other directions. The value of the gap has been greatly exaggerated. The excitations with $|k| > k_F$ are electron-like and they have group velocities parallel with the momentum, whereas the excitations with $|k| < k_F$ are hole-like and have group velocities opposite to the momentum direction.

[6]This can be seen from the fact that increasing energy leads to a decreasing momentum, see Fig. 5.2.

which can be verified by inserting them into eqn (5.14).

5.2.4 Energy gap and BCS divergence

It turns out from the BCS theory that the density of states for the quasiparticle excitations from the superconducting ground state is of the form (see Exercise 5.2)

$$N_S(E) = N_F \frac{|E|}{\sqrt{E^2 - |\Delta|^2}} \theta(|E| - |\Delta|). \tag{5.18}$$

Here E is measured with respect to the Fermi level E_F and N_F is the density of states at $E = E_F$ when the sample is in the normal state. This function is plotted in Fig. 5.3.

There are two main features in this expression that show up in the experiments on superconductors. First there is the *energy gap* $|\Delta|$: for $|E| < \Delta$ there are no available states for the quasiparticles. The pair potential is therefore often called the gap function. Thus, if the temperature is far below $|\Delta|$, quasiparticles do not affect the behaviour of the superconductors. The second notable feature is the *BCS divergence* of the density of states at $E = \Delta$. This shows up, for example, in the tunnelling current as discussed in Ch. 9.

The first experimental evidence of the energy gap was related to the electronic specific heat of the superconductors. In contrast to the linear dependence $C_{en} = \gamma T$ valid in the normal state (see below eqn (2.32)), well below T_c it was found to decay exponentially with the temperature, $C_{es} \propto T_c e^{-|\Delta|/(k_B T)}$. This is a direct consequence of eqn (5.18).

The energy gap is generally temperature dependent, and it vanishes at the critical temperature as any order parameter should do. In the BCS model the energy gap is related with the critical temperature by the relation

$$\Delta(T = 0) = 1.764 k_B T_c. \tag{5.19}$$

The critical temperature of Al, frequently used in mesoscopic experiments, is $T_c \approx 1.2$ K.[7] This means that the zero-temperature energy gap for Al is of the order of 180 μeV.

5.2.5 Coherence length

Apart from the penetration depth, there is another important characteristic length scale which shows up in phenomena related to superconductivity. This is the *coherence length* ξ_0, a characteristic scale that describes the position dependence of the changes in the superconducting order parameter. For example, at an interface between a superconductor and a normal metal where superconductivity is suppressed because of the presence of the normal metal, this is the length scale within which the order parameter regains its bulk value. The coherence length in a pure superconductor (when $\xi_0 \ll \ell_{\text{el}}$) is given by

$$\xi_0^{\text{clean}} = \frac{\hbar v_F}{\pi |\Delta|}, \tag{5.20}$$

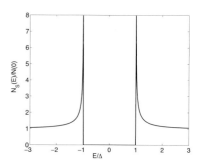

Fig. 5.3 BCS bulk superconducting density of states.

[7]In Al films, the critical temperature may be higher than this, even up to 2 K.

and in the dirty case $\xi_0 \gg \ell_{el}$ it is

$$\xi_0^{dirty} = \sqrt{\frac{\hbar D}{2\Delta}}. \tag{5.21}$$

A typical value for the elastic mean free path in mesoscopic metal films is $\ell_{el} \sim 30$ nm (note that this is just an order-of-magnitude estimate!). With this mean free path and with the Fermi velocity of Al, $v_F = 2.03 \times 10^6$ m/s, D would be roughly 0.02 m^2/s. With the above-mentioned energy gap this yields for aluminum $\xi_0 \approx 200$ nm.

The coherence length and the penetration depth indicate possible crossover lengths for mesoscopic phenomena: indeed, one of the interesting questions is how superconducting effects behave as the structure sizes are shorter than these lengths.

5.3   Josephson effect

The most important superconducting effect for mesoscopic systems is the Josephson effect, which arises if two superconductors are placed in a weak contact, such that the two systems are almost isolated, and the contact only allows the tunnelling of Cooper pairs across it.

In practice, there are three types of Josephson weak links: the most usually encountered weak link is the one where the two superconductors are connected through an insulating barrier (for example, an oxide layer formed between the two). In this case, the transmission probability of all the scattering channels of the contact are very small. Such a junction is called a superconductor–insulator–superconductor (SIS) junction. Other types of weak links are the superconductor–normal metal–superconductor (SNS) weak link and the superconductor-constriction-superconductor (ScS) weak link. In the latter, the weak link is also from the same superconducting material, but its transverse dimension is low.

For a phenomenological description of the Josephson effect, we follow the treatment proposed originally by Richard Feynman (Feynman *et al.*, 2005). Its idea is to start from the fact that the superconducting order parameter is a complex function, of the form $\psi = \sqrt{n}e^{i\varphi}$, where n refers to the amplitude of pairing ('density of Cooper pairs'), and φ is a phase of the order parameter. Now, let us assume that two such superconducting condensates, with order parameters $\psi_1 = \sqrt{n_1}e^{i\varphi_1}$ and $\psi_2 = \sqrt{n_2}e^{i\varphi_2}$ are brought into contact with each other, so that there is a coupling between the two systems. The two functions are thus related by

$$i\hbar\frac{\partial\psi_1}{\partial t} = U_1\psi_1 - K\psi_2 \tag{5.22a}$$

$$i\hbar\frac{\partial\psi_2}{\partial t} = U_2\psi_2 - K\psi_1. \tag{5.22b}$$

The constant K depends on the properties of the junction. For an isolated case, $K = 0$, and the remaining equations would simply describe the macroscopic coherent state of the superconductors. Now let us assume that both sides are connected to two sides of a voltage-biased tunnel barrier, such that there is a potential difference qV between the two, i.e., $U_1 - U_2 = qV$. Here q is the charge of the current carriers ($q = 2e$ for a Cooper pair). Defining the zero of energy halfway between U_1 and U_2 we get

$$i\hbar\frac{\partial\psi_1}{\partial t} = \frac{qV}{2}\psi_1 - K\psi_2 \tag{5.23a}$$

$$i\hbar\frac{\partial\psi_2}{\partial t} = -\frac{qV}{2}\psi_2 - K\psi_1. \tag{5.23b}$$

Let us use the definitions of ψ_i in terms of n_i and φ_i and rewrite the above equations as

$$\frac{\partial\psi_1}{\partial t} = \frac{1}{2\sqrt{n_1}}e^{i\varphi_1}\frac{dn_1}{dt} + i\sqrt{n_1}e^{i\varphi_1}\frac{d\varphi_1}{dt} = \frac{qV}{2i\hbar}\sqrt{n_1}e^{i\varphi_1} - \frac{K}{i\hbar}\sqrt{n_2}e^{i\varphi_2} \tag{5.24a}$$

$$\frac{\partial\psi_2}{\partial t} = \frac{1}{2\sqrt{n_2}}e^{i\varphi_2}\frac{dn_2}{dt} + i\sqrt{n_2}e^{i\varphi_2}\frac{d\varphi_2}{dt} = -\frac{qV}{2i\hbar}\sqrt{n_2}e^{i\varphi_2} - \frac{K}{i\hbar}\sqrt{n_1}e^{i\varphi_1}. \tag{5.24b}$$

Multiply eqn (5.24a) by $e^{-i\varphi_1}$ and eqn (5.24b) by $e^{-i\varphi_2}$, and then separate the real and imaginary parts of the resultant expressions, remembering that both n_i and φ_i are real. The real part yields

$$\frac{dn_1}{dt} = -2K\sqrt{n_1 n_2}\sin(\varphi) \tag{5.25a}$$

$$\frac{dn_2}{dt} = 2K\sqrt{n_1 n_2}\sin(\varphi), \tag{5.25b}$$

where $\varphi = \varphi_2 - \varphi_1$. The imaginary part can be written as

$$\frac{d\varphi_1}{dt} = -\frac{qV}{2\hbar} + K\sqrt{\frac{n_2}{n_1}}\cos(\varphi) \tag{5.26a}$$

$$\frac{d\varphi_2}{dt} = \frac{qV}{2\hbar} + K\sqrt{\frac{n_1}{n_2}}\cos(\varphi). \tag{5.26b}$$

From the real part we obtain an expression for the current from the left superconductor, $j = -q\frac{dn_1}{dt}$,

$$j = j_c\sin(\varphi_2 - \varphi_1), \tag{5.27}$$

where $j_c = 2qK\sqrt{n_1 n_2}$ is the critical current density for this junction. Multiplying this by the area A of the junction yields

$$I = I_C\sin(\varphi_2 - \varphi_1), \tag{5.28}$$

with $I_C = Aj_c$. Equation (5.28) is called the *dc Josephson relation*: it shows that even in the absence of a potential drop, there may be a current through the junction. Moreover, this current depends on the phase difference across the junction. If the current exceeds the critical current I_C, a voltage starts to appear. A more microscopic approach yields an equation for I_C, relating it to the junction resistance R_N in the normal state and the magnitude of the order parameter in the superconductors (see eqns (5.37) and (5.38) below).

Let us consider the difference of the imaginary parts of the above relations, eqn (5.26b), with similar superconductors. In this case $n_1 \approx n_2$ and we obtain

$$\frac{d\varphi}{dt} = \frac{d}{dt}(\varphi_2 - \varphi_1) = \frac{qV}{\hbar} = \frac{2eV}{\hbar}. \tag{5.29}$$

The latter form is obtained by using the charge $q = 2e$ of a Cooper pair. Solving this for a constant voltage yields $\varphi = \varphi(0) + 2eVt/\hbar$ and

$$I = I_C \sin(\varphi(0) + 2eVt/\hbar). \tag{5.30}$$

Therefore, the current through a Josephson junction with an applied dc voltage oscillates in time. This is the *ac Josephson effect*.[8]

The text above presents only a simplified model for a Josephson junction, which still captures the two most relevant effects. However, Josephson junctions are one of the most complicated systems (yet controllable) studied in physics, and they continue to be studied even today. Chapter 9 introduces few of the mesoscopic phenomena observed in Josephson junctions.

5.4   Main phenomena characteristic for mesoscopic systems

Apart from the phenomena related to the bulk properties of superconductors, two microscopic effects show up especially in small hybrid structures composed of normal metals and superconductors. These are the Andreev reflection and the proximity effect, which are in many ways related. To illustrate the microscopic theory of superconductivity, we calculate the Andreev reflection probabilities using the Bogoliubov–de Gennes equation and use these to provide a microscopic theory for the Josephson effect via the Andreev bound states.

5.4.1   Andreev reflection

An important phenomenon in mesoscopic superconducting systems is the Andreev reflection (Andreev, 1964), where an electron in a normal metal hits a superconductor and is converted into a hole. This is a result of the appearance of the energy gap in the superconductor. Consider a structure where a normal-metal wire (say, Cu) is brought in good contact with a superconductor (say, Al). Far in the normal metal, we see no effects of superconductivity, and the density of states is flat. On the other hand, far in the superconductor, there are no effects from the normal metal, and the density of states has a gap. Then, if an electron excitation in the normal metal with energy lower than the gap (i.e., $E < \Delta$)[9] hits a superconductor, it cannot enter it because there are no states for it

[8] The dc Josephson relation is not fundamental as the current-phase relation is not always sinusoidal (see Fig. 5.9). However, deviations from the ac Josephson relation are rarely encountered.

[9] Note that here energies are calculated with respect to the Fermi level.

below the gap. However, if the contact is clean,[10] there can also be no direct elastic scattering into another electron state traversing out of the contact. What happens is an *Andreev reflection*: the electron reflects as a hole (an excitation below the Fermi sea). Hole has a positive charge, and thus the net result is an extra Cooper pair (with a double electronic charge) inside the superconductor. This thus means that Andreev reflection carries charge current into or out of (in the inverse process, hole converting into an electron) the superconductor.[11]

The process can be exemplified with a simple system where a normal metal (with $\lambda(\vec{r}) = 0$) is in contact with a superconductor (with $\lambda(\vec{r}) = \lambda = $ const.), as in Fig. 5.4.[12] For simplicity, let us assume that the materials are clean, such that $U(\vec{r}) = 0$, and that there is no magnetic field. Moreover, we can neglect self-consistency by assuming a constant $\Delta(\vec{r}) = \Delta$ inside the superconductor.

The solution to the Bogoliubov–de Gennes equation is of the form $\psi = \psi_{\text{left}} + \psi_{\text{right}}$ (as in Fig. 5.4), where

$$\psi_{\text{left}} = \begin{pmatrix} 1 \\ 0 \end{pmatrix} e^{ik_N^+ x} + r_{ee} \begin{pmatrix} 1 \\ 0 \end{pmatrix} e^{-ik_N^+ x} + r_{he} \begin{pmatrix} 0 \\ 1 \end{pmatrix} e^{ik_N^- x} \quad (5.31)$$

is the wave function on the left-hand side of the interface, composed of an incident electron moving towards the right, a normally reflected electron, and a hole that has Andreev reflected from the electron state. The wave numbers are given in eqn (5.16) and in the normal state (denoted with subscript N) they are obtained by setting $\Delta = 0$. On the superconducting side, the electron can transmit into one of the two types of eigenstates of a bulk superconductor (see Example 5.1). Therefore, the solution in the superconductor is of the form

$$\psi_{\text{right}} = t_+ \begin{pmatrix} u_0 e^{i\phi_S} \\ v_0 \end{pmatrix} e^{ik_S^+ x} + t_- \begin{pmatrix} v_0 e^{i\phi_S} \\ u_0 \end{pmatrix} e^{-ik_S^- x}. \quad (5.32)$$

Here, the wave number is given by eqn (5.16) with a non-zero Δ. If the NS interface is clean, the transmission and reflection coefficients are obtained by requiring continuity of the functions and their derivatives at the surface point, $x = 0$. Such a calculation yields

$$r_{ee} = 0, t_{+/-} = 0, |r_{he}|^2 = 1, \quad |E| < \Delta$$

$$r_{ee} = t_- = 0, |r_{he}|^2 = \frac{E - \sqrt{E^2 - \Delta^2}}{E + \sqrt{E^2 - \Delta^2}}, |t_+|^2 = 1 - |r_{he}|^2, \quad |E| > \Delta.$$
$$(5.33)$$

For sub-gap energies Andreev reflection is thus the only allowed process.

If the interface is not clean, also normal reflection takes place. Figure 5.5 illustrates the probabilities $|r_{he}|^2$, $|r_{ee}|^2$, $|t_+|^2$ and $|t_-|^2$ of the various processes as functions of energy for a few values of the transmission probability T of the interface (this is the transmission probability one would obtain for $\Delta \to 0$). As the Andreev reflection entails the transmission of two particles through the interface (the initial electron and the reflected hole), its probability decays as T^2 for $E \ll \Delta$ and $T \ll 1$. In a tunnel structure where $T \ll 1$, it can therefore be disregarded.

[10] Strictly speaking, besides the absence of an extra scattering potential at the interface, this also requires that the Fermi velocities on the two sides of the interface are the same.

[11] However, it does not carry heat as the Cooper pairs do not carry entropy. The resultant poor heat conductance of normal-metal superconductor interfaces even in the presence of high charge conductance was the phenomenon Andreev explained.

ψ_{left} $\quad$ N $\mid$ S $\quad$ ψ_{right}

$\begin{pmatrix} 1 \\ 0 \end{pmatrix} e^{ik_N^+ x} \longrightarrow$

$\longleftarrow r_{ee} \begin{pmatrix} 1 \\ 0 \end{pmatrix} e^{-ik_N^+ x}$ $\quad t_+ \begin{pmatrix} u_0 e^{i\varphi_S} \\ v_0 \end{pmatrix} e^{ik_S^+ x} \longrightarrow$

$\longleftarrow r_{he} \begin{pmatrix} 0 \\ 1 \end{pmatrix} e^{ik_N^- x}$ $\quad t_- \begin{pmatrix} v_0 e^{i\varphi_S} \\ u_0 \end{pmatrix} e^{-ik_S^- x} \longrightarrow$

Fig. 5.4 Scattering states in the presence of Andreev reflection at a NS interface.

[12] This example is due to Blonder, Tinkham and Klapwijk (1982), frequently abbreviated as BTK.

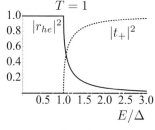

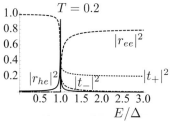

Fig. 5.5 Probabilities for Andreev reflection ($|r_{he}|^2$, solid lines), normal reflection ($|r_{ee}|^2$, dashed line), transmission without ($|t_+|^2$, dotted lines) and with ($|t_-|^2$, dash-dotted) branch crossing. Top: normal-state transmission of the interface $T = 1$. Bottom: normal-state transmission $T = 0.2$. For more details, see (Blonder *et al.*, 1982).

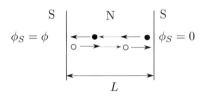

Fig. 5.6 Andreev bound states are formed from multiple Andreev reflections at the two NS interfaces, satisfying a resonant condition.

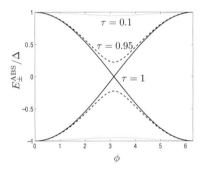

Fig. 5.7 Andreev bound state energies for a few values of the transmission probability.

[13]The general relation between the current I and the phase derivative of the total energy can be justified by noting that phase ϕ and charge Q are conjugate elements (see Sec. 9.4 below), satisfying the commutation relation (9.36). Then $I = -\dot{Q} = -(i/\hbar)[H, Q] = -(i/\hbar)\partial_\phi H[\phi, Q] = (2e/\hbar)\partial_\phi H$, where H is the Hamiltonian of the system. Strictly speaking, eqn (5.37) applies only for short junctions, whose length $L \ll \xi_0$: for longer junctions, there is also a contribution from states with $E \gtrsim \Delta$ (Wendin and Shumeiko, 1996).

5.4.2　Andreev bound states

Consider what happens if we separate two superconductors by a normal metal as depicted in Fig. 5.6. Now the hole that Andreev reflected from the left NS interface may find its way to the right interface, and Andreev reflect again into an electron state. Returning back to the left interface, this electron may again Andreev reflect, and so on. If the total phase acquired within a full cycle is a multiple of 2π, this results into a bound state. The phases acquired in an Andreev reflection of an electron into a hole (φ_{eh}) and of a hole into an electron (φ_{he}) are (see Exercise 5.5)

$$\varphi_{eh} = -\arccos(E/\Delta) + \phi_S, \quad \varphi_{he} = -\arccos(E/\Delta) - \phi_S, \quad (5.34)$$

where ϕ_S is the phase of the superconductor. Therefore, the total phase acquired in a full cycle is

$$\phi_{\text{tot}}^1 = (k_h + k_e)L + \phi + 2\arccos(E/\Delta),$$

where $(k_h+k_e)L \approx 2EL/v_F$ is the total dynamical phase acquired within traversing the normal metal. But there is also the opposite process, where a hole reflects into an electron from the left interface. This has the total phase

$$\phi_{\text{tot}}^2 = (k_e + k_h)L - \phi + 2\arccos(E/\Delta).$$

For a bound state these must satisfy the condition $\phi_{\text{tot}}^{1,2} = 2n\pi$, which give the bound state energies.

In the following, let us concentrate only on the limit of a short junction $L \ll \xi_0$, where we can disregard the dynamical phase. In this case the bound state energies are given by

$$E_{\pm}^{\text{ABS}} = \pm\Delta\cos(\phi/2). \quad (5.35)$$

These are Andreev bound states, which dictate the behaviour of superconducting weak links. This formula is valid for the case of a clean interface. In the presence of some scattering at the interface, the bound state energies become (see Example 5.2)

$$E_{\pm}^{\text{ABS}} = \pm\Delta[1 - \tau_p\sin^2(\phi/2)], \quad (5.36)$$

where τ_p is the transmission probability for a transmission eigenmode p through the system inside the normal metal. These energies are plotted in Fig. 5.7 and an example of their direct measurement is illustrated in Fig. 5.8.

As shown in the above example, the net result of each cycle of Andreev reflections is a transfer of a Cooper pair between the superconductors. Andreev bound states can hence carry supercurrent. This is actually the mechanism of the Josephson effect through the normal metal. The current carried by each pair of states is given by[13]

$$I_S = \frac{2e}{\hbar}\sum_{\pm}\frac{\partial E_{\pm}^{\text{ABS}}(\phi)}{\partial\phi}\tanh\left[\frac{E_{\pm}^{\text{ABS}}}{2k_BT}\right] = \frac{e\Delta^2}{2\hbar}\frac{\tau\sin(\phi)}{E_{+}^{\text{ABS}}}\tanh\left[\frac{E_{+}^{\text{ABS}}(\phi)}{2k_BT}\right].$$

$$(5.37)$$

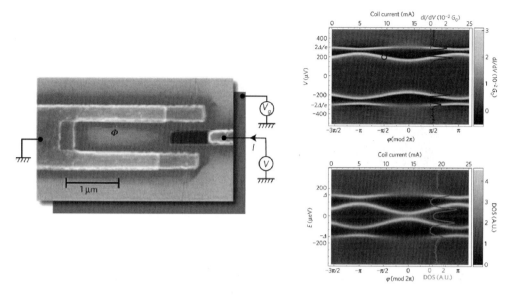

Fig. 5.8 Tunnel spectroscopy of Andreev bound states in a carbon nanotube quantum dot. Left: Measured system consisting of a superconducting fork with ends connected by a carbon nanotube (thin wire) and a superconducting tunnel probe (on the right). Right top: Measured differential conductance. Right bottom: the resultant density of states reveal the dependence of the ABS as functions of the superconducting phase difference. For an explanation of tunnel spectroscopy, see Example 1.1. (J.-D. Pillet *et al.*, *Nature Phys.* **6**, 965 (2010), Figs. 1 and 2.) Used by permission from Macmillan Publishers Ltd., © 2010.

This is plotted as a function of phase ϕ in Fig. 5.9. Note that the dc Josephson relation $I_S(\phi) = I_C \sin(\phi)$ is satisfied only for the case of small transmission or at high temperatures (see Exercise 5.6).

For low transmissions, $E_+^{\text{ABS}} \approx \Delta$. Summing over the transmission probabilities for different eigenchannels, we then get the Ambegaokar–Baratoff (1963) relation for the supercurrent

$$I_S = \frac{\pi \Delta \sin(\phi)}{2e R_N} \tanh\left(\frac{\Delta}{2k_B T}\right), \tag{5.38}$$

where $R_N^{-1} = 2e^2/h \sum_p \tau_p$ is the normal-state resistance of the junction. For example, for an Al Josephson junction with $R_N = 1\Omega$, we would obtain $I_C = 280 \ \mu\text{A}$ at $T = 0$.

The current-phase relation obtained from (5.37) was recently confirmed experimentally in atomic point contacts, see (Della Rocca *et al.*, 2007). Besides tunnel junctions and atomic contacts, the Josephson current has been observed in various systems, including carbon nanotubes (Kasumov *et al.*, 1999), semiconductor nanowires (Doh *et al.*, 2005), quantum dots (van Dam *et al.*, 2006) and graphene (Heersche *et al.*, 2007).

Besides providing a microscopic description of the Josephson effect, it has been suggested that the Andreev bound states may be used in realizing quantum bits; see for example (Despósito and Levy Yeyati, 2001; Zazunov *et al.*, 2003).

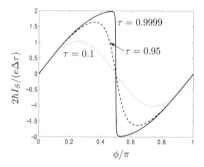

Fig. 5.9 Current carried by the Andreev bound states at $T = 0$.

Example 5.2 Andreev bound states with finite transmission

The effect of a finite transmission probability (i.e., $\tau_n \neq 1$) on the Andreev bound states of a short SNS junction can be calculated in a rather straightforward manner using the scattering theory (Beenakker, 1991*b*).

Consider the system sketched in Fig. 5.10. It can be divided into five parts (from left to right): a superconductor with an order parameter $\Delta e^{i\phi/2}$, the 'left' and 'right' parts of the normal metal and the scattering region connecting them, and another superconductor with the order parameter $\Delta e^{-i\phi/2}$. Besides the scattering region, these parts can be considered as 'leads' supporting propagating waves of the type

$$\psi_{n,\beta}^{\pm}(x,\vec{r}_\perp) = \begin{pmatrix} u_{n\beta}(\vec{r}_\perp) \\ v_{n\beta}(\vec{r}_\perp) \end{pmatrix} e^{\pm i k_n^\beta x}, \tag{5.39}$$

where the vector has been written in the Nambu space, and $u_{n\beta}(\vec{r}_\perp)$, $v_{n\beta}(\vec{r}_\perp)$ describe the local eigenfunctions of the Bogoliubov–de Gennes equation (see Example 5.1) for mode n in part β at the transverse coordinate $\vec{r}_\perp$. Inside the normal metal, $\Delta = 0$ and the eigenfunctions are electron-like ($\propto \begin{pmatrix} 1 & 0 \end{pmatrix}^T$) or hole-like ($\propto \begin{pmatrix} 0 & 1 \end{pmatrix}^T$).

As discussed in Ch. 3, the connections between the eigenfunctions in different parts of the system can be described using scattering matrices. Let us concentrate on the energy region $|E| < \Delta$ and assume that all normal scattering is concentrated within the region between the 'left' and 'right' parts of the normal metal. In this case, the waves inside the normal metal are not transmitted into the superconductor, and the absolute value of the Andreev reflection probability equals unity as discussed in Sec. 5.4.1. However, each Andreev reflection comes with a phase φ, which depends on the direction of the process, see eqn (5.34).

The waves in the 'left' and 'right' parts of the normal metal are connected via the scattering matrix of the type

$$s_N = \begin{pmatrix} s_0(E) & \hat{0} \\ \hat{0} & s_0^\dagger(-E) \end{pmatrix}, \qquad s_0 = \begin{pmatrix} r_L & t \\ t & r_R \end{pmatrix}, \tag{5.40}$$

where s_N is written within the Nambu electron–hole space, using the fact that holes can be viewed as time-reversed electrons and that normal scattering does not mix electrons and holes. Here and in the following, $\hat{0}$ refers to the matrix of zeros and $\hat{1}$ is the unit matrix, and the dimension of these matrices is clear from the context. Moreover, we may assume for simplicity the transmission matrices t for the scattering from the left to the right to equal the transmission matrix for the scattering from the right to left.[14] The usual unitarity constraints for s_N yield

$$r_{L/R}^\dagger r_{L/R} = \hat{1} - t^\dagger t, \quad r_L^\dagger t = -t^\dagger r_R, \quad r_{L/R} r_{L/R}^\dagger = \hat{1} - t t^\dagger. \tag{5.41}$$

The matrix s_N couples the incoming modes $\psi_{\text{in}} = \begin{pmatrix} \psi_{nL}^+ & \psi_{nR}^- \end{pmatrix}^T$ to the outgoing modes $\psi_{\text{out}} = \begin{pmatrix} \psi_{nL}^- & \psi_{nR}^+ \end{pmatrix}^T$. On the other hand, the Andreev reflections on the NS interfaces also couple these states, such that $\psi_{\text{in}} = s_A \psi_{\text{out}}$, where

$$s_A = \alpha \begin{pmatrix} \hat{0} & r_A \\ r_A^\dagger & \hat{0} \end{pmatrix}, \qquad r_A = \begin{pmatrix} e^{i\phi/2}\hat{1} & \hat{0} \\ \hat{0} & e^{-i\phi/2} \end{pmatrix}. \tag{5.42}$$

For the bound states, we hence get $\psi_{\text{out}} = s_N(E_p) s_A \psi_{\text{out}}$, which is satisfied at certain energies E_p. For these energies the matrix $\hat{1} - s_N(E_p) s_A$ is

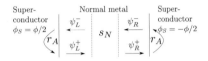

Super-conductor $\phi_S = \phi/2$ Normal metal   Super-conductor $\phi_S = -\phi/2$

ψ_L^- ψ_R^-

r_A s_N r_A

ψ_L^+ ψ_R^+

Fig. 5.10 Setup for the scattering theory description of the Andreev bound states.

[14]This corresponds to assuming an equal number of modes on each side and the time reversal symmetry. As the division to the 'left' and the 'right' is rather a mathematical model than an attempt to discuss the real 'left' and 'right' sides of the scatterer, such an assumption is fully justified.

non-invertible; in other words, $\det[\hat{1} - s_N(E_p)s_A] = 0$. Finding the zeros of this determinant hence establishes the Andreev bound state energies in the presence of scattering.

Let us start by using the property

$$\det\begin{pmatrix} a & b \\ c & d \end{pmatrix} = \det\left[\begin{pmatrix} a & 0 \\ c & \hat{1} \end{pmatrix}\begin{pmatrix} \hat{1} & a^{-1}b \\ 0 & d - ca^{-1}b \end{pmatrix}\right] = \det(a)\det(d - ca^{-1}b),$$

where the latter equality follows from $\det ab = \det a \det b$. This allows us to simplify the electron–hole part of the matrices, so that the bound state condition becomes

$$\det[\hat{1} - \alpha(E_p)^2 s_0(-E_p)^\dagger r_A^\dagger s_0(E_p)r_A] = 0. \tag{5.43}$$

In the following we assume that the normal metal is much shorter than the superconducting coherence length, or in other words $\Delta \ll \hbar/\tau_t$, where τ_t is the average time it takes for the electrons to traverse between the NS contacts. In this limit the scattering matrix s_0 is energy independent, and we may approximate $s_0(E_p) \approx s_0(0)$. Then a direct calculation for the bound state condition yields

$$\det\left[\hat{1} - \alpha(E_p)^2 \begin{pmatrix} r_L & t \\ t & r_R \end{pmatrix}^\dagger \begin{pmatrix} e^{-i\phi/2}\hat{1} & \hat{0} \\ \hat{0} & e^{i\phi/2}\hat{1} \end{pmatrix}\begin{pmatrix} r_L & t \\ t & r_R \end{pmatrix}\begin{pmatrix} e^{i\phi/2}\hat{1} & \hat{0} \\ \hat{0} & e^{-i\phi/2}\hat{1} \end{pmatrix}\right]$$

$$= \det\left[\hat{1} - \alpha(E_p)^2\begin{pmatrix} t^\dagger t(e^{i\phi}-1)+1 & r_L^\dagger t(e^{-i\phi}-1) \\ t^\dagger r_L(1-e^{i\phi}) & t^\dagger t(e^{-i\phi}-1)+1 \end{pmatrix}\right]$$

$$= \det\left[\hat{1}(1-\alpha^2)^2 - 4\alpha(E_p)^2 t^\dagger t \sin^2(\phi/2)\right]$$

$$= \frac{4[\Delta^2 + 2iE_p(iE_p + \sqrt{\Delta^2 - E_p^2})]}{\Delta^4}\det\left[(\Delta^2 - E_p^2)\hat{1} + \Delta^2 t^\dagger t \sin^2(\phi/2)\right] = 0.$$

This equation repeatedly uses the unitarity conditions (5.41). What is left is to diagonalize $t^\dagger t$, expressing it in terms of the transmission eigenvalues τ_p. The bound state condition then becomes

$$E_p = \pm\Delta[1 - \tau_p \sin^2(\phi/2)]. \tag{5.44}$$

Hence, for a system with N modes, there are $2N$ Andreev bound states with energy E_p.

In addition to the Andreev bound states, the SNS junction contains a continuum of states for energies above Δ and below $-\Delta$. However, in equilibrium for short junctions, these do not contribute to the supercurrent (Beenakker, 1991b).[15]

[15]For examples on the deviations from the short-junction limit $\Delta \ll \hbar/\tau_t$, see (Kopnin *et al.*, 2006) in the ballistic and (Heikkilä *et al.*, 2002) in the diffusive case, and from equilibrium, see (Bergeret *et al.*, 2010).

5.4.3 Proximity effect

Due to the Andreev reflection, Cooper pairs can 'pop in' the normal metal near the NS interface (see Fig. 5.11). This process induces a non-zero pairing amplitude $F(\vec{r})$ also inside the normal metal (note that as the coupling constant λ is assumed to vanish there, the pair potential stays zero). The inverse effect also exists: because the pairing amplitude vanishes far in the normal metal, that in the superconductor near the interface is diminished. The whole phenomenon is called the superconducting proximity effect.

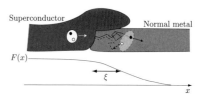

Fig. 5.11 Typical normal-superconductor contacts for the study of the proximity effect are deposited as overlap junctions of two materials, for example superconducting aluminum or niobium and normally conducting copper or silver. As a result, Cooper pairs leak from the superconductor to the normal-metal side. The 'penetration depth' of the proximity effect in diffusive systems is $\xi = \sqrt{\hbar D/(2\pi k_B T)}$. For typical metal samples at $T \approx 100$ mK, ξ is of the order of a few μm.

In most practical cases, the quantitative treatment of the proximity effect has to be carried out with the approach of (quasi-classical) Green's functions (Belzig *et al.*, 1999). As this is outside the scope of this book, we do not dwell on this subject. However, it is useful to discuss at least the basic properties of the proximity effect. It has been shown to slightly increase the local conductivity and decrease the thermal conductivity of the normal metal (Belzig *et al.*, 1999). In systems with multiple NS interfaces, the proximity effect can also give rise to thermoelectric effects; see (Virtanen and Heikkilä, 2004; Virtanen and Heikkilä, 2007). It also gives rise to variations in the local density of states near the N–S interface (for a recent measurement, see (le Sueur *et al.*, 2008)). However, the most important consequence of the proximity effect is the fact that a normal metal sandwiched between two superconductors can carry a supercurrent (Heikkilä *et al.*, 2002), and can hence realize a Josephson junction.

Further reading

- There are many books dedicated to superconductivity, such as (Tinkham, 1996). Also, many textbooks on solid state physics contain chapters on superconductivity.

- The properties of Josephson junctions are discussed, for example, in (Likharev, 1979) and

(Barone and Paterno, 1982).

- The superconducting proximity effect is discussed, for example, in (de Gennes, 1999), more recent reviews dealing with the electronic transport and supercurrent are for example (Belzig *et al.*, 1999) and (Golubov *et al.*, 2004).

Exercises

(5.1) Verify the eigenenergies and eigensolutions of the Bogoliubov–de Gennes equation (see Example 5.1).

(5.2) Consider a bulk clean superconductor ($\Delta(\vec{r}) =$ const.) at a vanishing magnetic field. Assuming that the normal-state density of states $\frac{d\xi_k}{dk} \approx 1/N_F =$ const. around the Fermi level, derive the BCS density of states, eqn (5.18). Hint: use the eigenenergies $E_{\vec{k}}$ from eqn (5.15) and the fact that in this case the density of states $N(E) = dk/dE = (dE/dk)^{-1}$.

(5.3) Show that the BCS density of states satisfies

$$\int_{-E^*}^{E^*} dE N_S(E) = 2N(0)E^*, \qquad (5.45)$$

for $E^* \gg \Delta$, i.e., there are the same total number of states in the superconducting state as in the normal state.

(5.4) **Shapiro steps.** Consider a Josephson junction driven with a time-dependent voltage, $V = V_0 + V_1 \cos(\omega_1 t)$. Calculate the corresponding time-dependent phase using the ac Josephson relation (5.29), and show that the time-averaged supercurrent,

$$\lim_{T \to \infty} \frac{1}{T} \int_0^T dt I(t), \qquad (5.46)$$

is finite for $V_0 = n\hbar\omega_1/2e$. Hint: You can use the expansion

$$e^{iz\sin\theta} = \sum_{k=\infty}^{\infty} e^{ik\theta} J_k(z), \qquad (5.47)$$

where $J_k(z)$ is the kth Bessel function. This effect in current-biased junctions defines the voltage V_0 so accurately that it is used for voltage metrology.

(5.5) Calculate the phases of the Andreev reflection probability amplitude, eqn (5.34), for $|E| < |\Delta|$ and for a clean NS interface where the superconductor has a pair potential $|\Delta|e^{i\phi_S}$. Hint: Assume the scattering setup depicted in Fig. 5.4, and use as boundary conditions the continuity of the wave function and its derivative at the NS interface. Express the wave functions inside the superconductor by using the BCS eigenfunctions from eqn (5.17).

(5.6) Show that the supercurrent through a weak link, eqn (5.37), becomes a sinusoidal function of the phase even with large transmission τ at high temperatures. What is the requirement for this temperature?

(5.7) **Question on a scientific paper.** Take the paper Kasumov *et al.*, *Science* **284**, 1508 (1999) presenting measurements of supercurrents through carbon nanotubes. There were two kinds of samples: nanotube ropes and single-wall nanotubes. Find out a) how the critical temperature of superconductivity (Fig. 2) and b) the critical supercurrent (Fig. 3) for these samples were obtained. Nanotubes are not intrinsically superconducting. Then, what can explain these superconducting characteristics?

6 Fluctuations and correlations

The above chapters frequently mention the fluctuations of currents or voltages. These fluctuations are an everyday phenomenon for any electronic engineer, or for example someone trying to analyze the resolution of electronic measuring equipment. In those cases the fluctuations act as 'noise', i.e., trying to hinder the actual operation based on the average quantities. This chapter describes such fluctuations, tells how they can be quantitatively characterized, and shows how they can sometimes be used to obtain further information about the studied objects.

6.1 Definition and main characteristics of noise

Consider a time-resolved current measurement with a constant bias voltage source:

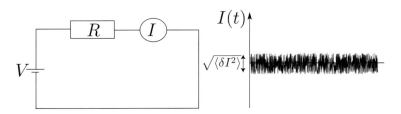

Fig. 6.1 Time-resolved current measurement with a constant voltage bias.

The magnitude of the current fluctuations around the average may be described by the correlation function

$$S_I(t, t') = 2\langle \delta I(t) \delta I(t') \rangle, \quad \text{'classical current'}$$
$$= \langle \{ \delta \hat{I}(t), \delta \hat{I}(t') \} \rangle, \quad \text{'quantum current'}, \tag{6.1}$$

where $\delta \hat{I} = \hat{I}(t) - \langle \hat{I}(t) \rangle$. Here the brackets $\langle \cdot \rangle$ denote an ensemble average over all the possible (microscopic) states of the system. Usually this is equivalent to an average over repeated measurements, or a time average.[1] In the quantum case, the current operators at different times do not necessarily commute. A typical way to define the correlation function is then to symmetrize it and to use the anticommutator $\{\hat{A}, \hat{B}\} \equiv \hat{A}\hat{B} + \hat{B}\hat{A}$. However, sometimes also the non-symmetrized correlator needs to be used, as discussed in Sec. 6.6.

[1] This is the ergodic hypothesis in statistical physics.

In a stationary system where the ensemble-averaged current $\langle I \rangle$ does not vary in time, $S_I(t,t')$ depends only on the time difference between the events, $S_I(t,t') = S_I(t'-t)$. In this case it is customary to describe the Fourier transform of $S_I(t'-t)$,

$$S_I(\omega) \equiv \int_{-\infty}^{\infty} d(t'-t) e^{i\omega(t'-t)} S_I(t'-t). \qquad (6.2)$$

In the following, we concentrate on the frequency domain, and thus there should be no confusion between the different functions $S_I(\omega)$ and $S_I(t'-t)$. The function $S_I(\omega)$ is often called the noise power spectral density, and as described by the Wiener–Khintchine theorem, it describes the squared deviation $\delta I(\omega)^2$ of the signal around frequency ω (see Exercise 6.1).

Example 6.1 Two-level current

To clarify the notion of the correlation function and its ensemble average, let us consider a fictitious system where the current may have only either the value $I_+ = +\delta I$ or $I_- = -\delta I$. Which value the current has at a given moment depends on the microscopic features of the system and the value of the current may change at any time, independent of what the states were at a prior time. Thus, the fluctuations are of the form shown in Fig. 6.2. This type of noise is called telegraph noise. Let us now calculate $S(t,t')$. For $t' \neq t$ we have

$$S_I \equiv 2\langle \delta I(t)\delta I(t') \rangle = 2 \sum_{n,n'=\pm 1} p_n p_{n'}(n\delta I)(n'\delta I)$$

$$= 2(\delta I)^2 \left(\frac{1}{2}\right)^2 (1\cdot 1 + (-1)\cdot 1 + 1\cdot(-1) + (-1)\cdot(-1)) = 0. \qquad (6.3)$$

Here the probability for state n is p_n, assumed equal for both states, $p_{+1} = p_{-1} = 1/2$. The sum over n and n' is due to the ensemble average, and it goes over all possible initial and final states at time instants t and t'.

For $t = t'$ we have to have $n = n'$ such that

$$S_I(t,t') = 2 \sum_{n=\pm 1} p_n(n\delta I)(n\delta I) = 2(\delta I)^2 \frac{1}{2}(1\cdot 1 + (-1)\cdot(-1)) = 2(\delta I)^2. \quad (6.4)$$

Thus, we obtain

$$S_I(t,t') = 2(\delta I)^2 \delta(t-t'). \qquad (6.5)$$

Fourier transforming, we get

$$S_I(\omega) = 2(\delta I)^2, \qquad (6.6)$$

independent of frequency ω. This kind of noise is called white, and it reflects the assumption that the correlations between the currents $I(t)$ and $I(t')$ disappear for any, however small time difference $t'-t$.[2] In practice, there always is a time scale τ during which such transitions between states with different currents may take place. This gives rise to a finite width $\sim \tau$ for $S_I(t'-t)$, and correspondingly, it makes $S_I(\omega)$ frequency dependent at frequencies $\omega \sim 1/\tau$ (see Fig. 6.3).

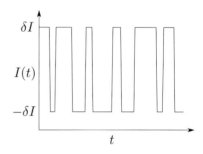

Fig. 6.2 Telegraph noise

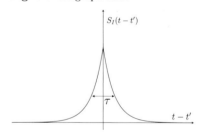

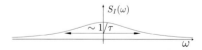

Fig. 6.3 Noise in a system with a correlation time scale τ. Top: time domain; bottom: frequency domain.

[2] Such a process without memory is called Markovian.

6.1.1   Motivations for the study of noise

Typically, noise is a feature that limits the accuracy of the detected average signal ($\langle I \rangle$), and hence it can be an *undesired property* that one tries to minimize. Attempts for such minimization are already of course a fair motivation for its study. For example, one can show that for any finite-frequency measurements, devices show an unavoidable noise source—quantum vacuum fluctuations—that does not vanish even at low temperatures. On the other hand, using low-frequency measurements eventually runs into trouble due to the so-called $1/f$-noise which is a consequence of the fact that all the microscopic features of the measurement setup are never fully known and controlled. However, the study of *shot noise* has shown that even the noise itself may *contain information on the measured system*. Moreover, the *descriptions of the interaction between small quantum systems and their environment*, for example an atom in a photon field, employ the concept of noise (although then, one often talks about correlations rather than noise). Such a description is especially important in quantum computing as noise provides the dephasing of the quantum systems, but also is the studied feature in the descriptions of quantum measurements. Examples of this are discussed in Sec. 6.6.

The main types of (current) fluctuations are[3]

- Thermal noise $S(\omega \approx 0) = 4\frac{k_B T}{R}$
- Shot noise $S(\omega \approx 0) = 2eFI$, where F is the *Fano factor*
- Quantum or vacuum fluctuations $S(\omega \gg \max(k_B T, eV)/\hbar) = 4\frac{\hbar\omega}{R}$
- $1/f$ or flicker noise due to uncontrollable and/or unknown changes in the system or the measurement equipment
- Correlations at some given frequency ω_0, reflecting some energy or time scale relevant for the studied system.

6.1.2   Fluctuation–dissipation theorem

Fluctuations of equilibrium linear systems can be quantified through the fluctuation–dissipation relation (Callen and Welton, 1951). Derivations of this relation are based on the theory of linear systems, linear response theory (Kubo, 1957). This theory is outlined in Appendix C, and it is used to derive the fluctuation–dissipation relation, which for the current fluctuations reads

$$S_I(\omega) = 2\hbar\omega \mathrm{Re}[Y(\omega)]\left[\coth\left(\frac{\hbar\omega}{2k_B T}\right) + 1\right]. \qquad (6.7)$$

Here $Y(\omega)$ is the impedance of the sample over which the fluctuations are quantified. The sample is assumed to be in equilibrium with temperature T. This is the theorem for the non-symmetrized current correlator; the symmetrized version does not have the second term.

The fluctuation–dissipation theorem is one of the most fundamental theorems of statistical physics: it states that a dissipative process is

[3]In this text we concentrate on the current fluctuations. However, the concept is very general, and we could use a similar type of description for the fluctuations of any observable.

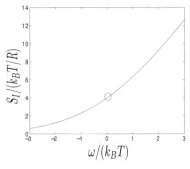

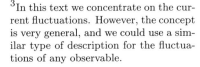

Fig. 6.4 Frequency dependence of equilibrium fluctuations with a frequency independent admittance $Y(\omega) = 1/R$. The circle points to the zero-frequency thermal fluctuations $S_I = 4k_B T/R$.

always accompanied by fluctuations. Moreover, it quantifies the fluctuations caused by the non-zero temperature and the quantum fluctuations due to the zero-point motion of the particles.

6.1.3 Thermal and vacuum fluctuations

Let us consider the limits of eqn (6.7). For $\hbar|\omega| \ll k_B T$,[4]

$$S(|\omega| \ll k_B T/\hbar) = 4\mathrm{Re}[Y(\omega)]k_B T. \tag{6.8}$$

Thus, in equilibrium and at low frequencies, the current through a resistance R fluctuates by the amount $\sim 4k_B T/R$. This is called the thermal noise.[5]

In the opposite limit $\hbar\omega \gg k_B T$, we get a temperature-independent noise,

$$S(\omega \gg k_B T/\hbar) = 4\mathrm{Re}[Y(\omega)]\hbar\omega. \tag{6.9}$$

These fluctuations are present even at $T = 0$, and they are called vacuum or quantum fluctuations. They may become the major noise 'source' in some of the present-day high-frequency, low-temperature experiments.[6]

In the limit $-\omega \gg k_B T/\hbar$, the non-symmetrized noise in eqn (6.7) vanishes. This shows that the vacuum fluctuations can only absorb energy, and hence they cannot be used, for example, as a power source.

6.1.4 Shot noise

Driven (non-equilibrium) systems contain an additional source of noise: shot noise. This was first discovered by Schottky (1918), who studied noise in vacuum tubes. Shot noise may be described with the following toy model.

Consider a conductor which transfers electrons in an uncorrelated fashion, such that the current can be described by a sequence of δ-peaks:

$$I(t) = e\sum_i \delta(t - t_i). \tag{6.10}$$

Here t_i are assumed random and uncorrelated.

The average current in a stationary system can then be expressed by

$$\langle I \rangle = \lim_{T \to \infty} \frac{1}{T}\int_0^T I(t)dt = \lim \frac{e}{T}\int_0^T \sum_i \delta(t - t_i)dt = \lim \frac{eN_T}{T} \equiv \frac{e}{\tau}, \tag{6.11}$$

where N_T is the number of electrons transmitted in time T, and $\tau = T/N_T$ is the average time between the transfers of the electrons.

[4] For $T = 1$ K, this would correspond to the angular frequency $|\omega| \ll 130$ GHz, or the frequency $f \ll 21$ GHz.

[5] Often the thermal current fluctuations of current are also called the Johnson–Nyquist noise (Johnson, 1928; Nyquist, 1928).

[6] 'Source' is in quotation marks because vacuum fluctuations cannot emit energy. However, they describe an intrinsic uncertainty in measurements due to the Heisenberg uncertainty principle.

The current noise is then

$$\frac{1}{2}S_I(t,t') = \langle I(t)I(t')\rangle - \langle I(t)\rangle\,\langle I(t')\rangle$$

$$= e^2 \left\langle \sum_{ij}\delta(t-t_i)\delta(t'-t_j)\right\rangle - \frac{e^2}{\tau^2}$$

$$= e^2 \left\langle \sum_{i}\delta(t-t_i)\delta(t'-t_i)\right\rangle + e^2 \left\langle \sum_{i\neq j}\delta(t-t_i)\delta(t'-t_j)\right\rangle - \frac{e^2}{\tau^2}$$

$$= e^2 \left\langle \sum_{i}\delta(t-t_i)\delta(t'-t_i)\right\rangle + e^2 \left\langle \sum_{i}\delta(t-t_i)\right\rangle\left\langle \sum_{j}\delta(t'-t_j)\right\rangle - \frac{e^2}{\tau^2}$$

$$= e^2 \left\langle \sum_{i}\delta(t-t_i)\delta(t'-t_i)\right\rangle = \frac{e^2}{\tau}\delta(t-t') = e\,\langle I\rangle\,\delta(t-t').$$

$$(6.12)$$

The second last equality holds provided that t_i and t_j are uncorrelated.

Thus, the noise equals $S_I(t,t') = 2e\langle I\rangle\delta(t-t')$ and its Fourier transform is

$$S_I(\omega) = 2e\langle I\rangle. \qquad (6.13)$$

What we can see from eqn (6.13) is that the current noise is proportional to the average current.[7] The origin of the noise is in the discreteness of electron charge, and it is therefore called the 'shot noise'.

6.2   Scattering approach to noise

Using the scattering approach we can make a more serious calculation of the noise correlator in a driven mesoscopic system. In Ch. 3 we discuss the expression for the current operator in lead α in terms of the scattering matrix,

$$\hat{I}_\alpha = \frac{2e}{h}\sum_{\beta\gamma}\sum_{m\in\beta,n\in\gamma}\int dEdE'\,e^{i(E-E')t/\hbar}\hat{a}^\dagger_{\beta m}(E)A^{\beta\gamma}_{mn}(\alpha,E,E')\hat{a}_{\gamma n}(E'),$$

$$(3.54)$$

where

$$A^{\beta\gamma}_{mn}(\alpha,E,E') = \delta_{mn}\delta_{\beta\alpha}\delta_{\gamma\alpha} - \sum_{j}(s^{\alpha\beta}_{jm})^*(E)s^{\alpha\gamma}_{jn}(E'). \qquad (3.52)$$

We may use this to evaluate the current correlator in the general non-equilibrium (but stationary) state of the system. For simplicity, let us concentrate on the symmetrized correlator defined in eqn (6.1), but with a generalization to describe also cross-correlations, i.e.,

$$S_{\alpha\beta}(t) \equiv \langle \delta\hat{I}_\alpha(t)\delta\hat{I}_\beta(t') + \delta\hat{I}_\beta(t')\delta\hat{I}_\alpha(t)\rangle. \qquad (6.14)$$

Both of the current operators are quadratic in the lead operators $\hat{a}$. To find an expression for their correlator, we need to calculate the statistical

[7] Note that this calculation does not include the possibility for currents in the opposite direction and thereby the thermal noise is not included.

expectation value of products of four operators $\hat{a}$. For a non-interacting Fermi (Bose) system at equilibrium this is

$$\langle \hat{a}^\dagger_{\alpha k}(E_1)\hat{a}_{\beta l}(E_2)\hat{a}^\dagger_{\gamma m}(E_3)\hat{a}_{\delta n}(E_4)\rangle - \langle \hat{a}^\dagger_{\alpha k}(E_1)\hat{a}_{\beta l}(E_2)\rangle\langle \hat{a}^\dagger_{\gamma m}(E_3)\hat{a}_{\delta n}(E_4)\rangle$$
$$= \delta_{\alpha\delta}\delta_{\beta\gamma}\delta_{kn}\delta_{ml}\delta(E_1 - E_4)\delta(E_2 - E_3)f_\alpha(E_1)[1 \mp f_\beta(E_2)].$$
$$(6.15)$$

The proof of these expressions requires the use of the Wick theorem, valid for macroscopic equilibrium quantum systems (see Exercise 6.3).[8] The upper sign in the above expression corresponds to Fermi and the lower to the Bose statistics, and f_α is the corresponding equilibrium distribution function in lead α.[9] Using this equation we obtain by a straightforward calculation

$$S_{\alpha\beta}(\omega) = \frac{e^2}{h} \sum_{\gamma\delta} \sum_{mn} \int dE A^{\gamma\delta}_{mn}(\alpha; E, E + \hbar\omega) A^{\delta\gamma}_{nm}(\beta; E + \hbar\omega, E)$$
$$\times \{f_\gamma(E)[1 \mp f_\delta(E + \hbar\omega)] + [1 \mp f_\gamma(E)]f_\delta(E + \hbar\omega)\}.$$
$$(6.16)$$

This equation includes spin in indices m, n.

Below, we mostly concentrate on the zero-frequency noise for fermions. This has a slightly simpler expression,

$$S_{\alpha\beta}(0) = \frac{e^2}{h} \sum_{\gamma\delta} \sum_{mn} \int dE A^{\gamma\delta}_{mn}(\alpha; E, E) A^{\delta\gamma}_{nm}(\beta; E, E)$$
$$\times \{f_\gamma(E)[1 - f_\delta(E)] + [1 - f_\gamma(E)]f_\delta(E)\}.$$
$$(6.17)$$

This expression holds in the presence of an arbitrary number of terminals. Below we look for a simplification in the case of only two terminals, and in Sec. 6.5 we use this expression to discuss cross-correlations in multi-terminal structures.

6.2.1   Two-terminal noise

In the two-terminal case current conservation implies $S_{LL} = S_{RR} = -S_{LR} = -S_{RL}$.[10] Expression (6.17) can be written down for example for the autocorrelation function where $\alpha = \beta = L$,

$$S_{LL} = \frac{2e^2}{h} \int dE \Big\{ \mathrm{Tr}[A^{LL}(L)A^{LL}(L)]f_L(E)(1 - f_L(E))$$
$$+ \mathrm{Tr}[A^{RR}(L)A^{RR}(L)]f_R(E)(1 - f_R(E))$$
$$+ \mathrm{Tr}[A^{LR}(L)A^{RL}(L)][f_L(E)(1 - f_R(E)) + f_R(E)(1 - f_L(E))] \Big\}.$$
$$(6.18)$$

Using the representation

$$S = \begin{pmatrix} r & t' \\ t & r' \end{pmatrix}$$
$$(3.42)$$

[8]The proof of Wick's theorem can be found in most books dealing with quantum many-body systems, such as (Negele and Orland, 1998), (Bruus and Flensberg, 2004) and (Fetter and Walecka, 2003).

[9]An analogous theory could be used to derive an expression for photon shot noise, for example. The reason for preserving the boson sign here is explained in Sec. 6.5.

[10]See Sec. 6.5

and definition (3.52), we get traces of the form $\mathrm{Tr}[r^\dagger r t^\dagger t]$ in the expression for the noise. From the unitarity of S it is straightforward to show that $r^\dagger r = I - t^\dagger t$, where I denotes the unit matrix of the same dimension as r. Diagonalizing the matrix $t^\dagger t$ and denoting the transmission eigenvalues by T_n allows to write the above trace as[11]

$$\mathrm{Tr}[r^\dagger r t^\dagger t] = \sum_n T_n(1 - T_n). \tag{6.19}$$

Now, after some algebra (see Exercise 6.4) we are ready to write the two-terminal noise power at zero frequency in the form

$$S = \frac{4e^2}{h} \sum_n \int dE \{ T_n(E)(\Lambda_L + \Lambda_R) + T_n(E)[1 - T_n(E)](f_L - f_R)^2 \},$$
$$\tag{6.20}$$

where $\Lambda_j = f_j(1 - f_j)$. If the transmission eigenvalues are energy independent, we may integrate this to find

$$S = \frac{4e^2}{h} \left[2k_BT \sum_n T_n^2 + eV \coth\left(\frac{eV}{2k_BT}\right) \sum_n T_n(1 - T_n) \right], \tag{6.21}$$

where V is the voltage applied between the two reservoirs. This is plotted in Fig. 6.5 for a tunnel junction.[12]

In the limit $eV \to 0$, this produces the thermal noise, $S = 4k_BTG$, where $G = 2e^2/h \sum_n T_n$. In the opposite, 'shot noise' limit, we get

$$S = 2eFG|V| = 2eF|\langle I \rangle|, \tag{6.22}$$

where

$$F = \frac{\sum_n T_n(1 - T_n)}{\sum_n T_n} \tag{6.23}$$

is called the Fano factor. In the tunnelling limit, $T_n \ll 1$ for all n, we obtain

$$S_I^{\mathrm{tunn}} = 2e|V|G = 2e|\langle I \rangle|, \tag{6.24}$$

which equals the Schottky result obtained above. One can also see that for such a non-interacting normal-metal system, the Fano factor is always between zero and one. For example, a ballistic conductor has $T_n = 1$ for all transmitting channels. In this case $F = 0$, i.e., the shot noise vanishes. Hence, through the shot noise, one may obtain information about the transmission distribution that may not be present in the average current.[13]

The calculation of the Fano factor for a given system requires in general the knowledge of, if not all the values T_n,[14] at least their probability distribution $\rho(T)$. As mentioned in Ch. 3, this can be found for some systems by applying the random matrix theory. Fano factors have been calculated in this way for[15]

- Diffusive wires for which $F = \frac{1}{3}$. See also Sec. 6.4 and Exercise 6.8.

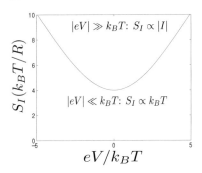

Fig. 6.5 Current noise vs. voltage in a tunnel junction for which all $T_n \ll 1$.

[11]See an example of this in Example 6.2.

[12]The crossover between thermal noise at $eV \ll k_BT$ and the shot noise at $eV \gg k_BT$ described by this equation can be used as a primary thermometer, see (Spietz *et al.*, 2003) and Exercise 6.16.

[13]This is often stated as a motivation to study shot noise. Personally, I think that usage of shot noise to learn about energy relaxation as described in Sec. 6.4 offers more valuable information than the information about transmission distribution, which is in most experiments already quite well known. Exceptions to this are the mechanical break junctions described in Fig. 1.7.

[14]In fact, these have been measured individually in mechanical break junctions, see for example (Scheer *et al.*, 1997).

[15]For details, see (Blanter and Büttiker, 2000).

- Chaotic cavities for which $F = \frac{N_L N_R}{(N_L + N_R)^2}$, where $N_{L/R}$ are the numbers of open channels in the leads coupled to the cavity. Such a chaotic cavity is a ballistic system with an irregular form and only surface scattering, connected to the leads via small point contacts. If the form of the cavity is irregular enough, electron dynamics inside it is chaotic, and its transmission properties are universal (i.e., independent of the microscopic details).

- Dirty two-dimensional interfaces, where the resistance arises from a 2d layer of randomly distributed scatterers (Schep and Bauer, 1997). In this case $F_2 = \frac{1}{2}$.

In normal-metal—superconductor contacts, Andreev reflection (see Ch. 5) needs to be taken into account when calculating the shot noise. At energies below the superconducting gap the electrons cannot enter the superconductor, and therefore the only contribution to the current is due to Andreev reflection with probability $R_{An} \in [0, 1]$ for channel n. In this case the conductance can be evaluated as (Beenakker, 1992)

$$G_{NS} = \frac{4e^2}{h} \sum_n R_{An} = \frac{4e^2}{h} \sum_n \frac{T_n^2}{(2 - T_n)^2}, \qquad (6.25)$$

where the latter equality is valid at the Fermi level. Here T_n are the transmission coefficients of the contact when the system is entirely in the normal state. Moreover, the zero-temperature shot noise is (de Jong and Beenakker, 1994)

$$S_{N-S} = 8e|V|\frac{e^2}{h} \sum_n R_{An}(1 - R_{A,n}) = \frac{64e^3|V|}{h} \sum_n \frac{T_n^2(1 - T_n)}{(2 - T_n)^4}. \qquad (6.26)$$

The Fano factor of a NIS tunnel contact (where $R_{An}, T_n \to 0$) is $F = 2$.

In the 'Poisson' limit of transport, where the probability T_n of transmission is low, the shot noise is directly proportional to the effective charge of the current carriers. Besides the tunnel contacts to superconductors described above, this has been shown to be the case for fractional quantum Hall samples (de Picciotto *et al.*, 1997).

Example 6.2 Transmission eigenvalues of a three-mode system
Let us calculate the transmission eigenvalues in a system containing two modes in the left lead and one mode in the right lead. The scattering matrix has the form

$$S = \begin{pmatrix} r_{11} & r_{12} & t_1' \\ r_{21} & r_{22} & t_2' \\ t_1 & t_2 & r_R \end{pmatrix}. \qquad (6.27)$$

Comparing to eqn (3.42) we identify $t' = \begin{pmatrix} t_1' & t_2' \end{pmatrix}^T$ and $t = \begin{pmatrix} t_1 & t_2 \end{pmatrix}$. Either of these can be used for the calculation of the transmission eigenvalues for the noise formula. Using the previous we get

$$t'^\dagger t' = |t_1'|^2 + |t_2'|^2 \equiv T_0 \qquad (6.28)$$

and with the latter we obtain

$$t^\dagger t = \begin{pmatrix} |t_1|^2 & t_1^* t_2 \\ t_2^* t_1 & |t_2|^2 \end{pmatrix}. \qquad (6.29)$$

This has the eigenvalues $|t_1|^2 + |t_2|^2 = T_0$ (by unitarity, see Exercise 3.2) and zero. Hence, the obtained conductance $G = 2e^2 T_0/h$ and shot noise $S = 4e^3/hT_0(1 - T_0)|V|$ are independent of which transmission matrix we use.

6.3   Langevin approach to noise in electric circuits

Once the noise properties of single mesoscopic elements are known, it is often desirable to figure out how these fluctuations show up in more complicated circuits. When many such noisy elements are coupled via macroscopic conductors,[16] a simple circuit theory with the noisy elements acting as Langevin[17] sources can be employed. An example of such a circuit is given in Fig. 6.6: in parallel to each dissipative element, we add an 'intrinsic' noise source δI_i. The different intrinsic noise sources are uncorrelated, $\langle \delta I_i \delta I_j \rangle = 0$ for $i \neq j$, and therefore only the autocorrelators $\langle \delta I_i^2 \rangle$ need to be taken into account. These intrinsic correlators on the other hand give rise to voltage fluctuations ΔV_j in the nodes of the system. When treating only fluctuations, the average voltages or currents do not need to be written out explicitly.

The total fluctuating current through element i is

$$\Delta I_i = \delta I_i + Y_i(\omega)(\Delta V_{i,L} - \Delta V_{i,R}), \tag{6.30}$$

where $Y_i(\omega)$ is the admittance of the element and the term in the brackets is the drop of the fluctuating voltage over it. For each node we can write the Kirchhoff law of current conservation (total current into each node vanishes) and through this solve for the voltages ΔV_j or the total fluctuating currents ΔI_i in terms of only the intrinsic fluctuations δI_i:

$$\Delta V_i = \sum_j c_{ij} \delta I_j \tag{6.31a}$$

$$\Delta I_i = \sum_j d_{ij} \delta I_j, \tag{6.31b}$$

where c_{ij} and d_{ij} are functions of the admittances/impedances of the circuit elements. The fluctuation correlators are then obtained via equations of the type (see Exercise 6.6 for the specific form of the frequency dependence)

$$S_{ij}(\omega) = \langle \Delta I_i(\omega) \Delta I_j(-\omega) \rangle \Delta f = \sum_{kl} d_{ik}(\omega) d_{jl}(\omega)^* \langle \delta I_k \delta I_l \rangle \Delta f. \tag{6.32}$$

Here Δf is the frequency bandwidth of the noise measurement, which cancels out from the final results. The above equation exploits the fact that for the admittances inverting the frequency corresponds to complex conjugation. The non-correlation of the intrinsic noise sources implies

$$S_{ij}(\omega) = \sum_k d_{ik} d_{jk}^* \langle \delta I_k \delta I_k \rangle \Delta f. \tag{6.33}$$

[16]'Macroscopic' in this context means that between the circuit elements, the electron systems can be treated as if they would be in a (local) equilibrium. In practice this typically means that the circuit elements are separated from each other with wires whose length exceeds the electron-phonon scattering length.

[17]A Langevin equation is an equation of motion containing a stochastic force whose average vanishes but whose correlation function is non-zero. See Example 11.1.

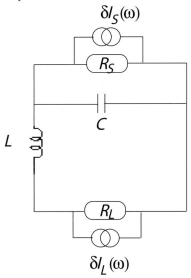

Fig. 6.6 Example of a transformer circuit with the noise current sources.

The intrinsic correlators $\langle(\delta I_k)^2\rangle$ correspond to shot noise, thermal noise, etc., depending on the parameter regime in which one is interested. This noise description is applied in Example 6.3 and Exercise 6.7.

Example 6.3 Measurement of noise

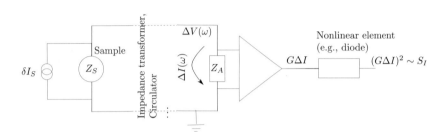

Fig. 6.7 Typical noise measurement scheme: Noise from the sample connects via some (optional) transformers and/or circulators to an amplifier, which amplifies either the voltage or current flowing through it. The amplified signal is then squared using a non-linear element, such as a diode. The output from this non-linear element may then be related to the noise in the sample. Many schemes also contain a bias tee (not shown), which separates the dc voltage driving the sample and the finite-frequency part of the noise, which is sent to the amplifier.

Current noise in nanoelectronics is typically measured by connecting the noisy current to an amplifier and detecting the square of the current by using some non-linear element, such as a Schottky diode. This measurement scheme can be analyzed with the Langevin circuit approach. Typically the relevant quantity is the noise power $S_P = \langle[\Delta I(\omega)\Delta V(-\omega) + \Delta I(-\omega)\Delta V(\omega)]/2\rangle$ induced by the noise source.

Let us calculate the power emitted by the noise source over the amplifier impedance Z_A utilizing the Langevin circuit scheme, disregarding the transformers and circulators. The sample noise δI_S induces a noise voltage ΔV across the amplifier impedance Z_A. Therefore the noise current across the latter is $\Delta I(\omega) = \Delta V(\omega)/Z_A(\omega)$. This equals the noise current across the sample, $\Delta I(\omega) = \delta I_S - \Delta V(\omega)/Z_S(\omega)$. Solving for the noise voltage yields

$$\Delta V = \frac{Z_A Z_S}{Z_A + Z_S}\delta I_S. \tag{6.34}$$

Using the fact that $Z(-\omega) = Z(\omega)^*$, the noise power across the amplifier is

$$S_P = \frac{\mathrm{Re}[Z_A(\omega)]|Z_S|^2}{|Z_A + Z_S|^2}S_I, \tag{6.35}$$

where S_I is the current noise correlator. Assuming for example that S_I describes thermal noise with some temperature T_S, and the frequencies of the measurement (say, the bandwidth of the amplifier) are much smaller than $k_B T_S$, we have $S_I(\omega) = \mathrm{Re}[4/Z_S(\omega)]k_B T_S$ and hence[18]

$$S_P = \frac{\mathrm{Re}[Z_A(\omega)]\mathrm{Re}[Z_S(\omega)]}{|Z_A + Z_S|^2}4k_B T_S. \tag{6.36}$$

[18]Here we use the fact that $\mathrm{Re}[1/Z_S(\omega)] = \mathrm{Re}[Z_S(\omega)]/|Z_S(\omega)|^2$.

This power (per unit bandwidth, giving it a dimension of energy) should be compared to the noise emitted by the amplifier, typically characterized by its noise temperature T_N: the smaller S_P/T_N, the longer the noise should be averaged for an accurate measurement. The best coupling of the source noise is obtained when the two impedances are the same, $Z_A \approx Z_S$. At high frequencies the amplifier impedance is given in terms of the characteristic impedance of the cables (see Appendix E) and typically $Z_A \approx 50\ \Omega$. Therefore noise measurements from high-impedance sources often require the use of transformer circuits (see Fig. 6.6 and Exercise 6.9), which on the other hand limit the frequency bandwidth over which the noise is coupled.

6.4 Boltzmann–Langevin approach

Besides the scattering approach, there are two other useful formulations from which one can calculate the non-equilibrium current noise in mesoscopic circuits. One relies on the non-equilibrium Keldysh Green's-function theory, introducing the fluctuation effects through a 'counting field' defined in the terminals.[19] This formulation is outside the scope of this book, but in the incoherent limit in a diffusive wire this theory, applied for the second-order correlator, produces the same noise formula as an incoherent semiclassical Boltzmann–Langevin approach.[20] The latter is an extension of the Boltzmann theory introduced in Ch. 2, incorporating the idea of Langevin noise sources into the Boltzmann theory. It assumes that due to the stochastic nature of the scattering inside the sample, the scattering currents, eqn (2.12), fluctuate in time, i.e., they are of the form

$$J_{\vec{p},\vec{p}'} = \bar{J}_{\vec{p},\vec{p}'} + \delta J_{\vec{p},\vec{p}'}. \tag{6.37}$$

Here $\bar{J}_{\vec{p},\vec{p}'}$ is the average scattering current, and $\delta J_{\vec{p},\vec{p}'}$ is the fluctuation. As a result, also the energy distribution function fluctuates, and we may denote

$$f(\vec{r},\hat{p},t) = \bar{f}(\vec{r},\hat{p},t) + \delta f(\vec{r},\hat{p},t). \tag{6.38}$$

Our aim is thus to derive an equation for δf and from this to find the fluctuations of the current density. These in turn can be related to the fluctuations of the total current. Below we concentrate only on the diffusive limit with much weaker inelastic than elastic scattering. In this case the fluctuations in the inelastic collision integrals can be disregarded, as their only effect is to deform the shape of the average distribution function $f(\vec{r})$ on which the fluctuations also depend.

The derivation of the noise formula in terms of the distribution functions is fairly lengthy and is presented in Appendix D. This is not the only possible type of a derivation of the semiclassical shot noise: see Exercise 6.10 for an alternative derivation utilizing eqn (6.20) and simple kinetic equations.

This derivation yields for the noise power spectral density

$$S_I(\omega \ll \omega^*) = \frac{4G_N}{L} \int_0^L dx \int dE\, \bar{f}_0(E,x)(1 - \bar{f}_0(E,x)), \tag{6.39}$$

where $G_N = A\sigma/L$ is the conductance of the wire.

Equation (6.39) is applied in Exercise 6.8 to show, among other things, that the Fano factor of a diffusive wire is 1/3, analogous to the result

[19]See for example (Belzig, 2003).

[20]This theory was first derived by Kogan and Shulman (1969). The derivation here follows the review by Blanter and Büttiker (2000).

obtained from the scattering theory. This result illustrates that shot noise is not a coherent effect, and therefore for example dephasing has no effect on it unless it also affects the average conductance.

The Boltzmann–Langevin formalism is especially useful when estimating the effect of inelastic scattering on the noise (see Exercise 6.8). Moreover, it can be used to show how noise can act as a local thermometer, probing the voltage-induced heating in the structure. As an example, let us consider what happens in the quasi-equilibrium regime where the distribution function becomes a Fermi function with a position dependent chemical potential $\mu(x)$ and temperature $T_e(x)$ (see Sec. 2.5.1). In that case, we may use eqn (A.64h) to obtain[21]

$$S_I = \frac{4G_N k_B}{L} \int_0^L dx T_e(x). \tag{6.40}$$

In this limit the current noise provides hence a direct measurement of the average electron temperature over the conductor. When the conductor is biased, this temperature depends on inelastic scattering effects within the conductor, and thereby S_I can be used to obtain information about the strength of inelastic scattering.[22]

The overall behaviour of shot noise Fano factor as a function of wire length, or voltage that can be used to tune the inelastic scattering lengths, is sketched in Fig. 6.8 (Steinbach *et al.*, 1996). When $L > \ell_{e-ph}$, the Fano factor is a direct measure of the average electron temperature inside the wire and therefore its voltage dependence reflects the temperature dependence of the dominant energy relaxation mechanism of electrons. In the macroscopic wires where electron–phonon scattering dominates, i.e., the electron temperature equals that of the surroundings, the only remaining noise is the thermal noise. Therefore, shot noise (voltage-induced extra noise) can only be measured in mesoscopic wires with long enough energy relaxation lengths.

6.5   Cross-correlations

The above discussion concentrates mostly on two-probe systems, where it is enough to consider the autocorrelation function of the current in one of the electrodes. However, cross-correlations in mesoscopic systems have recently attracted a lot of attention, as there are a few simple principles describing especially the sign of the cross-correlations[23] and because they yield different phase-sensitive information of current transport than the autocorrelators (Samuelsson *et al.*, 2004; Neder *et al.*, 2007).

The autocorrelators discussed above and the cross-correlators are not completely independent of each other. In the limit of low frequencies, not only the average current but also its fluctuations δI have to be conserved, i.e., the sum over the fluctuations in different leads α has to vanish, $\sum_\alpha \delta I_\alpha = 0$. Therefore, the fluctuations obey a sum rule,

$$\left\langle \left(\sum_\alpha \delta I_\alpha \right)^2 \right\rangle = \sum_\alpha \langle (\delta I_\alpha)^2 \rangle + \sum_{\alpha \neq \beta} \langle \delta I_\alpha \delta I_\beta \rangle = 0. \tag{6.41}$$

[21] Note that in the non-equilibrium limit the integrated shot noise differs from the quasi-equilibrium limit; see Exercise 6.8.

[22] For example, with this scheme energy relaxation has been studied in carbon nanotubes (Chaste *et al.*, 2010) and in graphene (Fay *et al.*, 2011; Betz *et al.*, 2012).

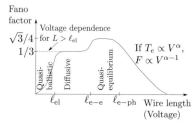

Fig. 6.8 Sketch of the Fano factor vs. wire length in different regimes of parameters (see also Exercise 6.8). As the inelastic scattering lengths depend on the energy of the electrons participating in the transport, this dependence can be studied by measuring the Fano factor at different voltages across the wire (except typically the part $L < \ell_{el}$ for diffusive wires).

[23] For a detailed scattering theory formulation of cross-correlations, see (Büttiker, 1992).

For example, for a two-probe system, this implies $S_{LL} = S_{RR} = -S_{LR} = -S_{RL}$, and cross-correlations yield no more information than the auto-correlation function studied above. Therefore cross-correlators are generally studied in multi-probe geometries. But there is something one can learn from this relation: As the autocorrelation functions are always positive, it seems that the cross-correlators should be negative. Thus, a current in one lead generally implies that there should be a correlated opposite current in another lead. It turns out that such a property holds for a general multi-probe non-interacting fermion system in the absence of inelastic scattering, but positive cross-correlations can be found, for example, in (partially) superconducting systems.

6.5.1   Equilibrium correlations

The general zero-frequency formula for the multi-terminal current–current correlator derived in Sec. 6.2 reads

$$
S_{\alpha\beta}(0) = \frac{e^2}{h} \sum_{\gamma\delta} \sum_{mn} \int dE A_{mn}^{\gamma\delta}(\alpha; E, E) A_{nm}^{\delta\gamma}(\beta; E, E)
$$
$$
\times \{f_\gamma(E)[1 - f_\delta(E)] + [1 - f_\gamma(E)]f_\delta(E)\}. \tag{6.17}
$$

Consider first the case when there is no voltage applied over the sample. In this case, assuming that the scattering matrix is energy independent within a few $k_B T$ around the Fermi energy, we have only simple integrals over $f^0(E)(1 - f^0(E))$ and thus

$$
S_{\alpha\beta}^{\mathrm{eq}} = \frac{2e^2}{h} k_B T \sum_{\gamma\delta} \mathrm{Tr}[A^{\gamma\delta}(\alpha)A^{\delta\gamma}(\beta)], \tag{6.42}
$$

independent of whether we consider fermions or bosons. Using the definition of $A^{\delta\gamma}$ we can obtain

$$
S_{\alpha\beta}^{\mathrm{eq}} = \frac{2e^2}{h} k_B T \mathrm{Tr}[2 \cdot 1_\alpha \delta_{\beta\alpha} - (s^{\alpha\beta})^\dagger s^{\alpha\beta} - (s^{\beta\alpha})^\dagger s^{\beta\alpha}]. \tag{6.43}
$$

Here 1_α is the unit (identity) matrix of size $M_\alpha \times M_\alpha$. For the autocorrelation function, i.e., $\alpha = \beta$, this yields the thermal noise $4k_B TG$. For the cross-correlators, we can define (positive definite) total transmission probabilities between α and β by $T_{\alpha\beta} \equiv \mathrm{Tr}[(s^{\alpha\beta})^\dagger s^{\alpha\beta}]$ and get

$$
S_{\alpha\neq\beta}^{\mathrm{eq}} = -\frac{2e^2}{h} k_B T[T_{\alpha\beta} + T_{\beta\alpha}]. \tag{6.44}
$$

Thus, equilibrium cross-correlations are always negative, both for bosons and fermions.

6.5.2   Finite-voltage cross-correlations

Consider the cross-correlations in the presence of a finite voltage and for an energy-independent scattering matrix within the interesting region of energies. The sum over lead indices γ and δ in eqn (6.17) can be separated in two parts: one where the terms contain $f_\gamma(1 \mp f_\gamma)$, and another with cross-terms of the form $f_\gamma(1 \mp f_\delta)$, $\delta \neq \gamma$. The previous terms yield a contribution proportional to the equilibrium fluctuations; they vanish at $T = 0$.[24] For the cross-terms, $A_{\gamma\delta}(\alpha) = -(s^{\alpha\gamma})^\dagger s^{\gamma\delta}$. These yield

[24] For bosons in this limit, one should also take into account the effects due to Bose condensation. However, bosons are presented here just to illustrate a point about the sign of the correlations.

$$S_{\alpha\neq\beta}^{\mathrm{cross}} = \frac{2e^2}{h} \sum_{\gamma\neq\delta} \int dE \mathrm{Tr}[(s^{\alpha\gamma})^\dagger s^{\alpha\delta}(s^{\beta\delta})^\dagger s^{\beta\gamma}] f_\gamma(E)[1 \mp f_\delta(E)]. \quad (6.45)$$

This equals[25]

[25] For details, see (Büttiker, 1992).

$$S_{\alpha\neq\beta}^{\mathrm{cross}} = \mp \frac{2e^2}{h} \int dE \mathrm{Tr}\left[\underbrace{\left(\sum_\gamma s^{\beta\gamma}(s^{\alpha\gamma})^\dagger f_\gamma(E)\right)}_{A^\dagger} \underbrace{\left(\sum_\delta s^{\alpha\delta}(s^{\beta\delta})^\dagger f_\delta(E)\right)}_{A}\right].$$

$$(6.46)$$

The upper sign is for fermions and the lower for bosons. This equation contains a trace of a product of a matrix A with its adjoint. Such a trace is always positive. Hence, for fermions, such transport cross-correlations are always negative, whereas for bosons they are positive. As the total correlations are a sum of these transport correlations and terms proportional to the equilibrium correlations, the sign of the cross-correlations for (non-interacting) fermions is always negative, but for bosons it can change the sign.[26]

The sign of the cross-correlations is related to the statistics of the current carriers. The negative correlations for fermion systems is a sign of 'anti-bunching' due to the Pauli principle that prohibits two electrons occupying the same state. On the contrary, bosons can 'bunch', i.e., occupy the same quantum state, and therefore boson systems can show positive cross-correlations.

[26] Negative cross-correlations for a non-interacting electron system have been measured, for example, by (Henny et al., 1999) and (Oliver et al., 1999). Some time ago it was predicted and measured that inelastic scattering may also lead to positive cross-correlations in electronic systems, see (Rychkov and Büttiker, 2006) for the theory and (Oberholzer et al., 2006) for the experiments.

It has been suggested that also electron systems could show bunching, i.e., positive cross-correlations, especially in certain interacting systems. For example, if one of the leads in a three-terminal system is superconducting, the cross-correlations between the other two leads may be positive (Börlin et al., 2002; Samuelsson and Büttiker, 2002). The source for such positive cross-correlations is due to Cooper pairs (see Ch. 5) emitted from the superconductors breaking into two excitations transmitting into different leads.[27]

Multi-probe cross-correlations have been recently studied, both theoretically and experimentally, especially in two-dimensional electron gases that are driven into the quantum Hall regime with a magnetic field. In this case the electrons transmit through the conductor within the Landau levels formed at the edges of the conductor. Transport within the Landau levels is ballistic, and the only scattering takes place in quantum point contacts defined separately with metallic gates. In this manner one can realize different types of electron interferometers (Ji et al., 2003) and realize 'electron optics' types of experiment. For example, cross correlation measurements in this type of an interferometer were used to show

[27] Alternatively, this can be viewed as a crossed Andreev reflection where the Andreev reflected particle ends up in a different electrode than the original particle.

interference between two electrons originating from separate independent sources (Neder *et al.*, 2007). The cross-correlations in this case are a sign of *entanglement* between the electrons (Samuelsson *et al.*, 2004). As a result, they can also be used for measuring the Bell inequalities (Chtchelkatchev *et al.*, 2002).

6.6 Effect of noise on quantum dynamics

The concept of noise and fluctuations is especially useful when describing the properties of small, spatially restricted quantum systems coupling to the outside world. In elementary quantum mechanics, we learn that the time dependence of the state (density matrix) of a given system described with the Hamiltonian H is given by

$$\rho(t) = U(t)\rho(0)U^\dagger(t), \tag{6.47}$$

where $U(t) = \mathcal{T}\exp(-\frac{i}{\hbar}\int_0^t H dt)$ is a unitary operator, i.e., $UU^\dagger = U^\dagger U = \mathbf{1}$. Because of unitarity, from the 'final' state $\rho(t)$ we can easily find the initial state $\rho(0)$:

$$\rho(0) = U^\dagger(t)\rho(t)U(t). \tag{6.48}$$

Thus, for quantum systems it does not greatly matter whether we describe things advancing forward or backward in time.

From any everyday phenomenon we know that for the classical world there is a big difference between the future and the past—for example, when one thinks about classically allowed physical processes. A window is easy to break with a snowball, but it is quite unlikely that a reverse process would take place, a crashed window turning back into a single piece and the snowball flying back into the thrower's hand. In the world of electronics, the essential concept differentiating the backward-in-time processes from the forward-moving processes is *dissipation*: when a current flows through a resistor, it heats up the lattice of the resistor, and this heat is carried away by the phonons, which then somewhere else heat up the rest of the universe. The reverse rarely takes place: phonons emerging from the measurement setup, and giving rise to a finite electric current.[28]

In isolated quantum systems the total energy of the system is conserved, which then leads to the unitarity of the time evolution. The way out is to describe the connection of this small system to the outside world via some *fluctuating* field.[29] That is, the Hamiltonian is of the form

$$H = H_0 + g\hat{A}_i\hat{f}(t), \tag{6.49}$$

where $\hat{f}(t)$ is a zero-average fluctuating field and g is a (scalar) coupling coefficient to this field. Generally $\hat{f}$ can still be an operator, but it is typically chosen such that it commutes with all the degrees of freedom of the interesting subsystem, i.e., all the operators $\hat{A}_i$. An example model for $\hat{f}$ is discussed in Sec. 11.4.2, but here I discuss the consequences of

[28] One way to understand the difference between time-reversible quantum-mechanical dynamics and the (seemingly) irreversible classical dynamics is to exploit the concept of *initial correlations*: In a complicated system the macrostate that we observe may consist of a huge number of different microstates (between which there may be correlations unknown to the observer) that we do not observe. Although the dynamics of the microstates may be fully reversible, the observed behavior of a given macrostate may seem irreversible: macrostates with a larger entropy (more microstates describing the one macrostate) very seldom tend towards macrostates with less entropy (fewer microstates), simply due to the fact that this is less probable.

[29] There are many different ways to make a quantitative analysis of dissipation in quantum-mechanical systems, although most of them rely on a 'system+environment' model similar to the one introduced here. One widely used reference in this context is the book by Weiss (1999).

the presence of such a fluctuating field. The main effects are that $\hat{f}$ can make the subsystem (described by H_0) *relax* towards an equilibrium state, and also lose its coherent features. The latter process is called *dephasing*. These two processes, along with their connection to noise, are discussed in Secs. 6.6.1 and 6.6.2.

6.6.1   Relaxation

Assume for simplicity that the Hamiltonian H_0 is time independent. Then we can describe the quantum system with the eigenvectors $|n\rangle$ of this H_0 with the eigenenergies E_n, satisfying

$$H_0|n\rangle = E_n|n\rangle. \tag{6.50}$$

Now assume that at some initial state $t = 0$, the occupation numbers of these states are described by the probabilities p_n. These p_n are the diagonal elements of the density matrix at time $t = 0$. As by definition $|n\rangle$ are the eigenenergies of the Hamiltonian, without the perturbation term in eqn (6.49) p_n would remain constant at later times as well. The fluctuating field can cause transitions in the system, making p_n time dependent. The most convenient way to describe this is using the interaction representation:[30] we make the transformation to the Heisenberg picture in the subsystem, i.e., $\hat{A}_i(t) = \exp(iH_0t/\hbar)\hat{A}_i(0)\exp(-iH_0t/\hbar)$, and include the interaction term $g\hat{f}\hat{A}_i(t)$ perturbatively. In the first order in g we then get the relation between the wave function at $t = 0$ and at some later time t (see also eqns (A.24) and (A.27)):[31]

$$|\psi_I(t)\rangle = |\psi_I(0)\rangle - \frac{i}{\hbar}\int_0^t dt'\, g\hat{f}(t')\hat{A}_i(t)|\psi_I(0)\rangle. \tag{6.51}$$

Assuming the system initially at state $|n\rangle$, the probability amplitude that it has made a transition to state $|m\rangle$, $m \neq n$ is then

$$\alpha_m = \langle m|\psi_I(t)\rangle = -\frac{ig}{\hbar}\int_0^t dt'\, \hat{f}(t')\langle m|\hat{A}_i(t)|n\rangle$$

$$= -\frac{ig}{\hbar}\int_0^t dt'\, \hat{f}(t')e^{i\omega_{mn}t}A_i^{mn}, \tag{6.52}$$

where $A_i^{mn} = \langle m|\hat{A}_i(0)|n\rangle$ and the dynamic phase with $\omega_{mn} = (E_m - E_n)/\hbar$ comes from the time evolution of the Heisenberg operator (or the states in the Schrödinger picture). The probability for the occupation of state m is then

$$p_m(t) = |\alpha_m|^2 = \frac{g^2}{\hbar^2}|A_i^{mn}|^2\int_0^t\int_0^t dt_1 dt_2 e^{-i\omega_{mn}(t_1-t_2)}\hat{f}(t_1)\hat{f}(t_2). \tag{6.53}$$

Let us now ensemble average this probability over all the fluctuations $\hat{f}$:

$$\bar{p}_m(t) = \frac{g^2}{\hbar^2}|A_i^{mn}|^2\int_0^t\int_0^t dt_1 dt_2 e^{-i\omega_{mn}(t_1-t_2)}\langle\hat{f}(t_1)\hat{f}(t_2)\rangle. \tag{6.54}$$

[30] See Appendix A.2.1.

[31] This derivation is quite similar to that for the Fermi golden rule in Appendix A.3. However, here we especially concentrate on the fact that the force is fluctuating.

Then, let us perform a change of variables in the integrals, $\tau = t_1 - t_2$, and $T = (t_1 + t_2)/2$. We get

$$\bar{p}_m(t) = \frac{g^2}{\hbar^2}|A_i^{mn}|^2 \int_0^t dT \int_{-B(T)}^{B(T)} d\tau e^{-i\omega_{mn}\tau}\langle \hat{f}(T+\tau/2)\hat{f}(T-\tau/2)\rangle,$$
(6.55)

where $B(T) = T$ if $T < t/2$ and $B(T) = t - T$ otherwise.

Assume that the system describing the fluctuating field $\hat{f}(t)$ is stationary and described by a correlation time τ_f. Then for $t \gg \tau_f$ we can take $B(T) \to \infty$ and get $\bar{p}_m(t) = \Gamma_{mn}t$ with

$$\Gamma_{mn} = \frac{g^2}{\hbar^2}|A_i^{mn}|^2 \int_{-\infty}^{\infty} d\tau e^{-i\omega_{mn}\tau}\langle \hat{f}(\tau)\hat{f}(0)\rangle = \frac{g^2}{\hbar^2}|A_i^{mn}|^2 S_f(-\omega_{mn}).$$
(6.56)

Here $S_f(\omega)$ is the (non-symmetrized) autocorrelator of $\hat{f}$, and Γ_{mn} is the *relaxation rate* of state n into state m. This example thus provides a natural interpretation for the positive- and negative-frequency noise. Excitation of the quantum system ($\omega_{mn} > 0$) is related to the negative-frequency noise whereas the down-relaxation ($\omega_{mn} < 0$) is related to the positive-frequency noise. Therefore $S(\omega > 0)$ describes *absorption* and $S(\omega < 0)$ *emission* of energy from the fluctuating force.

For a general initial density matrix p_n, we can construct a *master equation* to describe the occupations,

$$\frac{dp_n}{dt} = \sum_m (\Gamma_{nm}p_m - \Gamma_{mn}p_n).$$
(6.57)

This master equation can, for example, be used to show how a quantum system reaches an equilibrium state, and the temperature of this equilibrium state can be related to the properties of $S_f(\omega)$ (see Exercise 6.12).

In the long-time limit, assuming that the perturbation is sufficiently weak,[32] the noise properties of the 'outside world', described by $\hat{f}$, determine the occupation probabilities of state n.

In the case of two-level systems, the relaxation rate is often described with the relaxation time

$$T_1 = \Gamma_{01}^{-1}.$$
(6.58)

This notation is used especially in nuclear magnetic resonance.

6.6.2 Dephasing

For the full description of the properties of a quantum system, it is not sufficient to concentrate on the occupation numbers of the different energy levels—rather, all the quantum interference effects are contained in the interferences between different states. These are in turn described by the off-diagonal elements of ρ. The fluctuations $\hat{f}$ lead to a decay of these off-diagonal elements in time. For simplicity, I describe this process using a spin, i.e., a two-level system, as the small quantum system. The Hamiltonian for such a system can be specified by the Pauli spin matrices

[32] It can be shown that the next order in the perturbation parameter g is related to the third cumulant of the fluctuations in f; see (Ojanen and Heikkilä, 2006).

σ_i detailed in Appendix A.6.[33] The simplest way to treat such a system is to describe it in the energy eigenbasis, such that the Hamiltonian is

$$H_0 = \frac{\epsilon}{2}\sigma_z. \tag{6.59}$$

The most general coupling of this system to the external world is described by a term of the form

$$H_{\text{int}} = g_z \hat{f}_z \sigma_z + g_x \hat{f}_x \sigma_x. \tag{6.60}$$

A term proportional to σ_y could also be added, but it would not qualitatively change the outcome.

The term $\hat{f}_z \sigma_z$ commutes with the Hamiltonian, and does not lead to any relaxation (its matrix elements for $n \neq m$ would vanish). However, its presence does lead to dephasing. Let us simplify our problem a little more by assuming for a moment also $g_x = 0$. The time evolution of the density matrix can then be written as

$$\rho(t) = \exp(-iH_0 t/\hbar)\mathcal{T}\exp\left[-\frac{i}{\hbar}g_z\int_0^t dt' \hat{f}_z(t')\sigma_z\right]\rho(0)$$
$$\times \tilde{\mathcal{T}}\exp\left[\frac{i}{\hbar}g_z\int_0^t dt' \hat{f}_z(t')\sigma_z\right]\exp(iH_0 t/\hbar). \tag{6.61}$$

Here $\mathcal{T}$ is the time-ordering operator, and $\tilde{\mathcal{T}}$ is the anti-time-ordering operator. Assume the initial state is of the form[34]

$$\rho(0) = \begin{pmatrix} p & \alpha \\ \alpha & (1-p) \end{pmatrix} = \frac{1}{2}\sigma_0 + \left(p - \frac{1}{2}\right)\sigma_z + \alpha\sigma_x. \tag{6.62}$$

Here σ_0 is the 2×2 unit matrix. The term α thus describes the off-diagonal part containing the information on the interference of the two states. Plugging this into eqn (6.61) and noting that σ_0 and σ_z commute with the perturbation and H_0, we obtain

$$\rho(t) = \frac{1}{2}\sigma_0 + (p - \frac{1}{2})\sigma_z + \alpha\exp(-iH_0 t/\hbar)\mathcal{T}\exp\left[-\frac{i}{\hbar}g_z\int_0^t dt' \hat{f}_z(t')\sigma_z\right]$$
$$\times \sigma_x\tilde{\mathcal{T}}\exp\left[\frac{i}{\hbar}g_z\int_0^t dt' \hat{f}_z(t')\sigma_z\right]\exp(iH_0 t/\hbar). \tag{6.63}$$

Let us now concentrate on the σ_x-component of ρ, i.e., its off-diagonal part, and average it over the realizations of $\hat{f}_z$. We find

$$\rho_{od} = \frac{1}{2}\text{Tr}[\sigma_x\rho(t)] = \frac{\alpha}{2}\text{Tr}\left[\sigma_x\langle \mathcal{T}\exp(iA\sigma_z)\sigma_x\tilde{\mathcal{T}}\exp(-iA\sigma_z)\rangle\right], \tag{6.64}$$

where $A = -(\epsilon t/2 + g_z\int_0^t dt' \hat{f}_z(t'))/\hbar$. Using the fact that $\exp(a\sigma_z)\sigma_x = \sigma_x\exp(-a\sigma_z)$,[35] we finally obtain

$$\rho_{od} = \frac{\alpha}{2}\text{Tr}[\langle\exp(2iA\sigma_z)\rangle]$$
$$= \frac{\alpha}{2}\text{Tr}\left[\exp(-i\epsilon t\sigma_z/\hbar)\left\langle \mathcal{T}_K\exp\left(-\frac{i}{\hbar}g_z\int_{C_K} dt' \hat{f}_z(t')\sigma_z\right)\right\rangle\right]. \tag{6.65}$$

[33]Note that this discussion applies especially to qubits, quantum two-level systems.

[34]For example, the coherent state $\psi = \sqrt{p}|\downarrow\rangle + \sqrt{1-p}|\uparrow\rangle$ written in the basis of the Hamiltonian would be described by this $\rho(0)$ with $\alpha = \sqrt{p(1-p)}$.

[35]This can be proven by expanding the exponent and using the property $\sigma_z\sigma_x = -\sigma_x\sigma_z$.

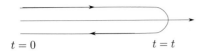

$t = 0$ $t = t$

Fig. 6.9 Keldysh Contour

[36]If $\hat{f}_z(t)$ commutes with itself at different time instants, this is simply twice the integral from $t = 0$ to t.

Here the contour C_K is called the Keldysh contour (see Fig. 6.9), which first runs from the initial time $t = 0$ to the time t and then backwards in time from t to $t = 0$. The operator $\mathcal{T}_K$ orders the operators $\hat{f}_z$ along this contour.[36] The bracketed expression is the characteristic function of the fluctuations of $\int_0^t dt'\,\hat{f}_z(t')$ introduced in the context of full counting statistics (see Sec. 6.7 below). Let us assume for simplicity that the noise is Gaussian, specified only by the variance which in this case is in the long-time limit (similar to what is done in eqns (6.54)–(6.56))

$$\int_0^t \int_0^t dt'\,dt'\,\langle f_z(t')f_z(t')\rangle \approx S_f(0)t.$$

Here $S_f(0)$ describes the fluctuations in f_z at frequencies $\omega \lesssim 1/t$.

Noting the fact that $\mathrm{Tr}[\exp(-i\epsilon t/\hbar\sigma_z)] = \mathrm{Tr}[\cos(\epsilon t/\hbar) - i\sin(\epsilon t/\hbar)\sigma_z] = 2\cos(\epsilon t/\hbar)$, we obtain for the off-diagonal element of the density matrix

$$\rho_{od} = \alpha\cos\left(\frac{\epsilon t}{\hbar}\right)\exp\left(-\frac{g_z^2 S_f(0)t}{\hbar^2}\right). \tag{6.66}$$

We hence find that the fluctuations lead to a decay of the off-diagonal terms in the density matrix with the dephasing rate given by[37]

$$\Gamma_\varphi^z = \frac{1}{\tau_\varphi} = g_z^2 S_f(0)/\hbar^2. \tag{6.67}$$

[37]This term is often called the 'pure dephasing rate', as the total dephasing rate may also include the $S_{f_x}(\omega)$ term.

Including the σ_x-term in the perturbation would give us another dephasing rate, which in the case of small g_x is (see Exercise 6.13)

$$\Gamma_\varphi^x = g_x^2 S_{f_x}(-\epsilon)/(2\hbar^2). \tag{6.68}$$

Here $S_x(\omega)$ is the noise power for the fluctuations of the field $f_x(t)$. The total dephasing rate is then the sum of these two contributions.[38]

In nuclear magnetic resonance, the dephasing time Γ_ϕ^{-1} is often denoted as T_2.

[38]As the dephasing and relaxation rates depend on the noise power of the environmental fluctuations, they can be used for noise measurements; see (Schoelkopf *et al.*, 2003) and (Astafiev *et al.*, 2004).

Complement 6.1 Dephasing time and length in electronic conductors

Chapter 4 discusses various quantum interference effects in electron transport through small conductors. There, the observables, such as conductance, depend on the phase ϕ acquired during the transmission of an electron through the conductor. At non-zero temperatures this phase information is lost at length scales exceeding the *dephasing* or the *phase coherence* length ℓ_φ due to the fact that different electrons traversing the conductor observe slightly different potential.

As shown in above sections, dephasing is closely connected with *energy relaxation* and the associated lengths are often of the same order of magnitude. The time scales for the latter can be estimated from the relaxation time approximations on the collision integrals presented in Ch. 2. The main contributors are electron–electron and electron–phonon scattering. As the electron–phonon scattering rate is strongly temperature dependent (see Complement 2.2), it dominates at high temperatures, typically above some tens of K. The dephasing and relaxation lengths at those temperatures typically become minuscule, and thus we concentrate here on temperatures of the order of 1 K, where interference effects are still visible. There the dominant mechanism in small samples arises from to the voltage fluctuations due to the other electrons, i.e., electron–electron scattering.

In the presence of fluctuations, the measured conductance is an average over the phase fluctuations occurring at the time an electron traverses the mesoscopic sample. It can typically be written as

$$G = G_0 + \delta G \langle \cos[(\phi(t_f) - \phi(t_i)]\rangle_\phi = G_0 + \delta G \mathrm{Re} \langle e^{i[\phi(t_f) - \phi(t_i)]} \rangle, \quad (6.69)$$

where G_0 is the phase-independent part of the conductance and δG the amplitude of the phase-dependent part. We are interested in the fluctuations of the phase at the times between the time t_i of the electron entering the phase coherent sample and the time t_f when it leaves it.

Assuming Gaussian statistics of the phase (see Sec. 6.7 below), we can evaluate the exponent as

$$\langle e^{i[\phi(t_f) - \phi(t_i)]} \rangle = e^{i\langle \phi(t_f) - \phi(t_i)\rangle} e^{-\langle (\delta\phi(t_f) - \delta\phi(t_i))^2\rangle/2} \equiv e^{i\langle \phi(t_f) - \phi(t_i)\rangle} e^{-(t_f - t_i)/\tau_\varphi}, \quad (6.70)$$

where $\delta\phi$ denotes the fluctuation of the phase from its average value $\langle \phi \rangle$. The first term gives the regular interference term, and the latter yields dephasing with the time scale τ_φ. The dephasing term can be connected to the voltage fluctuations via the vector potential $\vec{A}(t)$,

$$\phi(t_f) - \phi(t_i) = -\frac{e}{\hbar} \int_0^L [\vec{A}(t_f) - \vec{A}(t_i)] \cdot dl = \frac{e}{\hbar} \int_{t_i}^{t_f} V_L(t) dt. \quad (6.71)$$

Here $V_L(t)$ is the voltage across length L of the conductor. The dephasing rate can hence be related to the voltage fluctuations via

$$\frac{1}{\tau_\varphi} = \frac{\langle (\delta\phi(t_f) - \delta\phi(t_i))^2 \rangle}{2(t_f - t_i)} = \frac{e^2}{2\hbar^2(t_f - t_i)} \int_{t_i}^{t_f} dt \int_{t_i}^{t_f} dt' \langle \delta V_L(t) \delta V_L(t') \rangle. \quad (6.72)$$

Thermal voltage noise in the limit $\tau_\varphi \gg \hbar/(k_B T)$ is described by the white-noise correlator $\langle \delta V_L(t) \delta V_L(t') \rangle = 2k_B T R_L \delta(t - t')$ and therefore

$$\frac{1}{\tau_\varphi} = \frac{e^2}{\hbar^2} R_L k_B T = \frac{2\pi R_L k_B T}{\hbar R_K}, \quad (6.73)$$

where $R_K = h/e^2$.

Now the question is: which length L should be used in eqn (6.73)? In what follows, we concentrate on the typically most relevant, diffusive regime. There, a more refined calculation[39] taking into account the way correlations decay in a diffusive wire shows that the proper choice depends on the effective dimensionality of the wire (compared to the thermal length $\xi_T = \sqrt{\hbar D/k_B T}$). In a one-dimensional sample where $R_L = L/(\sigma A)$, this is the distance to which the phase information penetrates, i.e., the dephasing length $\ell_\varphi = \sqrt{D\tau_\varphi}$ — unless ℓ_φ exceeds the length of the conductor whose conductance is measured, in which case we should use that length. We hence have to solve eqn (6.73) self-consistently. For a one-dimensional wire we then obtain

$$\frac{1}{\tau_\varphi^{(1\mathrm{D})}} = \left(\frac{2\pi k_B T D^{1/2}}{\hbar A \sigma R_K} \right)^{2/3}. \quad (6.74)$$

Because the exponent is smaller than 1, at low enough temperatures the dephasing rate becomes of the order of temperature. At this point the resistance

[39]See (Dittrich *et al.*, 1998) or (Altshuler and Aronov, 1985).

R_L becomes of the order of R_K, and the wire undergoes a localization transition (see the discussion in Complement 4.3).

In two dimensions, for R_L we have to take into account the fact that an incoherent conductor of size $L \times L$ consists of phase-coherent pieces of size ℓ_φ. This leads to a logarithmic factor, i.e., $R_L \approx R_\square \ln(\ell_\varphi / L_T)$, where $R_\square$ is the sheet resistance and the scale $L_T = \sqrt{\hbar D / k_B T}$ comes from noting that the voltage correlator has a finite width $\sim \hbar / k_B T$. As a result, the dephasing rate becomes

$$\frac{1}{\tau_\varphi^{(2D)}} \simeq \frac{k_B T}{\hbar} \frac{R_\square}{R_K} \ln\left(\frac{k_B T \tau_\varphi}{\hbar}\right) \sim \frac{k_B T}{\hbar} \frac{R_\square}{R_K} \ln\left(\frac{R_K}{R_\square}\right), \tag{6.75}$$

provided $R_\square \ll R_K$.

For three dimensions, assuming $R_L = 1/(\sigma \ell_\varphi)$ would yield a dephasing rate $1/\tau_\varphi \propto T^2$. However, it turns out (Altshuler and Aronov, 1985) that one should instead use $R_L = 1/(\sigma \xi_T) = [k_B T/(\hbar D \sigma^2)]^{1/2}$, i.e., the voltage correlations affecting the decoherence come from smaller length scales $L_T \ll \ell_\varphi$. In this case the dephasing rate is

$$\frac{1}{\tau_\varphi^{(3D)}} \simeq \frac{2\pi}{R_K \sigma D^{1/2}} \left(\frac{k_B T}{\hbar}\right)^{3/2}. \tag{6.76}$$

In a three-dimensional conductor the dephasing rate is essentially the same as the relaxation rate. However, for not extremely dirty three-dimensional wires the dephasing is dominated by electron–phonon scattering (see Exercise 6.14).

6.7   Full counting statistics

In the above sections we characterize the current fluctuations through their variance, i.e., the second moment. One of the findings is that studying the non-equilibrium shot noise reveals information on the studied system not present in the average current. In the absence of inelastic collisions, this information is encoded in the Fano factor. Further information can be obtained by studying the higher moments (cumulants) of charge transfer, or in general the full counting statistics (FCS).

6.7.1   Basic statistics

Let us briefly go through a few definitions concerning the statistical description of a stochastic phenomenon. Consider a process indexed by number n, taking place with a certain probability $P(n)$. For example, this could be the probability that n charges have passed through a conductor within a given time. A *characteristic function* $\chi(\lambda)$ is defined by the (discrete or continuous) Fourier transform of this probability:

$$\chi(\lambda) \equiv \langle e^{i\lambda n} \rangle \equiv \sum_n e^{i\lambda n} P_n. \tag{6.77}$$

The mth derivative of the characteristic function is proportional to the mth *raw moment* $\langle n^m \rangle$,

$$\left(\frac{d^m}{d\lambda^m} \chi(\lambda)\right)_{\lambda=0} = i^m \langle n^m \rangle. \tag{6.78}$$

Taking first the logarithm and then differentiating $\chi(\lambda)$, we obtain the *cumulants* C_m,

$$C_m = \langle\langle n^m \rangle\rangle \equiv (-i)^m \left(\frac{d^m}{d\lambda^m} \ln \chi(\lambda) \right)_{\lambda=0}. \tag{6.79}$$

The lowest cumulants can be expressed through the *central moments* $M_m \equiv \langle (n - \langle n \rangle)^m \rangle$ as

$$\begin{aligned}
C_1 &= \langle n \rangle = \text{average} \\
C_2 &= M_2 = \text{variance} \\
C_3 &= M_3 \\
C_4 &= M_4 - 3M_2^2 \\
C_5 &= M_5 - 10M_2 M_3.
\end{aligned} \tag{6.80}$$

The point in describing a probability distribution via its cumulants is that they generally show how a given distribution deviates from Gaussian. Namely, for a Gaussian distribution, $C_m = 0$ for $m > 2$.

Besides Gaussian, two important distributions in terms of full counting statistics are the Poisson and the binomial distributions, which have the characteristic functions $\chi_{\text{Poisson}}(\lambda) = e^{M_1(e^{i\lambda}-1)}$, and $\chi_{\text{Binomial}} = (1 - p + pe^{i\lambda})^n$, where M_1, p and n are numbers parameterizing the distributions. Binomial distribution describes the probability distribution of the number of events when the probability for each event is p, and the Poisson distribution is obtained in the case of rare events ($p \to 0$).

6.7.2 Full counting statistics of charge transfer

The natural observable to study in mesoscopic systems is the statistics of the transferred charge within a measurement time t_0. In terms of the fluctuating current, the cumulants of the number $n(t_0)$ of transmitted charges are expressed as

$$\langle\langle n^m(t_0) \rangle\rangle = \frac{1}{e^m} \int_0^{t_0} dt_1 \ldots dt_m \langle\langle I(t_1)I(t_2)\ldots I(t_m) \rangle\rangle. \tag{6.81}$$

The first cumulant is $\langle\langle n(t_0) \rangle\rangle = \langle I \rangle t_0/e$, and the second in the limit of a long t_0 is proportional to the noise power,

$$\langle\langle n^2(t_0) \rangle\rangle = \frac{S t_0}{2e^2}. \tag{6.82}$$

The quantum generalization of the electron counting statistics was first proposed by Levitov and Lesovik (1993) and Levitov *et al.* (1996). They suggested, as an idealized scheme of current measurement, using a spin $1/2$ coupled to the current so that it precesses at the rate proportional to the current (see Fig. 6.10). This is totally analogous to the study of the dephasing in a two-level system discussed in Sec. 6.6.2 above. This idea is to define ρ_{od} from eqn (6.65) in the limit $\epsilon = 0$ as the quantum version of the generating function for full electron counting

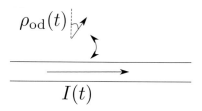

Fig. 6.10 Idealized scheme for a current (statistics) measurement based on a spin $1/2$ coupled to the current.

[40]For a summary of the research on
FCS in mesoscopic systems, see for example (Belzig, 2003).

statistics. Later it was shown by Kindermann and Nazarov (2003) that
the same definition can be directly used for a direct charge counting
measurement as well.

The characteristic function has been calculated for many types of
mesoscopic systems, but here we only concentrate on the simplest: a
normal-metal conductor between two normal-metal probes. In this case,
the characteristic function obtained from a quantum-mechanical calculation is of the form $\chi(\lambda) = \exp(\mathcal{S}(\lambda))$ with[40]

$$\mathcal{S}(\lambda) = \frac{2t_0}{h} \sum_j \int dE \ln \left[1 + T_{jLR}(E)(e^{i\lambda} - 1) + T_{jRL}(E)(e^{-i\lambda} - 1) \right].$$
(6.83)

Here $T_{jLR} = T_j f_L(E)(1 - f_R(E))$ is the probability for tunnelling from
the left to the right of the contact for the eigenchannel j, and T_{jRL} is the
corresponding probability from right to left. The argument of the logarithm summarizes the possible processes: no transmission, transmission
from left to right, and transmission from right to left. In this expression, the *counting factors* $e^{\pm i\lambda} - 1$ correspond to single-charge transfers
between the contacts. For example, for Andreev reflection, which is a
two-electron effect, the corresponding counting factor would be of the
form $e^{\pm 2i\lambda} - 1$.

At equilibrium and for energy-independent transmission probabilities,
the counting statistics is described by

$$\mathcal{S}(\lambda) = -\frac{8t_0 k_B T}{h} \sum_j \arcsin^2 \left(\sqrt{T_j} \sin \left(\frac{\lambda}{2} \right) \right).$$
(6.84)

Thus, even at equilibrium the fluctuations are not Gaussian, except for
a ballistic contact which has $T_j = 1$ for all propagating channels j, in
which case $\mathcal{S}(\lambda) = -t_0 k_B T \lambda^2 / h$.

At a vanishing temperature the term in eqn (6.83) describing transport
opposite to the voltage vanishes, and the resultant characteristic function
describes a series of binomial processes,

$$\chi(\lambda) = \prod_j \left(1 + T_j(e^{i\lambda} - 1) \right)^N,$$
(6.85)

where $N = 2et_0|V|/h$ is the *number of attempts* in a given time t_0. This
has an easy interpretation of a particle in channel j trying to transmit
through the scatterer for N times, the probability of success being T_j.

We can expand eqn (6.85) in the cumulants of the charge transfer.
The lowest cumulants are

$$\langle\langle n \rangle\rangle = N \sum_j T_j$$

$$\langle\langle n^2 \rangle\rangle = N \sum_j T_j(1 - T_j)$$

$$\langle\langle n^3 \rangle\rangle = N \sum_j T_j(1 - T_j)(1 - 2T_j)$$
(6.86)

$$\langle\langle n^4 \rangle\rangle = N \sum_j T_j(1 - T_j)(1 - 6T_j(1 - T_j)).$$

The first two are thus consistent with the formulae for the average current and noise, eqns (3.49) and (6.22).

Note that for a tunnel junction with $T_j \ll 1$, the counting distribution turns into a Poissonian, for which all the cumulants are of the form $\langle\langle n^m \rangle\rangle = N \sum_j T_j$.

Whereas already a lot is known about the properties of the full counting statistics, or higher cumulants of different types of systems, there are only a few measurements of the higher (than second) cumulants.[41]

6.8 Heat current noise

Apart from the charge current noise, also the heat current $\dot{Q}$ in mesoscopic systems fluctuates in time. These fluctuations are described with the power–power correlator

$$S_Q(\omega) = \int dt e^{i\omega t} \langle \dot{Q}(t)\dot{Q}(0)\rangle. \tag{6.87}$$

There is one relevant difference between the charge and heat current fluctuations, similar to the difference between the charge and heat currents: whereas charge current flows only within the electronic system, there may be heat current between any connected subsystems, for example between electrons and phonons. At equilibrium (no temperature gradients) the heat current is described by an analogue of the thermal noise for the charge current. At zero frequency it is

$$S_Q = 4k_B T^2 G_{\text{th}}, \tag{6.88}$$

where G_{th} is the heat conductance for the link where the heat flows. However, as the heat current is not strictly speaking an observable in quantum mechanics, it does not have to satisfy the fluctuation dissipation theorem (Averin and Pekola, 2010).

Out of equilibrium, heat current statistics in electronic contacts has been studied for example by employing the counting field technique (Kindermann and Pilgram, 2004).

In small systems the heat current fluctuations lead to the fluctuations of the internal energy. In the quasi-equilibrium limit (strong internal relaxation) these show up as fluctuations of the electron temperature.[42] The connection between the two in the free-electron model is given by

$$E_I = \frac{C_h T_e}{2} = \frac{\pi^2 k_B^2 T_e^2}{6\delta_I}, \tag{6.89}$$

where $C_h = \pi^2 k_B^2 T_e/(3\delta_I)$ is the electron specific heat and $\delta_I = 1/(N_F \Omega)$ is the average energy level spacing for a system with volume Ω.

At equilibrium, the temperature fluctuations are (close to) Gaussian.[43] The instantaneous (in contrast to zero-frequency) equilibrium temperature fluctuations are described by a Maxwell–Boltzmann distribution of energy,[44]

[41] The third cumulant of a tunnel junction was measured by Reulet *et al.* (2003) and Bomze *et al.* (2005). The attention has turned to using other mesoscopic systems for detecting the non-Gaussian noise; see for example (Gustavsson *et al.*, 2006) and (Timofeev *et al.*, 2007).

[42] For recent work, see (Heikkilä and Nazarov, 2009; Laakso *et al.*, 2010; Laakso *et al.*, 2012).

[43] 'Close to' since the probability becomes zero for $T < 0$.

[44] See for example (Landau and Lifshitz, 1985).

$$P(T_e) \propto \exp\left[-\frac{E_I - E_a}{k_B T_a}\right] = \exp\left[-\frac{C_h(T_e - T_a)^2}{2k_B T_a^2}\right], \qquad (6.90)$$

where T_a is the average temperature and E_a the average internal energy.

Heat current noise is especially relevant for defining the resolution of temperature measurements in small systems, as the accuracy of such measurements is limited by the heat fluctuations. Such accurate and fast thermometry is required, for example, in thermal radiation detectors where the radiation coupled to the electron system heats it up. From the change in the temperature one can then measure the amount of radiation.[45]

[45] For a detailed description of thermal radiation detectors, see Ch. IV in (Giazotto *et al.*, 2006).

Further reading

- Fluctuations in different types of systems are treated in (Kogan, 1996).

- The standard reference for shot noise is the review by (Blanter and Büttiker, 2000).

- Dissipative quantum systems are described in (Weiss, 1999).

- A sample of topics describing the quantum characteristics of fluctuations is discussed in (Nazarov, 2003).

- Moreover, the basic phenomena, such as equilibrium fluctuations described by fluctuation–dissipation theorem, are described in most books on advanced-level statistical mechanics.

Exercises

(6.1) **Wiener–Khintchine theorem.** Assume we measure random noise signal $\delta x(t)$ in a stationary system over a long time $t \to \infty$. As $\delta x(t)$ is a real quantity, its Fourier transform satisfies $\delta x(-\omega) = \delta x^*(\omega)$. Then we can write

$$\delta x(t) = \int_0^\infty \frac{d\omega}{2\pi}[\delta x(\omega)e^{-i\omega t} + \delta x^*(\omega)e^{i\omega t}]. \quad (6.91)$$

Consider a spectrum analyzer which measures this signal. It contains a narrow-band-pass filter which leaves only the (angular) frequencies $\omega \in [\omega_0 - \Delta\omega/2, \omega_0 + \Delta\omega/2]$, and an output detector that measures the mean square of this signal (corresponding to the signal power). Show that the averaged squared signal from the filtered band equals the spectral density of noise, $S(\omega_0)$ times the band width $\Delta f = \Delta\omega/2\pi$ (assuming that $S(\omega_0)$ does

not essentially change within this band), i.e.,

$$\langle \delta x^f(t)\delta x^f(t)\rangle = S(\omega_0)\frac{\Delta\omega}{2\pi} = S(\omega_0)\Delta f. \quad (6.92)$$

Here $\delta x^f(t)$ is the signal after the band-pass filtering,

$$\delta x^f(t) = \int_{\omega_0-\Delta\omega/2}^{\omega_0+\Delta\omega/2} \frac{d\omega}{2\pi}[\delta x(\omega)e^{-i\omega t} + \delta x^*(\omega)e^{i\omega t}]. \quad (6.93)$$

Hint: Write down the spectral density of noise for two times, t and t', and use the fact that in a stationary setup it depends only on $t - t'$.

(6.2) Prove the fluctuation–dissipation theorem between the susceptibility χ_{ij} and correlator S_{ij} with $i \neq j$.

(6.3) Prove eqn (6.15) for fermions by using the Wick theorem, valid for macroscopic systems. According to this theorem, state averages of multiple products of fermion or boson operators may be calculated as a sum of averages of all possible pairs of such operators. When taking the different pairings, the fermion/boson (anti)commutation property has to be used. For example, for fermion operators $\hat{a}_i$

$$\langle\hat{a}_1\hat{a}_2\hat{a}_3\hat{a}_4\rangle$$
$$=\langle\hat{a}_1\hat{a}_2\rangle\langle\hat{a}_3\hat{a}_4\rangle - \langle\hat{a}_1\hat{a}_3\rangle\langle\hat{a}_2\hat{a}_4\rangle + \langle\hat{a}_1\hat{a}_4\rangle\langle\hat{a}_2\hat{a}_3\rangle.$$

Hint: Use the commutation relations for Heisenberg operators,

$$\{\hat{a}_{\alpha n}(E), \hat{a}^\dagger_{\alpha m}(E')\} = \delta_{nm}\delta(E-E')$$
$$\{\hat{a}_{\alpha n}(E), \hat{a}_{\alpha m}(E')\} = 0$$
$$\{\hat{a}^\dagger_{\alpha n}(E), \hat{a}^\dagger_{\alpha m}(E')\} = 0$$

and the property $\langle\hat{a}^\dagger_{\alpha l}(E)\hat{a}_{\alpha' l'}(E')\rangle = \delta_{\alpha\alpha'}\delta_{ll'}\delta(E-E')f_\alpha(E_l)$, where $f_\alpha(E)$ is the distribution function for lead α. In normal systems 'anomalous' correlators such as $\langle\hat{a}\hat{a}\rangle$ vanish.

(6.4) Prove eqns (6.20) and (6.21).

(6.5) Starting from eqn (6.16), derive an expression for the frequency-dependent zero-temperature two-terminal noise at a finite voltage V in the case when the scattering matrix is energy independent.

(6.6) Show that for classical current fluctuations $\delta I(\omega)$ in the presence of time-independent driving, the frequency dependence of the stationary current–current correlator can be obtained from $S_I(\omega)\Delta f = \langle\delta I(\omega)\delta I(-\omega)\rangle$, where $\Delta f = \Delta\omega/(2\pi)$ is a bandwidth of the noise measurement. Hint: use the properties of the Fourier transform.

(6.7) Consider two tunnel barriers, with resistances R_1 and R_2 in series, connected to an ideal voltage source (see Fig. 6.11) with voltage V_{ext}. Assume that the intrinsic current fluctuations through R_1 and R_2 are uncorrelated. In this case the noise in the entire system can be analyzed via the Langevin circuit scheme. a) By using the fact $S_{\text{int}}(\omega)\Delta f = \langle\delta I(\omega)\delta I(-\omega)\rangle$, find the total current noise $S_I\Delta f = \langle\Delta I(\omega)\Delta I(-\omega)\rangle$. The bandwidth Δf cancels out from the final result. b) Assuming $R_1 = R_2$, show that adding tunnel barriers but keeping the same total resistance decreases the shot noise (noise at $T = 0$, i.e., $S_{\text{int}} = 2e\langle I\rangle$), but that the thermal noise $S^i_{\text{int}} = 4k_BT/R_i$ remains unaffected. c) Calculate the voltage noise $S_V = \langle\Delta V\Delta V\rangle/\Delta f$ at point A.

(6.8) Using the Boltzmann–Langevin formalism and eqn (6.39), consider the shot-noise limit where the temperature vanishes, $T = 0$, and there is a voltage V over the wire. Calculate S_I for a) non-equilibrium diffusive system where $L \ll l_{ee}, l_{eph}$ b) quasi-equilibrium diffusive system with $l_{ee} \ll L \ll l_{eph}$ and c) equilibrium diffusive system with $l_{ee}, l_{eph} \ll L$. In each case, compare this to the average current,

$$I = \frac{A\sigma_N}{L}V \tag{6.94}$$

and find the corresponding Fano factor (see also Fig. 6.8). Use the distribution functions calculated in Examples 2.2 and 2.3. Hint: Assuming the integrals converge, you may interchange the order of integration in eqn (6.39). Besides the integrals in eqn (A.64), the following integral may be useful:

$$\int_0^1 \sqrt{x(1-x)} = \frac{\pi}{8}.$$

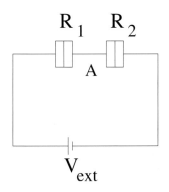

Fig. 6.11 Two tunnel barriers in series.

(6.9) Consider the setup shown in Fig. 6.6. Calculate the total finite-frequency current noise across the 'load' resistance R_L, by assuming that the intrinsic noise sources are uncorrelated and the auto-correlation function is frequency independent, i.e., $\langle\delta I_{S/L}(\omega)\delta I_{S/L}(-\omega)\rangle = S_{L/R}$. At the resonance frequency $\omega_0 = 1/\sqrt{LC}$, find L and C that 'match' the resistance R_S of the source with the resistance R_L of the load, i.e., maximize the coupling between the total current noise across the load and the intrinsic noise across the source.

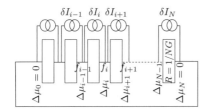

Fig. 6.12 Circuit for the derivation of the Boltzmann–Langevin equation.

(6.10) An alternative derivation of the Boltzmann–Langevin equation, eqn (6.39) for a diffusive wire. Model the diffusive wire by a series of N tunnel barriers ($T_n \ll 1$), each with resistance $R = 1/(NG_N)$, and with an intrinsic noise source δI_i ($i = 1, \ldots N$) whose correlator $\langle \delta I_i \delta I_j \rangle \propto \delta_{ij}$ is of the form of eqn (6.20), with f_L (f_R) given by the distribution functions at the left (right) of the junction (see Fig. 6.12). i) Find a difference equation for the voltage fluctuations following a similar scheme as in Exercise 6.7. ii) Convert this into a differential equation in the limit $N \to \infty$ and solve it by using the boundary conditions that the fluctuating potential vanishes at the ends of the wire. iii) Find an expression for the total noise current ΔI consisting of the intrinsic noise and the induced noise due to potential fluctuations. iv) Calculate the correlator $\langle \Delta I^2 \rangle$ by noting that in the limit $N \to \infty$ the distribution function becomes smooth, i.e., $f_i \approx f_{i+1}$.

(6.11) Assume that a two-level system described by the Hamiltonian

$$H_0 = \epsilon \sigma_z + \Delta \sigma_x \qquad (6.95)$$

is coupled to the environment through a force $gf\sigma_z$. The spectrum of the fluctuations of f is given through the thermal noise in a resistor with resistance R_N:

$$S_f(\omega) = 2\frac{R_N}{R_K}\hbar^2 \omega \coth\left(\frac{\hbar\omega}{2k_B T}\right), \qquad (6.96)$$

where $R_K = h/e$. Calculate the dephasing and relaxation times of the two-level system, assuming that this is the only source of noise. Then, try to obtain an estimate of these times for $\Delta = 0.01\epsilon$, $\epsilon/k_B = 1$K, $g^2 = 0.01$ and $R_N = 50\Omega$ at temperature $T = 50$mK.

(6.12) Consider a quantum system coupled to a 'bath' with equilibrium fluctuations obeying the fluctuation–dissipation theorem with temperature T. Show that the steady-state occupation numbers p_n of energy levels E_n satisfy detailed balance,

$$\frac{p_m}{p_n} = \exp\left(\frac{E_n - E_m}{k_B T}\right). \qquad (6.97)$$

What is the resultant steady-state energy distribution function of the quantum system? Why is it of this form?

(6.13) Using a similar approach as in Sec. 6.6.1, but now for the off-diagonal matrix element $\langle 0|\rho|1\rangle$ of the density matrix of a two-level system, show that the force proportional to σ_x produces a dephasing rate given by eqn (6.68).

(6.14) Using the results in eqns (6.74) and (6.76), estimate the dephasing lengths $\ell_\varphi = \sqrt{D\tau_\varphi}$ of one-dimensional and three-dimensional conductors at $T = 1$ K and $T = 300$ K resulting from voltage fluctuations in a diffusive wire having a mean free path $\ell_{el} = 50$ nm, Fermi wavelength $\lambda_F = 0.46$ nm and Fermi velocity $v_F = 1.6 \times 10^6$ m/s. In the one-dimensional case, use the cross-section $A = (100 \text{ nm})^2$. Use the Drude formula for the conductivity σ and the free-electron density of states N_F from eqn (1.5). Compare these to the electron–phonon scattering lengths in Complement 2.2.

(6.15) Find the four lowest cumulants of the equilibrium fluctuations described by eqn (6.84).

(6.16) **Question on a scientific paper.** Take the paper Spietz *et al.*, *Science* **300**, 1929 (2003), which presents the use of shot noise for primary thermometry. a) Analyze the measurement setting in Fig. 2. What is the role of the capacitances separating the sample from the noise measurement amplifier? Explain how these capacitances limit the measured frequency band for the noise. Explain also why the bare thermal noise cannot be used for primary thermometry. b) Using Fig. 3, explain how the shot noise can be used for measuring the temperature without calibration.

Single-electron effects

7

When a junction between two conductors is made very small, the energy scale required to transfer a single electron through it becomes observable. If two junctions are placed in series such that there is a conducting island between them, or a single junction is placed in a high-resistance environment, this *charging energy* E_C (Fig. 7.1) limits the current transport through the system. In this case no current can flow at voltages and temperatures lower than E_C/e. This is the regime of *Coulomb blockade*.

Coulomb blockade shows up in various types of small structures. In a *single-electron transistor* the island is typically made of a metal with a high density of states (average level spacing much less than E_C, k_BT or eV), and the charge transport is controlled by an extra *gate voltage* coupled capacitively to the junction. For certain values of gate voltages, Coulomb blockade is lifted and current can flow. This is why the system can be used as a transistor and as an extremely sensitive detector of charge. In semiconductor or graphene quantum dots and molecular structures, the typical energy level spacing $\Delta\epsilon$ is large (up to the order of 1 eV in molecules). In this case, the transport properties sensitively depend on both energy scales, E_C and $\Delta\epsilon$. The latter systems are described in Ch. 8.

In Exercise 1.1 you estimated the charging energy for a typical mesoscopic tunnel barrier. With $C = 0.89 \times 10^{-15}$ F$= 0.89$ fF, the charging energy $E_C = 86$ μeV$= 1$ K$\cdot k_B$.

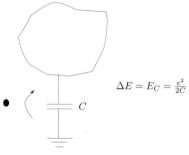

$$\Delta E = E_C = \frac{e^2}{2C}$$

Fig. 7.1 Adding an electron to an island costs an interaction energy E_ϵ, which can be quantified by the capacitance C.

7.1 Charging energy

Let us study the amount of energy required to charge an initially neutral capacitor with capacitance C by the charge q. The instantaneous charge Q induces a voltage $V = Q/C$ over the capacitor. On the other hand, a small change dQ in the charge changes the (grand canonical) energy of the capacitor by $dE = dQV = QdQ/C$. Integrating from the neutral state $Q = 0$ to some finite charge $Q = q$ results into an energy change $\Delta E = \frac{q^2}{2C}$.

Let us consider an event where an electron tunnels through a contact with capacitance C (with a constant voltage bias) as depicted in Fig. 7.2. Before tunnelling, the charge across the capacitor is $q = CV$ and the electrostatic energy stored on the capacitor is $E_i = q^2/(2C)$. After an electron has tunnelled through, the charge is $q - e$ and the energy of the

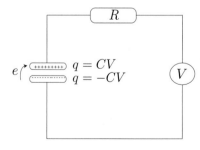

Fig. 7.2 Electron tunnelling through a contact with capacitance C.

final state is

$$E_f = \frac{(q-e)^2}{2C} = \frac{q^2}{2C} - \frac{e}{C}q + \frac{e^2}{2C} = E_i - eV + E_C. \quad (7.1)$$

At a vanishing temperature, tunnelling thus seems to be possible only provided that $eV > E_C$.

However, in the case of a single junction, the quantum fluctuations in the environment of the junction may enable the electron tunnelling even at lower voltages (Ingold and Nazarov, 1992). An estimate of this effect is provided by the uncertainty principle: Let us denote the time it takes for the electron charge to relax into the environment of the junction by τ. Within this time, the environment is in a non-equilibrium state, and tunnelling may take place provided that the energy uncertainty $\Delta E = \hbar/(2\tau)$ is greater than E_C, i.e.,

$$E_C \tau \lesssim \hbar/2. \quad (7.2)$$

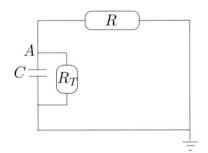

Fig. 7.3 Single junction with capacitance C in an environment with resistance R.

But what is the time τ? Consider the circuit in Fig. 7.3: Here R_T is the resistance of the junction, and R the resistance of the environment. It is easy to derive an equation for the charge q on the capacitor by requiring that the voltage drop across the resistors and the capacitor is the same. We obtain

$$\frac{dq}{dt} = -\frac{q}{R_{||}C}, \quad (7.3)$$

where $R_{||} = (1/R + 1/R_T)^{-1}$. The solution to this is $q(t) = q_0 e^{-t/(R_{||}C)}$, from which we can identify the time $\tau = R_{||}C$. Substituting this in eqn (7.2) we obtain the condition for Coulomb blockade (i.e., the inverse of the condition in eqn (7.2) that states when tunnelling is made possible by uncertainty):

$$E_C \tau = \frac{e^2}{2C} R_{||} C = \frac{e^2 R_{||}}{2} > \hbar/2. \quad (7.4)$$

As $R_{||}$ is dominated by the smaller of the two impedances, this imposes two conditions for the observation of Coulomb blockade in a single junction:[1]

- The resistance R_T of the tunnel barrier must exceed the (modified) quantum resistance $\hbar/e^2 = 4108\ \Omega$. For low tunnelling resistances, higher-order tunnelling effects such as cotunnelling (Sec. 7.4) lift the Coulomb blockade.

- The environment of the junction has to have a resistance that exceeds $\hbar/e^2$.

The current suppression in the case of a single junction is called *dynamical Coulomb blockade* or *environmental Coulomb blockade*, and is discussed more in Sec. 7.5.

The description of the Coulomb blockade theory presented in this chapter explicitly assumes that the electrons transmit through tunnel barriers. Also, other types of contact with a transmission probability close to unity have been considered (Flensberg, 1993; Matveev, 1995; Nazarov, 1999a). It turns out that Coulomb blockade features can still be found, but with an effective charging energy that is much smaller than the bare charging energy $E_C = e^2/(2C)$.

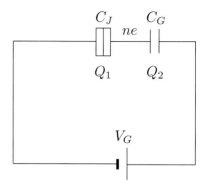

Fig. 7.4 Single-electron box. The circuit element denoted with capacitance C_J is a tunnel junction (in contrast to the pure capacitor C_G through which no tunnelling is assumed).

7.1.1 Single-electron box

Consider the system depicted in Fig. 7.4. It shows a metal island with a tunnel contact (the circuit element left from the island denotes a normal tunnel junction), and a capacitive contact to a gate voltage V_G. For $V_G = 0$, the ground state of the island is neutral, i.e., the number of electrons on the island equals the number of positive ions. The higher-energy states may involve $n = \pm 1, \pm 2, \ldots$ extra (deficit) electrons on the island. If V_G is turned on, the ground state of the system changes in discrete steps from $n = 0$ to another integer value of n. The system reacts to this change of the ground state by letting extra electrons tunnel through the tunnel contact.

The gate voltage induces continuous displacement charges Q_1 and Q_2 on the capacitors. This is due to the spatial distribution of charge on the island. However, the total charge $Q_1 + Q_2 = -ne$ still remains an integer multiple of the electron charge. The charging energy of the system is now obtained in terms of this n and the gate voltage V_G by noting that $V_G = Q_1/C_J - Q_2/C_G$ and eliminating Q_1 and Q_2 in favor of n and V_G. This gives

$$Q_1 = C_J U, \quad Q_2 = -C_G(V_G - U),$$

where $U = -(ne + C_G V_G)/(C_J + C_G)$ is the potential of the island. The sum of the individual charging energies of the junctions constitute the internal energy of the single-electron box. In the discussion of charge dynamics, the relevant grand canonical free energy has to involve the work $\mu_G N_2 = V_G Q_2$ done by the voltage V_G in polarizing the capacitor. This free energy is thus

$$E_{\rm ch}(n, Q_G) = \frac{Q_1^2}{2C_J} + \frac{Q_2^2}{2C_G} + V_G Q_2 = \frac{(ne - Q_G)^2}{2C} - \frac{Q_G V_G}{2}. \quad (7.5)$$

The latter term is often disregarded as it is independent of the charge state of the island. Here $C = C_J + C_G$ is the total capacitance of the island. The effect of the gate voltage is to define a continuous offset charge $Q_G = C_G V_G$. This is often called the *gate charge*.

The charging energy for different values of n as a function of Q_G is a series of parabola as illustrated in Fig. 7.5. The single-electron box has degeneracy points whenever $Q_G/e = n + 1/2$. At these points, the island may change its charge state. Changing the gate V_G slowly, the charge n on the island stays constant until a degeneracy point is reached, and an extra electron tunnels in or out of the island, depending on the direction of the gate charge change. As a result, the average charge on the island behaves as depicted in Fig. 7.6.

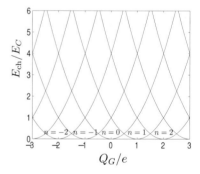

Fig. 7.5 Charging energy for different values of the excess number n of electrons in the single-electron box.

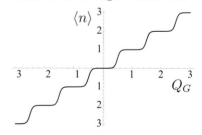

Fig. 7.6 Average number of excess charges in a single-electron box for different values of the gate charge. Here finite temperature $k_B T = 0.1 E_C$ rounds the charge steps.

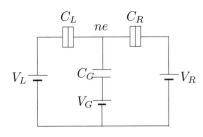

Fig. 7.7 Single-electron transistor.

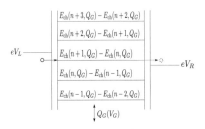

Fig. 7.8 Sequential tunnelling: electrons may tunnel through the single-electron transistor in the lowest-order tunnelling process if charging energy difference $E_{\text{ch}}(n+1, Q_G) - E_{\text{ch}}(n, Q_G)$ falls within the bias window defined by V_L and V_R.

[2]Here $\text{mod}(Q_G, e)$ gives the remainder of Q_G minus an integer number of elementary charges.

[3]However, the first theories on Coulomb blockade date back to the 1970s, see (Shekter, 1973; Kulik and Shekter, 1975).

7.1.2   Single-electron transistor (SET)

The most basic yet practically useful system exhibiting single-electron effects in a direct current is a single-electron transistor (SET) shown in Fig. 7.7. Here tunnelling through both contacts to the island is allowed, such that there may be a net current flowing through the system. The current is controlled by the transport or bias voltage $V_L - V_R$, and the state of the island can be controlled by the gate voltage. Similar considerations as for the single-electron box (see Exercise 7.1) show that we again get $E_{\text{ch}}(n, Q_G) = (ne - Q_G)^2/C$ with $C = C_L + C_R + C_G$. In this case the gate charge $Q_G = C_G V_G + C_L V_L + C_R V_R$ consists of contributions from all voltage sources.

Assume that we apply a bias voltage such that the potential on the left is higher than that on the right. In this case, at $T = 0$ tunnelling from the left electrode (i.e., changing the charge state from n to $n + 1$) to the island is possible if

$$eV_L > E_{\text{ch}}(n+1, Q_G) - E_{\text{ch}}(n, Q_G) = \left(n + \frac{1}{2} - \frac{Q_G}{e}\right) e^2/C \equiv \delta E_{\text{ch}}(n).$$
(7.6)

Similarly tunnelling from the island to the right lead ($n + 1 \to n$) is possible if

$$eV_R < E_{\text{ch}}(n+1, Q_G) - E_{\text{ch}}(n, Q_G).$$
(7.7)

Both conditions have to be satisfied simultaneously in order for the current to flow (see Fig. 7.8).

Let us consider a simple example where we have a symmetric bias, $V_L = -V_R = V/2$. Then for a given Q_G, we have to apply a bias $V > (e - 2\text{mod}(Q_G, e))/C$ in order to lift the Coulomb blockade.[2] Thus, changing the gate voltage, the current of the island at some finite bias oscillates between almost zero ($Q_G < (e - VC)/2$) and some finite value which depends on the tunnelling resistances. These are *Coulomb oscillations* (see Fig. 7.12), which can be understood from the stability diagram of the SET, but the quantitative calculation of the current requires the use of *orthodox theory* outlined in the next two sections.

7.2   Tunnel Hamiltonian and tunnelling rates

The orthodox theory of single-electron tunnelling was constructed by Averin and Likharev (1986).[3] It is based on calculating the tunnelling rates through single junctions that are part of a single-electron transistor or a multi-island system. This is done using the Fermi golden rule, i.e., the lowest-order perturbation theory in tunnelling. In connection with single electronics, this is the *sequential tunnelling approximation*. With the knowledge of these rates, we may construct a master equation that describes the probabilities of the charge states, and from the solution of this master equation we then obtain the observables, such as the current through the system.

The analysis starts from the Hamiltonian of the system. For a SET it can be written as

$$H = H_L + H_R + H_I + H_{\text{ch}} + H_t. \tag{7.8}$$

Here the first three terms describe the non-interacting electrons in the left and right electrode and the island.[4] They can thus be written in the second-quantized notation. For example, the Hamiltonian for the left electrode is[5]

$$H_L = \sum_{k,\sigma} \epsilon_k c^\dagger_{k,\sigma} c_{k,\sigma}, \tag{7.9}$$

where $c_{k\sigma}$ annihilates an electron in momentum, spin state k, σ in the left electrode and $c^\dagger_{k\sigma}$ is the corresponding creation operator. The Hamiltonian H_R for the right electrode is analogous, and that for the island reads

$$H_I = \sum_{q,\sigma} \epsilon_q b^\dagger_{q,\sigma} b_{q,\sigma}, \tag{7.10}$$

$b_{q,\sigma}$ being the annihilation operator for the electrons on the island. The charging energy term is assumed to depend only on the total charge on the island,

$$H_{\text{ch}} = \frac{(\hat{n}e - Q_G)^2}{2C}, \tag{7.11}$$

where $\hat{n} = \sum_{q,\sigma} b^\dagger_{q,\sigma} b_{q,\sigma} - N_+$ is the number operator describing the number of total charge on the island. We subtract the number N_+ of the positive ions for simplicity. Finally, the tunnelling Hamiltonian $H_t = H_{t,L} + H_{t,R}$ describes a process where an electron is annihilated in one of the electrodes and created on the island, and vice versa,[6]

$$H_{t,L} = \sum_{k,q,\sigma} T_{k,q} b^\dagger_{q,\sigma} c_{k,\sigma} + \text{h.c.} \tag{7.12}$$

We thus assume that tunnelling conserves the spin. A similar Hamiltonian can be written for the contact to the right electrode. This tunnelling Hamiltonian allows us to calculate the tunnelling rate using the Fermi golden rule. The transition rate induced by operator H' between an initial state $|i\rangle$ and the final state $|f\rangle$ can be found from (see Sec. A.3)

$$\Gamma_{i \to f} = \frac{2\pi}{\hbar} \langle |\langle f|H'|i\rangle|^2 \delta(E_i - E_f)\rangle. \tag{7.13}$$

The outer brackets denote the statistical averaging. Here the initial and final states are the different charge states on the island. We assume that the electrodes quickly relax into their equilibrium charge state and hence we do not consider the differences in the charge states of the electrodes. This assumption is valid when the system is in a low-resistance environment, and thus the dynamical Coulomb blockade effects (see Sec. 7.5) can be disregarded.

If the final or initial states can be achieved through various configurations of the internal degrees of freedom, we must sum over these and

[4] In fact, there the interactions are described by a mean field model, such that they only alter the dispersion relation ϵ_k, which is not necessarily of the free-electron form.

[5] Note that the momenta k, q are usually vectors, but we omit the vector sign in these formulae for simplicity.

[6] The letters h.c. denote the Hermitian conjugate of the previous expression. Here this term denotes the tunnelling into the opposite direction.

weigh the initial states with their occupation probabilities W_{i_α}:

$$\Gamma_{\beta\alpha} = \frac{2\pi}{\hbar} \sum_{f_\beta i_\alpha} |\langle f_\beta | H' | i_\alpha \rangle|^2 W_{i_\alpha} \delta(E_{f_\beta} - E_{i_\alpha}).$$

Let us first calculate the tunnelling rate through the left junction, so that $H' = H_{t,L}$, and the presence of other junctions being first disregarded in this calculation. We also concentrate only on the tunnelling *into* the island, so that the charge state changes from n to $n+1$. The final state is thus of the form $|f_{n+1}\rangle = \hat{b}_{q\sigma}^\dagger \hat{c}_{k\sigma} |i_n\rangle$, which can be achieved for any combination of $\{k, q, \sigma\}$, and we should therefore sum over them. We have

$$\Gamma_{LI}(n) = \frac{2\pi}{\hbar} \sum_{k,q,\sigma} \sum_{i_n} \left| \langle i_n | \hat{c}_{k\sigma}^\dagger \hat{b}_{q\sigma} H_{TL} | i_n \rangle \right|^2 W_{i_n} \delta(\delta E_{\text{ch}} + \varepsilon_q - \varepsilon_k),$$

$$= \frac{2\pi}{\hbar} \sum_{k,q,\sigma} \sum_{i_n} |T_{k,q}|^2 \left| \langle i_n | \hat{c}_{k\sigma}^\dagger \hat{b}_{q\sigma} \hat{b}_{q\sigma}^\dagger \hat{c}_{k\sigma} | i_n \rangle \right|^2 W_{i_n} \delta(\delta E_{\text{ch}} + \varepsilon_q - \varepsilon_k).$$

The initial state can be taken to be a tensor product of the initial states of the lead and the island, i.e., $|i_n\rangle = |i_L\rangle \otimes |i_{In}\rangle$. Then

$$\Gamma_{LI}(n) = \frac{2\pi}{\hbar} \sum_{k,q,\sigma} \sum_{i_L, i_{In}} |T_{k,q}|^2 \tag{7.14}$$

$$\times \left| \langle i_L | \hat{c}_{k\sigma}^\dagger \hat{c}_{k\sigma} | i_L \rangle \langle i_{In} | \hat{b}_{q\sigma} \hat{b}_{q\sigma}^\dagger | i_{In} \rangle \right|^2 W_{i_L} W_{i_{In}} \delta(\delta E_{\text{ch}} + \varepsilon_q - \varepsilon_k).$$

Objects of the type $\langle i_L | \hat{c}_{k\sigma}^\dagger \hat{c}_{k\sigma} | i_L \rangle$ are occupation numbers and hence either zero or one (for fermions). We can then omit the square and use

$$\sum_{i_L} W_{i_L} \langle i_L | \hat{c}_{k\sigma}^\dagger \hat{c}_{k\sigma} | i_L \rangle = f_L(\varepsilon_k),$$

$$\sum_{i_{In}} W_{i_{In}} \langle i_{In} | \hat{b}_{q\sigma} \hat{b}_{q\sigma}^\dagger | i_{In} \rangle = 1 - f_I(\varepsilon_q),$$

where $f_{L/I}(E)$ are the distribution functions of the electrons in the left reservoir and the island, respectively. Next we make the conversion from the momentum sums to energy integrals, assuming that the tunnelling matrix element is independent of momentum direction,

$$\sum_{k,q,\sigma} |T_{k,q}|^2 \mapsto \Omega_L \Omega_I \int d\epsilon_k d\epsilon_q |T|^2 N_L N_I.$$

Here the volumes Ω_L and Ω_I refer to the size of the region in space where the coupling between the electrode and the island takes place.[7] This also assumes that the transmission probability $|T_{k,q}|^2 = |T|^2$ through the junction is independent of the channel indices k, q with energies within the relevant interval.[8] We hence obtain

---

[7] Here Ω_I is typically the volume of the island, whereas Ω_L is of the order of the cross section of the contact times λ_F. Note that this volume is often omitted. Such an omission leads to expressions where the dimension of the density of states is 1/energy, not 1/(energy × volume). However, as these factors are eventually absorbed in the tunnel resistance, this omission does not change any relevant physics.

[8] In practice, this assumption means that the transmission probability is independent of the momentum angle, at least approximatively satisfied in diffusive systems.

$$\Gamma_{LI}(n)$$

$$= \frac{2\pi}{\hbar} \Omega_L \Omega_I \int d\varepsilon_L d\varepsilon_I |T|^2 N_L N_I f_L(\varepsilon_L)(1 - f_R(\varepsilon_I))\delta(\delta E_{\text{ch}} + \varepsilon_I - \varepsilon_L),$$

$$= \frac{1}{e^2 R_L} \int d\varepsilon_L d\varepsilon_I f(\varepsilon_L)(1 - f(\varepsilon_I))\delta(\delta E_{\text{ch}} + \varepsilon_I - \varepsilon_L).$$

$$(7.15)$$

The second equality assumes that the tunnelling matrix element $|T|^2$ and the densities of states N_L and N_I are constant within the transport window (see also the discussion in Example 1.1).

The energy conservation includes the change in the charging energy of the island, $\delta E_{\text{ch}} = \left(n + \frac{1}{2} - \frac{Q_G}{e}\right)e^2/C$. The effect of the voltage is included in the different potentials for the left electrode and the island (the potential difference equalling eV_L) in the distribution functions. Above expression introduces the tunnel conductance

$$R_L^{-1} = 4\pi e^2 N_I(0) N_L(0) \Omega_I \Omega_L |T|^2/\hbar,$$

whose expression is analogous to that used in Example 1.1.

Similar considerations apply for the reverse process $\Gamma_{IL}(n+1)$, and for the tunnelling between the island the the right reservoir. In these cases, the only changes that are needed are the change of the distribution functions corresponding to the initial and final states and the change of the tunnelling resistance to that of the right junction.

Now assume that the electron–phonon relaxation time is short, such that the electrodes and the island can be described by Fermi functions with the lattice temperature for each potential.[9] In this case, we can use identities (A.64a) and (A.64h). The resultant single-electron tunnelling rate is

$$\Gamma_{LI}(n) = \frac{R_K}{hR_L} \frac{\delta E_{\text{ch}}(n) - eV_L}{\exp[(\delta E_{\text{ch}}(n) - eV_L)/(k_B T)] - 1}, \qquad (7.16)$$

where $R_K = h/e^2$ is the resistance quantum for a single spin. Note that the sequential tunnelling term is linear in the factor R_K/R_L. For $(\delta E_{\text{ch}} - eV_L)/(k_B T) \gg 1$ the tunnelling is suppressed because of the large exponential factor in the denominator of $\Gamma_{LI}(n)$. This corresponds to the Coulomb blockade situation depicted in the previous section. At low temperatures we thus obtain

$$\Gamma_{LI}(n) \xrightarrow{T \to 0} \frac{1}{e^2 R_L} |\delta E_{\text{ch}}(n) - eV_L|\theta(eV_L - \delta E_{\text{ch}}(n)), \qquad (7.17)$$

where the Heaviside function $\theta(x) = 1$ for $x > 0$ and 0 otherwise. At non-zero temperatures all processes are allowed, but their rate depends strongly on the ratio $(\delta E_{\text{ch}} - eV_L)/k_B T$.

7.3 Master equation

To the lowest-order in the tunnel coupling, the charge state of the island can change at most by one electron in a single process. The knowledge

[9] When this is not the case, tunnelling heats up the island and the temperature is determined self-consistently, see for example (Laakso *et al.*, 2010).

of the tunnelling rates allows us to describe the time dependence of the probability $P(n,t)$ for the system occupying a given charge state n at time t. This probability satisfies the master equation (for an illustration, see Fig. 7.9)

$$\frac{dP(n,t)}{dt} = -[\Gamma_{LI}(n) + \Gamma_{IL}(n) + \Gamma_{RI}(n) + \Gamma_{IR}(n)]P(n,t)$$
$$+ [\Gamma_{LI}(n-1) + \Gamma_{RI}(n-1)]P(n-1,t) \qquad (7.18)$$
$$+ [\Gamma_{IL}(n+1) + \Gamma_{IR}(n+1)]P(n+1,t).$$

In other words, the rate of change of the probability $P(n,t)$ is equal to the difference of rates for tunnelling into state n (two lower lines) and the tunnelling out of state n (upper line).

We can also determine the average current from the rates and the probabilities. In the left contact it is

$$I_L(t) = e\sum_n [\Gamma_{LI}(n) - \Gamma_{IL}(n)]P(n). \qquad (7.19)$$

In the stationary state, where the left-hand side of eqn (7.18) vanishes, it is straightforward to show that $I_L = -I_R$.

In a stationary system the master equation can be solved with a probability matrix satisfying the detailed balance condition

$$[\Gamma_{LI}(n) + \Gamma_{RI}(n)]P(n) = [\Gamma_{IL}(n+1) + \Gamma_{IR}(n+1)]P(n+1). \qquad (7.20)$$

This has to be appended by the normalization $\sum_n P(n) = 1$.

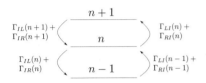

Fig. 7.9 Sequential tunnelling-induced transitions between charge states.

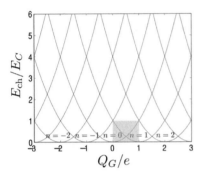

Fig. 7.10 Regime of energies and gate charges where the two-state approximation is valid, indicated by the grey box.

[10] For simplicity, we include here the potential shifts into the charging energy changes.

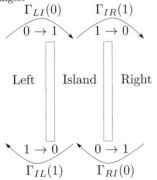

Fig. 7.11 Charge tunnelling processes involved in the two charge state approximation.

Example 7.1 Approximation with two charge states

Let us consider an example of solving the master equation. To specify the system, assume a symmetric bias $eV_L = -eV_R = eV/2$, and Q_G between 0 and e (see Fig. 7.10). At low temperatures and bias voltages, only the charge states 0 and 1 are occupied and hence contribute to the current. The energy changes $\pm\delta E_{\text{ch}}^L$ determining the rates $\Gamma_{LI}(0)$ and $\Gamma_{IL}(1)$ are[10]

$$\pm\delta E_{\text{ch}}^L = \pm\left[\left(\frac{1}{2} - \frac{Q_G}{e}\right)\frac{e^2}{C} - \frac{eV}{2}\right], \qquad (7.21)$$

where the upper sign denotes the transition from the electrode to the island and the lower from the island to the electrode. The corresponding energy change δE_{ch}^R in the right junction is obtained by replacing $eV/2$ by $-eV/2$. The charge tunnelling processes involved in the two charge state approximation are indicated in Fig. 7.11

Using the the fact that $\sum_n P(n) = P(0) + P(1) = 1$ we then get

$$P(0) = \frac{\Gamma_{IL}(1) + \Gamma_{IR}(1)}{\Gamma}, \text{ and}$$
$$P(1) = \frac{\Gamma_{LI}(0) + \Gamma_{RI}(0)}{\Gamma} \qquad (7.22)$$

where $\Gamma = \Gamma_{IL}(1) + \Gamma_{LI}(0) + \Gamma_{IR}(1) + \Gamma_{RI}(0)$ is the total rate. The current is given by

$$I = e[\Gamma_{LI}(0)P(0) - \Gamma_{IL}(1)P(1)]$$
$$= e\frac{\Gamma_{LI}(0)\Gamma_{IR}(1) - \Gamma_{RI}(0)\Gamma_{IL}(1)}{\Gamma}. \qquad (7.23)$$

This can be interpreted as the difference of the rates for going into the island from the left and then out to the right and the opposite process. At low temperatures with $k_B T \ll |eV|$ only one of these directions is possible. This equation can be readily analyzed by inspecting eqn (7.21). At low temperatures the tunnelling through the left junction is allowed when $Q_G - e/2 \geq -VC/2$. The rate for tunnelling into the right is non-zero when $Q_G - e/2 \leq VC/2$. Thus, both processes are present when $|Q_G - e/2| \leq |V|C/2$ with $V > 0$. The opposite processes would be allowed in a similar gate charge regime when $V < 0$. The total current in a symmetric setup $R_L = R_R = R_T$ at $T = 0$ is then

$$I = \frac{1}{4R_T} \left[V - \frac{4e^2}{C^2 V} \left(\frac{Q_G}{e} - \frac{1}{2} \right)^2 \right] \theta \left(|Q_G - e/2| - \frac{1}{2} C|V| \right). \quad (7.24)$$

The Heaviside function thus specifies the transport window, i.e., the window of bias/gate voltages where the current is finite.

Figures 7.12 and 7.13 show the current as a function of the transport voltage and the gate charge solved numerically from the master equation. As a function of the latter, the current shows the e-periodic Coulomb oscillations. When $Q_G = n + 1/2$, states n and $n + 1$ are degenerate and at $V = 0$ both charge states can coexist. In this case a finite current is possible even for a very small bias.

Due to the sensitivity of the current to the value of the gate charge, the single-electron transistor can be used as a very accurate electrometer.

7.4 Cotunnelling

The orthodox theory outlined above works mostly in the regime where the tunnelling can be treated with the lowest-order perturbation theory. This is the regime of sequential tunnelling, i.e., an electron tunnels into the island in a single process, and in another process another electron tunnels out. There is no coherence between these processes. The higher-order effects, such as the coherent cotunnelling through several junctions, are especially important in the Coulomb-blockade regime, where they give rise to a finite current.

The second-order tunnelling rates can be calculated from the standard second-order theory,[11]

$$\Gamma_{i \to f} = \frac{2\pi}{\hbar} \left| \sum_{\text{int}} \frac{\langle i | H_t | \text{int} \rangle \langle \text{int} | H_t | f \rangle}{E_{\text{int}} - E_i} \right|^2 \delta(E_i - E_f). \quad (7.25)$$

Here the sum goes over all intermediate charge states $|\text{int}\rangle$. The cotunnelling process thus corresponds to an electron tunnelling from the first electrode into a virtual charge state (above or much below the Fermi level) and from there to the second electrode. As the rate is inversely proportional to the distance $E_{\text{int}} - E_i$ between the virtual state energy and the initial state energy, the process is the weaker the further the closest allowed charge states are. But this is not as strong suppression

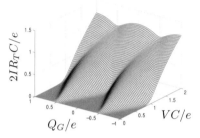

Fig. 7.12 Current vs. bias voltage and gate charge through a symmetric single-electron transistor in the sequential tunnelling approximation at $T = 0.05 E_C / k_B$. Calculated with a code written by Matti Laakso.

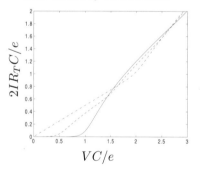

Fig. 7.13 Current–voltage characteristic of a symmetric SET for $Q_G = 0$ (solid line, Coulomb blockade), $Q_G = 0.25$ (dashed line, intermediate) and $Q_G = 0.5$ (dash-dotted line, linear $I - V$ due to the degeneracy of $n = 0$ and $n = 1$). Broadening of the edges is due to the finite temperature $T = 0.05 E_C / k_B$. Calculated with a code written by Matti Laakso.

[11] See Appendix A.3.1.

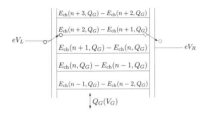

Fig. 7.14 Inelastic cotunnelling: two electrons tunnel through the two contacts in a single coherent process. The energy is conserved only between the initial and final states.

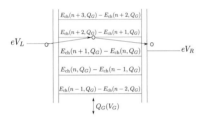

Fig. 7.15 Elastic cotunnelling: an electron tunnels through the two contacts in a single coherent process.

as in the first-order theory where the rate is exponentially suppressed ($\sim \exp[-(E_{\mathrm{int}} - E_i)/(k_B T)]$).

In a single-electron transistor, the level spacing is small compared to the charging energy scale. Hence, with a very high probability the electron tunnelling out of the island is not the same as the electron that tunnelled into the island in the same coherent process. Therefore, after the process an electron-hole excitation is left on the island. This process is often called *inelastic cotunnelling* (Fig. 7.14) in contrast to the *elastic cotunnelling*, where the same electron tunnels through the whole system in a single coherent process (Fig. 7.15).

Another point that one has to take into account is that two different types of events have to be added coherently: for example, in tunnelling from left to right, either (a) first an electron tunnels into the island from the left electrode, and then another electron tunnels into the right electrode, or (b) first an electron tunnels out of the island, and only then the first electron tunnels in. In the previous case the energy difference between the initial and the virtual states is $\delta E_L = E_{\mathrm{ch}}(n+1, Q_G) - E_{\mathrm{ch}}(n, Q_G) - eV$, whereas in the latter case it is $\delta E_R = E_{\mathrm{ch}}(n-1, Q_G) - eV - E_{\mathrm{ch}}(n, Q_G)$. These processes have to be added coherently, i.e., before taking the square of the matrix element.

The resultant rate for the inelastic cotunnelling in a normal-metal single-electron transistor is (Averin and Nazarov, 1990)

$$\Gamma_{\mathrm{cot}} = \frac{1}{4\pi^2 h} \frac{R_K^2}{R_L R_R} \int_{k \in L, q, q' \in I, k' \in R} dE_k dE_q dE_{q'} dE_{k'}$$

$$\times f(E_k)[1 - f(E_q)]f(E_{q'})[1 - f(E_{k'})]$$

$$\times \left[\underbrace{\frac{1}{E_q + \delta E_L - E_k}}_{\text{first (a) and then (b)}} + \underbrace{\frac{1}{E_{k'} + \delta E_R - E_q'}}_{\text{first (b), then (a)}} \right]^2 \delta(eV + E_k - E_q + E_{q'} - E_{k'}).$$

$$(7.26)$$

Here the potentials of the island and the electrodes are assumed equal and the voltages are included in the energy shifts. At $T = 0$, $eV \ll \delta E_L, \delta E_R$ the integrals can be performed analytically, with the result (see exercises)

$$\Gamma_{\mathrm{cot}} = \frac{e^3}{24\pi^2 h} \frac{R_K^2}{R_L R_R} \left(\frac{1}{\delta E_L} + \frac{1}{\delta E_R} \right)^2 V^3. \qquad (7.27)$$

Hence, we see that the rate for cotunnelling in a left–right symmetric SET ($R_R = R_L$) is proportional to $(R_K/R_L)^2$. Cotunnelling is thus the more important the lower the junction resistance. In systems with N junctions in series, the relevant higher-order effects can be found from the Nth order perturbation theory, and in general they lead to a current-voltage dependence $I \propto V^{2N-1}$ (Jensen and Martinis, 1992). These cotunnelling processes are important because they limit the accuracy of the single-electron transistor and other single-electron devices, examples of which are discussed below in Sec. 7.6.

7.5   Dynamical Coulomb blockade

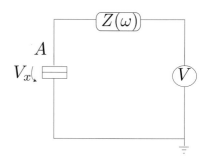

Section 7.1 shows that the rate of charge relaxation is important for the occurrence of Coulomb blockade in a single tunnel junction. When relaxation is slow, the tunnel junction is in the regime of *dynamical* or *environmental Coulomb blockade*.

Let us now lift the assumption that the charge on the reservoirs relaxes quickly. This is generally due to the fact that the junction resides in an environment characterized by an impedance $Z(\omega)$ (like that in Fig. 7.16). The presence of $Z(\omega)$ gives rise to two effects. First, the average voltage drop at the junction depends on the current,

Fig. 7.16 Tunnel junction in an environment characterized by the impedance $Z(\omega)$.

$$V_x = V - I(V_x)Z(\omega). \tag{7.28}$$

Since the current may depend on the voltage drop in a non-trivial manner, this problem has to be solved self-consistently. In addition to this, the resistor gives rise to voltage fluctuations, δV over the junction. As described in Ch. 6, at equilibrium or for macroscopic systems, the correlation function of the voltage fluctuations is described by the fluctuation-dissipation theorem, which states

$$S_V(\omega) \equiv \int_{-\infty}^{\infty} dt e^{i\omega t} \langle \delta\hat{V}(t)\delta\hat{V}(0)\rangle = \hbar\omega\mathrm{Re}[Z_t(\omega)]\left(\coth\left(\frac{\hbar\omega}{2k_BT}\right)+1\right). \tag{7.29}$$

For the description of the dynamical Coulomb blockade, it is necessary to use this non-symmetrized correlator, as the quantum fluctuations described by the antisymmetric part $S_V(\omega) - S_V(-\omega)$ play an important role in determining the tunnelling current.

By eqn (7.29) we see that the fluctuations depend on the real part of the total impedance $Z_t(\omega)$ between point A in Fig. 7.16 and the ground. In the case when $Z(\omega) \ll R_T$, we can ignore the tunnel resistance and use only $Z(\omega)$ and the capacitance C of the tunnel junction,

$$Z_t(\omega) = \frac{1}{i\omega C + Z^{-1}(\omega)}. \tag{7.30}$$

The charge relaxation time of the circuit can now be related to $Z(\omega)$ and C.

Because of voltage fluctuations, the electrochemical potential of one of the electrodes, say left, fluctuates in time. Thus, the state of the system is described by the Hamiltonian

$$H = H'_L + H_R + H_T + H_{\mathrm{env}} = \sum_{k,\sigma}(\epsilon_k + eV_x + e\delta\hat{V}(t))c^{\dagger}_{k\sigma}c_{k\sigma} +$$
$$\sum_{q,\sigma}\epsilon_q c^{\dagger}_{q\sigma}c_{q\sigma} + \sum_{k,q,\sigma}t_{k,q}c^{\dagger}_{k,\sigma}c_{q,\sigma} + \mathrm{h.c.} + H_{\mathrm{env}}, \tag{7.31}$$

where H_{env} describes the environment and gives rise to the fluctuations $\delta\hat{V}$. To be able to make the pertubation theory on the tunnelling, it is simpler to transform the time dependence of the electrode Hamiltonian

[12]The form of the transformation can be seen by requiring that the transformed wave function $|\psi'(t)\rangle = U|\psi(t)\rangle$ satisfies the Schrödinger equation with the transformed Hamiltonian H': $i\hbar\partial_t|\psi'(t)\rangle = i\hbar\dot{U}|\psi(t)\rangle + Ui\hbar\partial_t|\psi(t)\rangle = (i\hbar\dot{U} + UH)|\psi(t)\rangle = (i\hbar\dot{U}U^\dagger + UHU^\dagger)|\psi'(t)\rangle \equiv H'|\psi'(t)\rangle$.

[13]This can be seen as follows. First we note that for any operators A, B, C, we have $[A, BC] = B[A, C] + [A, B]C = \{A, B\}C - B\{A, C\}$. In particular, $[c_j^\dagger, c_k^\dagger c_k] = \{c_j^\dagger, c_k^\dagger\}c_k - c_k^\dagger\{c_j^\dagger, c_k\} = \delta_{jk}c_k^\dagger$ whereas $[d, c_k^\dagger c_k] = 0$. Then, define operator $O = e^{Ac_k^\dagger c_k}c_k^\dagger c_q e^{-Ac_k^\dagger c_k}$ and calculate $\partial_A O = e^{Ac_k^\dagger c_k}[c_k^\dagger c_k, c_k^\dagger c_q]e^{-Ac_k^\dagger c_k} = e^{Ac_k^\dagger c_k}(c_k^\dagger[c_k^\dagger c_k, c_q] + [c_k^\dagger c_k, c_k^\dagger]c_q) \times \times e^{-Ac_k^\dagger c_k} = e^{Ac_k^\dagger c_k}c_k^\dagger c_q e^{-Ac_k^\dagger c_k} = O$. Solving this equation, we get $O = \exp(A)c_k^\dagger c_q$. Substituting $A = i\delta\hat{\phi}$ and taking into account the fact that the other terms in the sum over k commute with the tunnelling operator, we obtain eqn (7.33).

to the tunnelling Hamiltonian. In other words, we look for a transformation that puts us in a frame that fluctuates along with the voltage. This can be realized with the transformation $H' = \hat{U}H\hat{U}^\dagger + i\hbar(\partial_t\hat{U})\hat{U}^\dagger$ where[12]

$$\hat{U} = \exp\left[\frac{i}{\hbar}\int^t dt' e\delta\hat{V}(t')\sum_{k,\sigma}c_{k,\sigma}^\dagger c_{k,\sigma}\right]. \tag{7.32}$$

Note that $\hat{U}$ commutes with the Hamiltonians of the electrodes, but the second part of the transformation, proportional to $\partial_t\hat{U}$, removes the time-dependent part from H_L'. The tunnelling part of the Hamiltonian transforms to[13]

$$H_T' = \sum_{k,q,\sigma}t_{k,q}e^{i\delta\hat{\phi}}c_{k\sigma}^\dagger c_{q\sigma} + \text{h.c.}, \tag{7.33}$$

where

$$\delta\hat{\phi} = \frac{e}{\hbar}\int^t \delta\hat{V}(t')dt'. \tag{7.34}$$

The total Hamiltonian resulting from this transformation is thus of the form

$$H' = H_L + H_R + H_{\text{env}} + H_T', \tag{7.35}$$

where H_L and H_R are the electrode Hamiltonian operators defined above, except that the energies of the quasiparticle states in the left electrode are shifted by the average potential V_x.

It remains to make the Fermi golden rule calculation of the tunnelling rates, similarly to Sec. 7.2. The initial and final states in the calculation of the transmission probabilities are $|i\rangle = |i\rangle_e \otimes |i\rangle_{\text{env}}$ and $|f\rangle = |f\rangle_e \otimes |f\rangle_{\text{env}}$, where $|\cdot\rangle_e$ refers to the electronic excitations and $|\cdot\rangle_{\text{env}}$ to the state of the electromagnetic environment. For example, the tunnelling rate from the left to the right is given by (see eqn (7.15))

$$\Gamma_{LR}(V_x) = \frac{1}{e^2 R_T}\int d\varepsilon_k d\varepsilon_q f_L(\varepsilon_k)[1 - f_R(\varepsilon_q)]$$
$$\times \sum_{f,i} P_i^{\text{env}}|\langle f|_{\text{env}}e^{i\delta\hat{\phi}}|i\rangle_{\text{env}}|^2\delta(\varepsilon_q - \varepsilon_k + E_f^{\text{env}} - E_i^{\text{env}}). \tag{7.36}$$

Here P_i^{env} is the probability of the initial state of the environment, and E_f^{env} and E_i^{env} are the energies of the final and initial environment states.

Let us use the following representation for the δ-function,

$$\delta(\varepsilon_q - \varepsilon_k + E_f^{\text{env}} - E_i^{\text{env}}) = \int\frac{dt}{2\pi\hbar}e^{\frac{i}{\hbar}(\varepsilon_k - \varepsilon_q + E_i^{\text{env}} - E_f^{\text{env}})t}. \tag{7.37}$$

Inserting this into eqn (7.36) yields

$$\Gamma_{LR}(V_x) = \frac{1}{e^2 R_T}\int d\varepsilon_k d\varepsilon_q f_L(\varepsilon_k)[1 - f_R(\varepsilon_q)]\int\frac{dt}{2\pi\hbar}e^{i(\varepsilon_k - \varepsilon_q)t/\hbar}$$
$$\times \sum_{f,i} P_i^{\text{env}}\underbrace{\langle i|_{\text{env}}e^{iE_i^{\text{env}}t/\hbar}}_{\langle i|_{\text{env}}e^{iH_{\text{env}}t/\hbar}}e^{-i\delta\hat{\phi}}\underbrace{e^{-iE_f^{\text{env}}t/\hbar}|f\rangle_{\text{env}}}_{e^{-iH_{\text{env}}t/\hbar}|f\rangle_{\text{env}}}\langle f|_{\text{env}}e^{i\delta\hat{\phi}}|i\rangle_{\text{env}}.$$

$$\tag{7.38}$$

Assume the states $|i\rangle$ and $|f\rangle$ are eigenstates of energy. In this case we can make the replacement $e^{-iE_f^{\text{env}}t/\hbar}|f\rangle_{\text{env}} = e^{-iH_{\text{env}}t/\hbar}|f\rangle_{\text{env}}$. This amounts to a transformation to the Heisenberg picture, $e^{-i\delta\hat{\phi}(t)} = e^{iH_{\text{env}}t/\hbar}e^{-i\delta\hat{\phi}}e^{-iH_{\text{env}}t/\hbar}$. We may then use the resolution of unity, $\mathbf{1}_{\text{env}} = \sum_f |f\rangle_{\text{env}}\langle f|_{\text{env}}$, and obtain

$$\Gamma_{LR}(V_x)$$
$$= \frac{1}{e^2 R_T} \int d\varepsilon_k d\varepsilon_q f_L(\varepsilon_k)[1 - f_R(\varepsilon_q)] \int \frac{dt}{2\pi\hbar} e^{i(\varepsilon_k - \varepsilon_q)t/\hbar} \langle e^{-i\delta\hat{\phi}(t)} e^{i\delta\hat{\phi}(0)} \rangle,$$
$$(7.39)$$

where $\langle \cdot \rangle = \sum_i p_i \langle i| \cdot |i\rangle$ denotes the statistical average.

Let us define the function

$$P(E) = \int \frac{dt}{2\pi\hbar} e^{iEt/\hbar} \langle e^{-i\delta\hat{\phi}(t)} e^{i\delta\hat{\phi}(0)} \rangle. \qquad (7.40)$$

Using this definition, the tunnelling rate can be expressed as

$$\Gamma_{LR}(V_x) = \frac{1}{e^2 R_T} \int d\varepsilon_k d\varepsilon_q f_L(\varepsilon_k)[1 - f_R(\varepsilon_q)] P(\varepsilon_k - \varepsilon_q)$$
$$= \frac{1}{e^2 R_T} \int dE \frac{E}{1 - \exp(-E/(k_B T))} P(eV_x - E), \qquad (7.41)$$

which takes into account the potential difference eV_x between the electrodes. The corresponding rate for the backward tunnelling (from right to left) satisfies the symmetry $\Gamma_{RL}(V_x) = \Gamma_{LR}(-V_x)$. Therefore, the current is given by

$$I(V_x) = e(\Gamma_{LR} - \Gamma_{RL})$$
$$= \frac{1}{e R_T} \int dE \frac{E}{1 - \exp(-E/(k_B T))} (P(eV_x - E) - P(-eV_x - E)). \qquad (7.42)$$

The function $P(E)$ has a probability interpretation (see Exercise 7.8): it describes the probability that the environment can absorb energy E during the tunnelling process.

7.5.1 Phase fluctuations

The function $\langle e^{-i\delta\hat{\phi}(t)} e^{i\delta\hat{\phi}(0)} \rangle$ can be interpreted as the characteristic function of the fluctuating quantity $\delta\hat{\phi}(t) - \delta\hat{\phi}(0)$, ordered such that $\delta\hat{\phi}(t)$ precedes $\delta\hat{\phi}(0)$.[14] As discussed in Sec. 6.7.1, the characteristic function can be expanded in the cumulants of the distribution,

$$\chi(\lambda) = \langle e^{i\lambda x} \rangle = \exp\left(\sum_{n=0}^{\infty} \kappa_n \frac{(i\lambda)^n}{n!} \right). \qquad (7.43)$$

Now let us assume that the phase fluctuations are Gaussian.[15] This is the case if the environment of the junction consists of macroscopic resistors much larger than the energy relaxation length. In this case, only the

[14]Note that as long as we never commute the operators $\hat{\phi}(t)$ and $\hat{\phi}(0)$, we can treat them as scalars. For any operators A and B, we could define $e^A e^B = \mathcal{T}_{AB}(e^{A+B})$, where $\mathcal{T}_{AB}$ orders the operators such that A precedes B.

[15]For an example of a non-Gaussian $P(E)$ theory, see (Heikkilä *et al.*, 2004).

second cumulant of the phase fluctuations is non-vanishing, and we may write

$$\langle e^{-i\delta\hat{\phi}(t)}e^{i\delta\hat{\phi}(0)}\rangle = e^{i^2\langle\delta\hat{\phi}(t)^2 - 2\delta\hat{\phi}(t)\delta\hat{\phi}(0) + \delta\hat{\phi}(0)^2\rangle/2}$$
$$= e^{\langle[\delta\hat{\phi}(t) - \delta\hat{\phi}(0)]\delta\hat{\phi}(0)\rangle} \equiv e^{J(t)}. \tag{7.44}$$

The latter expression uses the fact that in a stationary situation $\langle\delta\hat{\phi}(t)^2\rangle = \langle\delta\hat{\phi}(0)^2\rangle$.

Using the definition (7.34), we may now relate the spectrum of phase fluctuations to the voltage fluctuations over the junction. These are given by

$$S_\phi(\omega) = \frac{e^2}{\omega^2\hbar^2}S_V(\omega). \tag{7.45}$$

Substituting eqn (7.29) to eqns (7.45), (7.44) and (7.42) allows us to calculate the lowest-order current–voltage characteristics for low-capacitance tunnel junctions placed in an arbitrary macroscopic environment.

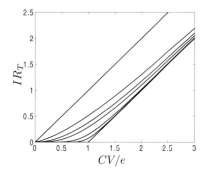

Fig. 7.17 Current–voltage curves of a tunnel junction in a resistive environment. From top to bottom: $R/R_K = 0, 0.2, 0.5, 1, 5, 25$ and ∞.

Example 7.2 Dynamical Coulomb blockade with a resistive environment

Let us consider the simplest example of a purely resistive environment ($Z = R$ is frequency independent) of the tunnel junction.[16] In this case, using eqn (7.29), we have

$$J(t) = \int \frac{d\omega}{\omega} \frac{R}{R_K} \frac{e^{-i\omega t} - 1}{1 + \omega^2(RC)^2}\left[1 + \coth\left(\frac{\hbar\omega}{2k_BT}\right)\right]. \tag{7.46}$$

This $J(t)$ has two rather simple limits. For $R \ll R_K$, $J(t) \to 0$, and $P(E) \approx \delta(E)$. In this case, we recover eqn (1.15), and thus a linear $I-V$ curve for a single tunnel junction. For the opposite limit of a high-impedance environment, $R \gg R_K$, $\text{Re}[Z_t] \to \pi/C\delta(\omega)$, and we get[17]

$$J(t) \approx -\frac{\pi}{CR_K}\left(it + \frac{k_BTt^2}{\hbar}\right). \tag{7.47}$$

With this $J(t)$, the $P(E)$ function becomes a Gaussian,

$$P(E) = \frac{1}{\sqrt{4\pi E_C k_B T}}\exp\left[-\frac{(E - E_C)^2}{4E_C k_B T}\right]. \tag{7.48}$$

At $k_BT \ll E_C$, this $P(E) \to \delta(E - E_C)$. In this limit, the expression for the current simplifies to

$$I(V_x) = \frac{eV_x - E_C}{eR_T}\theta(eV_x - E_C). \tag{7.49}$$

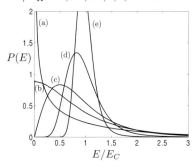

Fig. 7.18 Example $P(E)$ functions for an environment characterized by a capacitance C and resistance R. From (a) to (e): $R/R_K = 0.2, 0.5, 1, 5$ and 25.

[16]For a description of other types of environments, see (Ingold and Nazarov, 1992).

[17]This result can be obtained by expanding the term $e^{-i\omega t} - 1$ in a Taylor series, multiplying by the series expansion of the remaining term in the integrand of eqn (7.46), and retaining the constant terms as $\omega \to 0$.

That is, we get Coulomb blockade, which is washed out by increasing the temperature or decreasing R/R_K. The $I(V)$ curve for a few environment resistances is shown in Fig. 7.17 and the corresponding $P(E)$ functions are in Fig. 7.18. In the case of a general resistance R and at non-zero temperature T, the curves have to be calculated numerically. However, the function $J(t) = J_R(t) + iJ_I(t)$ can be evaluated analytically. Its real part in the low-temperature limit $k_BT \lesssim E_C$ is

$$J_R(t) = \frac{R}{R_K}\left\{e^{-t/\tau}E_1(-t/\tau) + e^{t/\tau}E_1(t/\tau)\right.$$
$$\left. - 2\ln\left[\frac{E_CR_K\sinh[k_BT|t|/(E_C\tau)]}{\pi Rk_BT}\right] - 2\gamma\right\}, \tag{7.50}$$

where $\tau = RC$, $E_1(x) = \int_1^\infty (e^{-zt}/t)dt$ is the exponential integral, and $\gamma \approx 0.577$ is the Euler constant. The imaginary part is

$$J_I(t) = -\pi \frac{R}{R_K}[1 - \exp(-|t|/\tau)]\text{sgn}(t). \qquad (7.51)$$

Using the facts that $J_R(-t) = J_R(t)$ and $J_I(-t) = -J_I(t)$ we can then write $P(E)$ in the form

$$P(E) = 2\int_0^\infty \frac{dt}{2\pi}e^{J_R(t)}\cos(J_I(t) + Et/\hbar). \qquad (7.52)$$

This allows for a numerical evaluation of $P(E)$.

7.6 Single-electron devices

Besides the single-electron transistor, there are many closely related devices that are based on a similar setting, or improve on the SET concept. We introduce here a few of the most relevant ones. A Coulomb blockade thermometer (CBT) can be used as an accurate primary electron thermometer, the radio frequency SET improves the charge measurement resolution of the SET by employing high frequencies, and a single-electron pump is an extremely accurate current source.[18]

7.6.1 Coulomb blockade thermometer

A Coulomb blockade thermometer (CBT) works in the regime $k_B T \gg E_C$ where the thermal energy almost overwhelms the Coulomb blockade. There, we can start from the infinite temperature limit of the current–voltage curves and calculate the lowest-order correction from Coulomb blockade. This correction turns out to be a universal function of the temperature with two operating modes: primary (no calibration required) and secondary (easier operation, but the junction parameters need to be measured separately).[19]

In the limit $k_B T \gg E_C$, the first-order correction from the charging effects to the linear current–voltage curve can be used as a thermometric quantity. In the case of a symmetric SET at $Q_G = 0$, $V_L = -V_R = V/2$, the differential conductance $G = dI/dV$ is of the form (number 2 comes from the fact that we have two junctions; see also Exercise 7.10)

$$2GR_T = 1 - ug(v), \qquad (7.53)$$

where $u = 2E_C/k_B T$, $v = eV/(2k_B T)$ and

$$g(v) = \frac{v\cosh\left(\frac{v}{2}\right) - 2\sinh\left(\frac{v}{2}\right)}{4\sinh^3\left(\frac{v}{2}\right)}. \qquad (7.54)$$

Measuring the conductance correction yields the electronic temperature: the conductance at $v = 0$ is $G = 1/(2R_T) - \Delta G$ with $2\Delta GR_T = E_C/(3k_B T)$. Finding the precise temperature from here requires calibration: knowledge of the tunnel resistance R_T and the charging energy E_C. However, the voltage V for which $g(v) = 1/12$, i.e., the

[18]When some or all of the metallic parts of a SET are made superconducting, there is a wealth of new device concepts: quantum bits, microcoolers, heat transistors, more efficient pumps, and more accurate electrometers, to name a few. Some of these devices are discussed in Ch. 9.

[19]See (Pekola *et al.*, 1994). A brief review of the main CBT properties is given in (Giazotto *et al.*, 2006). Recently, an improved concept, dealing with the measurement of the conductance of a single junction under high-impedance environment, was introduced in (Pekola *et al.*, 2008).

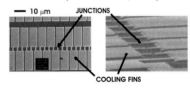

Fig. 7.19 Typical CBT sensor for the temperature range 20 mK–1 K. The structure has been fabricated by electron-beam lithography, combined with aluminium and copper vacuum evaporation. Both top view and a view at an oblique angle are shown; the scale indicated refers to the top view. (M. Meschke *et al.*, *J. Low Temp. Phys.* **134**, 1119 (2004), Fig. 2.) With kind permission from Springer Science and Business Media.

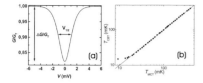

Fig. 7.20 (a) Measured conductance-voltage curve of a CBT thermometer. $G(V)/G_T$ is the differential conductance scaled by its asymptotic value at large positive and negative voltages, plotted as a function of bias voltage V. $V_{1/2}$ indicates the full width at half minimum of the characteristics. The full depth of the line, $\Delta G/G_T$, is another parameter to determine the temperature. In (b), the temperature deduced by CBT has been compared to that obtained by a ^{3}He melting curve thermometer. The latter is used as the official reference thermometer for low temperatures. However, it is not primary. Saturation of CBT below 20 mK indicates typical thermal decoupling between electrons and phonons. (M. Meschke *et al.*, *J. Low Temp. Phys.* **134**, 1119 (2004), Fig. 5.) With kind permission from Springer Science and Business Media.

conductance change is half from its maximum change, is a primary thermometric quantity: this position is reached with $v = 2.7196$, i.e., $eV = 5.4392k_BT$—independent of R_T or E_C. Thus, the measurement of the width of the $G(V)$ curve yields an ideal measurement of the electronic temperature.

In practice, cotunnelling limits the sensitivity of CBT devices. This can be cured by placing several junctions in a row. For N junctions in series, the conductance is (Hirvi *et al.*, 1995)

$$2GR_T = 1 - u_N g(v_N), \tag{7.55}$$

where $v_N = eV/(Nk_BT)$ and $u_N = 2(N-1)E_C/(Nk_BT)$ with $E_C = e^2/(2C)$ calculated for a single-junction capacitance C.

Figures 7.19 shows an example of a CBT and Fig. 7.20 its conductance-voltage curve, together with a comparison to another frequently used thermometer.

7.6.2   Radio frequency SET

The normal-state single-electron transistor is in principle a very sensitive detector of charge. However, it is typically quite slow to operate due to the long RC times of the measuring circuit (C is in such a case given by the capacitive load from cabling, and is typically between 0.1 to 1 nF (Schoelkopf *et al.*, 1998)). Therefore, the speed is limited to below some 1 MHz, i.e., obtaining the charge information requires measurement of the state at least for a μs. In this regime, the single-electron devices show large $1/f$ fluctuations—noise that scales as one over frequency of operation—due to the presence of dynamical offset charges in the SET environment giving rise to slow fluctuations of the gate charge.

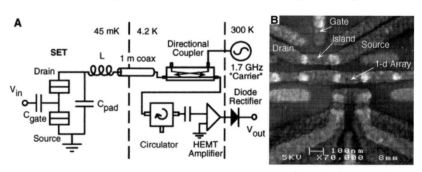

Fig. 7.21 A. Scheme of a radio-frequency SET and B. Electron micrograph of a practical device. In A, the resonant circuit is formed from the inductor L and the parasitic capacitance C_{pad} of the contact pad to the SET. (R. Schoelkopf *et al.*, *Science* **280**, 1238 (1998), Fig. 1.) Reprinted with permission from AAAS.

A radio-frequency SET (rf-SET) cures this problem by placing the SET in a resonant circuit (see Fig. 7.21). Instead of measuring the current–voltage curve of the SET, one monitors the reflection of a sinusoidal high-frequency carrier from the resonant circuit. The reflection coefficient is given by[20]

$$\Gamma = \frac{Z - Z_0}{Z + Z_0}, \tag{7.56}$$

where Z is the impedance of the load (the resonant circuit with the SET) and Z_0 is the characteristic impedance of the waveguide (the coaxial

[20]See Appendix E.

cable). These impedances should be measured at the carrier frequency, which can be taken up to a few GHz. At these frequencies, $Z_0 \approx 50\Omega$ and the resonant circuit can be used to transform the SET impedance downwards. On resonance, impedance Z depends strongly on the exact value of the SET impedance, which can be controlled by the gate charge.

This scheme allows to use SET for charge detection at frequencies of the order of hundreds of MHz (depends on the bandwidth of the amplifiers, and not so much on the SET itself). In the first demonstration of this device (Schoelkopf *et al.*, 1998), charge sensitivity of order 10^{-5} $e/\sqrt{\text{Hz}}$ was obtained (i.e., within 1 μs, one could measure the charge with the accuracy of 0.01 e).

As the charge detection can be made fast in an rf-SET, the same device can be used also for an extremely accurate measurement of small currents by directly counting electrons flowing in another conductor, by coupling the SET to this conductor (see Exercise 7.11). This way, the charges flowing through the conductor give rise to pulsed shifts of the SET gate charge, which then can be read off with the rf-SET.

7.6.3 Single-electron pump

Fabricating two or more junctions in series and controlling each of the islands with separate gate voltages allows for realizing a current pump. Let us consider the charging energy of the device shown in Fig. 7.22. Carrying out a similar calculation as done for the single-electron box above, assuming a vanishing bias voltage and $C_{Gi} \ll C$, we get

$$E_{\text{ch}} = \frac{2}{3}E_C[(n_1 - n_{G1})^2 + (n_2 - n_{G2})^2 + (n_1 - n_{G1})(n_2 - n_{G2})], \quad (7.57)$$

where $E_C = e^2/2C$, n_i is the charge on the ith island ($i = 1, 2$), and $n_{Gi} = V_{Gi}C_{Gi}/e$ is the corresponding gate charge. The ground state charge configuration (n_1, n_2) for given n_{Gi} is plotted in Fig. 7.23.

Consider what happens if the gate voltages are changed in time according to the circle drawn in Fig. 7.23 around the states $(0,0)$, $(1,0)$ and $(0,1)$, such that the system follows the gates adiabatically (see Fig. 7.24). When $n_{Gi} < 1/3$, the ground state of the system corresponds to no extra electrons on the islands. Increasing n_{G1} but keeping n_{G2} constant changes the ground state to $(1,0)$, i.e., one extra electron tunnels from the left electrode into the left island. Now increasing n_{G2} and keeping n_{G1} between $1/3$ and $2/3$ changes the ground state to $(0,1)$, and one electron tunnels from the left island to the right island. After this decreasing first n_{G1} and then n_{G2} below $1/3$ leads to the charge state $(0,0)$, i.e., one extra electron tunnels from the right island to the right electrode. As a result of this process, one electron has tunnelled through the whole system, resulting into a current equal to e/τ, where τ is the time in which the sequence was carried out. Repeating this sequence with frequency f leads to a current

$$I_p = ef \quad (7.58)$$

pumped through the device. As the frequency f can be controlled very precisely, this is a very precisely defined current.

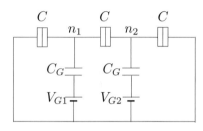

Fig. 7.22 Symmetric two-island single-electron device which can be used for pumping.

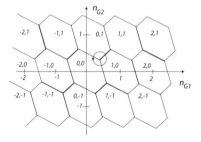

Fig. 7.23 Ground-state configuration of the two-island system as a function of the two gate charges n_{Gi}. The circle drawn around the edge between the $(0,0)$, $(1,0)$ and $(0,1)$ states represents a typical pumping cycle.

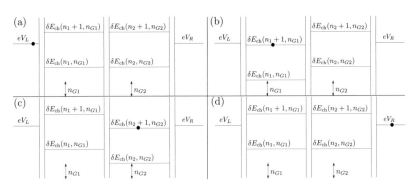

Fig. 7.24 Pumping cycle in a two-island system. The system starts from state (n_1, n_2), and proceeds via the states $(n_1 + 1, n_2)$ and $(n_1, n_2 + 1)$ back to the initial charge configuration. As a result, an electron is transferred from the left to the right.

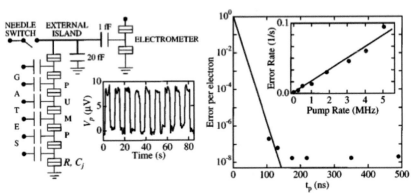

Fig. 7.25 Left: Schematics of a charge pump based on a single-electron transistor with many islands in series. Right: Error rate per electron vs. time $t_p \sim 1/f$ to pump a single electron. (M.W. Keller *et al.*, *Appl. Phys. Lett.* **69**, 1804 (1996), Figs. 2 and 4.) © 1996, American Institute of Physics.

[21]There are also other approaches slightly modified from the basic single-electron pumping scheme, for example, based on superconducting junctions or modulating the barriers rather than the gate to the island, see for example (Blumenthal *et al.*, 2007) and (Pekola *et al.*, 2008). The aim with these systems is to produce higher currents than in the normal pumps. However, their reported pumping accuracy is so far still too limited for use as current standards.

In practice, cotunnelling limits the accuracy of the current, as it leads to the possibility of stochastic processes where electrons are carried back and forth through the whole device and not following the pumping sequence. Fabricating more junctions in series reduces the cotunnelling current, and leads to an increasing accuracy. Figure 7.25 shows a practical device consisting of seven junctions (i.e., six islands) where an accuracy of 10^{-8} was reached. This is the most accurate known way to produce currents, and it has been proposed for use as a current standard. However, the method is not suitable for this purpose yet, as the adiabaticity requirement sets the limiting frequency to a few MHz, corresponding to currents of the order of $0.1 \ldots 1$ pA, which is too low for real-world applications.[21]

Further reading

- A somewhat more detailed description of Coulomb blockade effects is presented in Ch. 3 of (Dittrich *et al.*, 1998).

- For a thorough account of the dynamical Coulomb blockade, see (Ingold and Nazarov, 1992).

Exercises

(7.1) Find the charging energy for a single-electron transistor in terms of n, $V_{L/R/G}$ starting from $E_{\mathrm{ch}} = Q_L^2/(2C_L) + Q_R^2/(2C_R) + Q_G'^2/(2C_G) + V_L Q_L + V_R Q_R + V_G Q_G'$. Here Q_L and Q_R are the polarization charges on the left and right junctions, and Q_G' is the charge on the gate capacitor.

(7.2) Show that the forward and backward tunnelling rates satisfy the detailed balance criterion,

$$\frac{\Gamma_{LI}(n)}{\Gamma_{IL}(n+1)} = e^{-\delta E_{\mathrm{ch}}/(k_B T)}. \qquad (7.59)$$

(7.3) **Coulomb diamond.** Construct the stability diagram of the SET at $T = 0$: for given gate charges Q_G and bias voltages $V_L = -V_R = V/2$, find the charge states n whose occupation number is finite, i.e., which can contribute to the transport. Plot this in a two-dimensional plane, varying the gate charge between $-2e$ and $2e$, and the bias voltage between $-2e/C$ and $2e/C$. You do not need to evaluate the tunnelling rates. The figure you obtain is a characteristic conductance plot measured for all single-electron devices. Hint: find the $\{V, Q_G\}$ pairs where the conditions (7.6) and (7.7) apply as an equality. Above and below these lines the allowed charge states are different.

(7.4) Solve the master equation in the finite-temperature equilibrium ($V_L = V_R = 0$) case and calculate the average charge on the island for a given gate charge. You do not need to evaluate the sums analytically, but plot the results using some numerical software.

(7.5) Calculate the linear conductance I/V of a left–right symmetric SET in the limit $|eV| \ll k_B T$. Hint: in the linear regime you may assume the charge distribution to be the same as in equilibrium (see Exercise 7.4), and the only voltage dependence comes from the tunnelling rates. There you can expand $f(E + eV/2) \approx f(E) + f'(E)eV/2$, where $f(E) = f^0(E)$ is the Fermi function.

(7.6) Calculate the cotunnelling rate, eqn (7.27), in the limit $eV \ll \delta E_L, \delta E_R$, $T = 0$.

(7.7) Justify the Hamiltonian in eqn (7.35) by making the transformation defined in and above eqn (7.32).

(7.8) Show that the function $P(E)$ satisfies the sum rules

$$\int dE P(E) = 1 \qquad (7.60)$$

$$\int dE E P(E) = E_C. \qquad (7.61)$$

The first equation is a requirement for the probability interpretation of $P(E)$. In the latter equation, use the Gaussian result with eqns (7.29) and (7.30) with a frequency-independent Z.

(7.9) Using the sum rules of the previous exercise, show that at very large voltages $eV_x \gg k_B T, E_C$, the current through a single tunnel junction satisfies

$$I(V_x) = \frac{V_x - E_C/e}{R_T}. \qquad (7.62)$$

Note that the result $I = V_x/R_T$ obtained in the limit of a vanishing environmental impedance does not satisfy the second sum rule. Assuming a small but finite impedance Z satisfies the sum rule and leads to the above offset, but yields an (almost) linear $I - V$ curve.

(7.10) Derive the conductance change ΔG of a CBT in the case of two junctions. Hint: i) Write the tunnelling rates $\Gamma_{LI}(n)$ and $\Gamma_{IL}(n)$ for a vanishing gate voltage, $Q_G = 0$ and for a symmetric CBT, with $V_L = -V_R = V/2$. ii) Expand the difference $\Gamma_{LI}(n) - \Gamma_{IL}(n)$ in $u \equiv E_C/k_B T$ and find the zeroth- and first-order terms in u. iii) Write the current $I = e \sum_{n=-\infty}^{\infty} P(n)(\Gamma_{LI}(n) - \Gamma_{IL}(n))$ using this expansion, and use the symmetries of $P(n)$ at $Q_G = 0$ to eliminate the sums. iv) Differentiate your results to obtain the differential conductance.

(7.11) **Question on a scientific paper.** Take the paper Bylander *et al.*, *Nature* **434**, 361 (2005) and answer the following questions: a) Take Fig. 1 and try to understand how the measurement was conducted. How does one measure the SET impedance? What does this impedance tell about the charge flowing in the junction array? b) Figure 2 shows the $I - V$ curve of the device, and the curve labelled with N shows what happens when the system is in the normal state. How was the charging energy measured? c) Explain how Fig. 3a tells about real-time electron counting, and which quantity in Fig. 3b corresponds to the average current.

8 Quantum dots

[1]Sometimes also metallic islands with Coulomb blockade are referred to as quantum dots. Here we reserve the term quantum dot explicitly for structures where the finite energy-level spacing shows up.

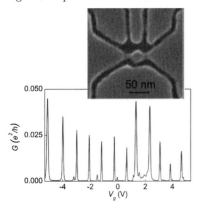

Fig. 8.1 Top: Micrograph of a graphene quantum dot. Bottom: Conductance through a graphene quantum dot as a function of the gate voltage. (L.A. Ponomarenko *et al.*, *Science* **320**, 356 (2008), Figs. 1 and 4.) Reprinted with permission from AAAS.

In the previous chapter we assume implicitly that once an electron tunnels into the island there are many available kinetic energy states where it can enter. This is the case in large metallic islands, where the energy-level spacing is small. In a two-dimensional electron gas formed in semiconductor heterostructures, metallic gate electrodes can be used to define a small island whose energy-level spacing is large compared to the externally controllable energy scales of temperature and voltage (for an example, see Fig. 1.6). Such a system is a 0-dimensional quantum dot (abbreviated below as QD).[1] Other systems where the finite energy level spacing shows up in the transport properties are, for example, molecules connected to two electrodes via different types of methods. This includes carbon nanotubes connected to metal electrodes, but also smaller molecules, such as fullerenes or even hydrogen molecules. Also, in graphene experimentalists have fabricated systems forming quantum dots (see Fig. 8.1). You will estimate the different energy-level spacings in these systems in Exercise 8.1.

The microscopic Hamiltonians of the different systems which show quantum-dot-type behaviour can be vastly different. However, typically at energies of the order of the level spacing all of these can be described by considering one or a few possibly degenerate electron energy levels of the isolated quantum dot, lying close to the Fermi levels of the leads, and in some way coupled to the electrodes. In the case of molecules these are the molecular orbital energy levels, and, for example, in a carbon nanotube they are typically the energy levels related to the longitudinal dynamics along the tube.[2] The relevant effective Hamiltonian hence has to include the description of those energy levels (which are assumed to be eigenstates of the isolated QD Hamiltonian[3]), of the coupling to the electrodes, and naturally the electrodes themselves. In addition, as discussed in the previous chapter, we need to describe interaction effects upon changing the number of electrons on the quantum dot. The latter are described by the charging energy E_C, which in small molecules also

[2]The energy levels of a few-electron semiconductor quantum dots have many similarities to those of atoms: for example, filling of the shells follows Hund's rules (Kouwenhoven *et al.*, 2001). Because of this similarity, semiconductor quantum dots are sometimes referred as the *artificial atoms*. As discussed in this chapter, the positions of these energy levels can be directly measured via the measurement of the current through the quantum dots.

[3]What to include in the 'isolated QD Hamiltonian' is rather a matter of choice. Generally, the included effects should be those that are stronger than the tunnel coupling. This then typically leaves out additional degrees of freedom besides the electronic ones, i.e., phonons, hyperfine states of the nuclei, etc.

depends on the actual number of electrons on the molecule.

A sketch of the QD energy level diagram and the transport setting through the QD is illustrated in Fig. 8.2. The coupling to the electrodes leads to a finite lifetime of the energy levels. This lifetime can be described by the inverse of the rate Γ of escaping from the quantum dot to the electrodes. Transport through the quantum dot then sensitively depends on the relation between Γ and the charging energy E_C. When $\Gamma \ll E_C, \Delta\epsilon, k_B T$, where $\Delta\epsilon$ is the distance from the closest QD level to the Fermi level in the leads, the system can be described via a tunnelling approximation very similar to that applied in the previous chapter. This limit is discussed in Sec. 8.3. In the limit $\Gamma \gg E_C$, the charge fluctuations on the electrodes mask the Coulomb blockade. In this case we may consider the non-interacting limit of a quantum dot (see Sec. 8.2) and obtain the Fabry–Perot-type description similar to that derived in Sec. 3.4. In the intermediate case of $\Gamma \sim k_B T, \Delta\epsilon \ll E_C$ the QD shows very rich physics. An important phenomenon found in this limit is the Kondo effect, which was first suggested and accepted as a theory of the effect of magnetic impurities on electrical conduction in bulk conductors. This effect is sketched in Sec. 8.4. Section 8.5 changes the perspective and discusses some of the basic properties of double quantum dots. In that case the discussion concentrates more on the stability diagram and its associated effects on transport than the exact formulas for the transmission. Double dots are especially interesting because of their suggested use as quantum bits—basic building blocks of a quantum computer. These are explained briefly towards the end of the section.

8.1   Electronic states in quantum dots

The general Hamiltonian of a quantum dot (QD) includes the electrodes, the QD single-particle states, the QD interaction energy and the tunnelling between the electrodes and the quantum dot. In the most general case we thus have

$$
\begin{aligned}
H =& H_L + H_R + \sum_{l\sigma} \epsilon_{l\sigma}(\{n_{l\sigma}\})\hat{n}_{l\sigma} \\
&+ \sum_{l\sigma, l'\sigma', l''\sigma'', l'''\sigma'''} U_{ll'l''l'''\sigma\sigma'\sigma''\sigma'''} c_{l\sigma}^{\dagger} c_{l'\sigma'}^{\dagger} c_{l''\sigma''} c_{l'''\sigma'''} + \text{h.c.} \\
&+ \sum_{k,\sigma,\beta=L,R} t_{k\beta l\sigma\sigma'} \hat{d}_{l\sigma}^{\dagger} \hat{c}_{k\beta\sigma'} + \text{h.c.},
\end{aligned} \tag{8.1}
$$

where H_L and H_R describe the electrodes and are similar to those presented in Sec. 7.2, $\epsilon_{l\sigma}$ denote the single-particle states with level index l and spin σ, and $U_{ll'l''l'''\sigma\sigma'\sigma''\sigma'''}$ is the interaction energy between the electrons on different levels.[4]

The Hamiltonian (8.1) includes the possible dependence of the single-particle spectrum on the number of electrons on the island. However, this model is seldom used in practical cases, but one may resort to

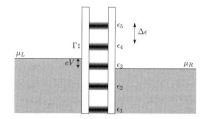

Fig. 8.2 Transport through a quantum dot (QD) can be visualized using the energy-level diagram shown here. The energy levels are in general not equispaced, and their broadening Γ (inverse lifetime) is finite due to the coupling to the electrodes.

[4]Typically the interaction energy is constrained by the total momentum and angular momentum (spin) conservation, such that for example the momenta of the states l, l', l'' and l''' have to add up to zero. However, as the rest of the chapter uses simpler forms of the interaction energy, it does not need to be specified further.

either of the two most commonly used approximations: (i) constant interaction (CI) model (Kouwenhoven *et al.*, 2001), or the (ii) Anderson (1961) model. The former is typically used when more than one single-electron energy level is involved in the transport, and the latter is a model for a single spin-degenerate model. In addition to making one of these approximations, within this chapter we assume spin-independent tunnelling amplitude coupling only equal-spin states.

In the constant interaction model, the single-particle states are assumed independent of the number of electrons, and the interaction energy is parametrized by a constant capacitance C, i.e., $U = e^2/(2C) \equiv E_C$.[5] In that case the interaction energy is of the form of eqn (7.5).

Often it is enough to consider transport through only a single spin-degenerate single-particle level ϵ_l. This is the case when the level spacing in the system is large compared to other characteristic scales of the process (temperature, voltage, level broadening). Such a system may be described by

$$H = H_L + H_R + \sum_\sigma \epsilon_l \hat{n}_\sigma + U \hat{n}_\uparrow \hat{n}_\downarrow + \sum_{k,\sigma,\beta=L,R} t_{k\beta} d_\sigma^\dagger c_{k\beta\sigma} + \text{h.c.} \quad (8.2)$$

where $\hat{n}_\sigma = \hat{d}_\sigma^\dagger \hat{d}_\sigma$ is the number operator. This type of a Hamiltonian was considered by Anderson (1961)[6] to describe the appearance of magnetic moments of certain magnetic ions embedded in a metal. Therefore, it is frequently called the Anderson Hamiltonian.

[5]This chapter uses the symbols U and E_C interchangably. The former is typically used when referring to the interaction energy between two levels, i.e., the Anderson model, whereas the latter refers to the general interaction energy related to the interaction of an excess electron with all other electrons on the dot.

[6]In Anderson's paper there was only a single reservoir coupled to the system, and hence no 'bias' was possible.

8.1.1 Spectral function

Upon coupling of the quantum dot to the electrodes, the QD eigenstates become 'quasi-eigenstates': they are not eigenstates of the whole coupled system, but typically still describe the properties of the quantum dot fairly well. The coupling to the surroundings causes a finite lifetime τ for the electrons on the dot. This lifetime can be described as an imaginary part of the eigenenergies:

$$\epsilon_l \to \epsilon_l - i\Gamma/2 \equiv \epsilon_l^*,$$

where $\Gamma = 2\hbar/\tau$ describes the coupling of the energy level to the environment. The model with the imaginary part can be argued by considering the time evolution of the electron wave function on a state with energy ϵ_l^*. This is

$$\psi(t) = \psi_0 e^{-i\epsilon_l^* t/\hbar}$$

Due to the finite broadening, this wave function is not normalizable. Rather, the norm, describing the occupation probability of the state ϵ_l^* is

$$n_l(t) = \psi(t)\psi(t)^* = |\psi_0|^2 e^{-\Gamma t/(2\hbar)} = |\psi_0|^2 e^{-t/\tau}. \quad (8.3)$$

The imaginary part of the energy hence leads to a decay of the occupation probability of state ϵ_l^*.

A useful quantity in describing the properties of a quantum dot coupled to the electrodes is the *spectral function* $A_{l\sigma}(\epsilon)$. This function describes the positions and lifetimes of the electron energy levels inside the quantum dot. The precise definition is typically made via the retarded Green's function,[7] but here we just give its definition in terms of the energy level ϵ_l:

$$A_{l\sigma}(\epsilon) \equiv \lim_{\eta \to 0^+} -2\mathrm{Im}\left[\frac{1}{\epsilon - \epsilon_l^* + i\eta}\right]. \qquad (8.4)$$

[7] See for example Ch. 8 in (Bruus and Flensberg, 2004).

The definition of the spectral function includes the spin index σ, although for a spin-degenerate level it does not show up in the actual form of $A_{l\sigma}(\epsilon)$.

The spectral function has a probability interpretation: it describes the probability of an electron in level l to have the energy ϵ. For example, it satisfies (see Exercise 8.2)

$$\int \frac{d\epsilon}{2\pi} A_{l\sigma}(\epsilon) = 1. \qquad (8.5)$$

Moreover, it turns out that the spectral function can be used to find the average occupation of the energy level ϵ_l. This can be found from

$$\langle n_\sigma \rangle = \int \frac{d\epsilon}{2\pi} A_{l\sigma}(\epsilon) f_D(\epsilon), \qquad (8.6)$$

where f_D is the energy distribution function of the quantum dot.

In the case of an isolated quantum dot, $\Gamma = 0$. Then the spectral function is

$$A_{l\sigma}(\epsilon) = \lim_{\eta \to 0^+} -2\mathrm{Im}\left[\frac{\epsilon - \epsilon_l - i\eta}{(\epsilon - \epsilon_l)^2 + \eta^2}\right] = 2 \lim_{\eta \to 0^+} \frac{\eta}{(\epsilon - \epsilon_l)^2 + \eta^2}. \qquad (8.7)$$

One can see that $A_{l\sigma}(\epsilon \neq \epsilon_l) = 0$. However, for $\epsilon = \epsilon_l$, the function diverges. This is similar to the behaviour of the Dirac δ-function. Using eqn (8.5) we hence get that for the isolated quantum dot

$$A(\epsilon) = 2\pi\delta(\epsilon - \epsilon_l).$$

In the presence of a finite lifetime the spectral function is a Lorentzian function

$$A_{l\sigma}(\epsilon) = \frac{\Gamma}{(\epsilon - \epsilon_l)^2 + (\Gamma/2)^2}. \qquad (8.8)$$

The escape rate Γ can hence be described as a 'broadening' of the energy level ϵ_l as in Fig. 8.2.

In equilibrium, the average level occupation can be found by using the Fermi distribution function in eqn (8.6). At $T = 0$ in the presence of the broadening Γ, this is

$$\langle n_\sigma \rangle = \frac{1}{2} - \frac{1}{\pi} \arctan\left(\frac{2(\epsilon_l - \mu)}{\Gamma}\right), \qquad (8.9)$$

where μ is the chemical potential of the quantum dot, typically imposed by the surroundings. This function is plotted in Fig. 8.3 as a function of

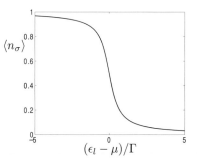

Fig. 8.3 Occupation of a quantum dot with broadening Γ as a function of the level energy ϵ_l.

[9]The coupling to the environment often leads also to a 'renormalization' of the energy level position ϵ_l. Typically, such a renormalization is very weakly energy dependent (except perhaps when dealing with small molecules), and one can simply include it in the definition of ϵ_l.

the potential μ. When $\mu \ll \epsilon_l$ the level is unoccupied, and when $\mu \gg \epsilon_l$ it is occupied, as can be expected from the simple picture in Fig. 8.2.

When the couplings $t_{k\beta}$ in the quantum dot Hamiltonian (8.1) are independent of the level index k, the transmission rate to an electrode with density of states N_β and volume Ω_β of the coupling region[8] can be described with $\Gamma_\beta = \pi N_\beta \Omega_\beta |t_\beta|^2$, $\beta \in L, R$.[9] The total broadening is the sum of the broadenings due to the coupling to the electrodes and that due to, for example, inelastic scattering acting as a further 'environment' to the quantum dot, i.e., $\Gamma = \Gamma_L + \Gamma_R + \Gamma_{\text{env}}$.

Example 8.1 Calculating the coupling energy from the Hamiltonian
The calculation of the tunnelling rates in the quantum dot case proceeds almost similarly to that in the metallic island (Sec. 7.2). The only difference is that we are concerned with the tunnelling rate into a single level l in the dot. Therefore, the only change we make in eqn (7.14) is that we do not sum over the island momentum l or spin σ. We thus obtain

$$
\begin{aligned}
\Gamma_{LD,l}(n) &= 2\pi \sum_k |t_{kl,\sigma}|^2 f(\epsilon_k)[1 - f(\epsilon_l)]\delta[\delta E_{\text{ch}}(n) + \epsilon_l - \epsilon_k] \\
&= \pi |t_L|^2 N_L \Omega_L \int d\epsilon_k f(\epsilon_k)[1 - f(\epsilon_l)]\delta[\delta E_{\text{ch}}(n) + \epsilon_l - \epsilon_k] \\
&= \Gamma_L f(\epsilon_l + \delta E_{\text{ch}}(n))[1 - f(\epsilon_l)].
\end{aligned} \tag{8.10}
$$

where on the second line the factor $1/2$ comes from the absence of spin summation usually included in the density of states. The transition rate from the island is obtained by interchanging the energies ϵ_l and ϵ_k. We can thus see that whenever ϵ_l and $\epsilon_l + \delta E_{\text{ch}}$ are around the Fermi level, the coupling energy is of the order of Γ_L.

8.2   Weakly interacting limit

[10]A detailed discussion of the mean-field description for quantum dots can be found, for example, in (Datta, 2004).

When the broadening Γ of the energy levels due to the coupling to the electrodes exceeds the interaction energy U in the Anderson Hamiltonian, Coulomb blockade-type effects are suppressed, and it is enough to consider a mean field model for the interactions.[10] In such a model, the term $U n_\uparrow n_\downarrow$ quadratic in the number operator in eqn (8.2) is replaced by

$$
U n_\uparrow n_\downarrow \to \frac{U}{2}(n_\uparrow \langle n_\downarrow \rangle + n_\downarrow \langle n_\uparrow \rangle). \tag{8.11}
$$

If the level position and the coupling to the levels are spin independent, the average occupation numbers must also be equal: $\langle n_\uparrow \rangle = \langle n_\downarrow \rangle \equiv \langle n_d \rangle$. In this case, the interaction term can be included in the definition of ϵ_l:

$$
\epsilon_d \equiv \epsilon_l + \frac{U}{2}\langle n_d \rangle. \tag{8.12}
$$

In this mean-field model we can use the expressions for the non-interacting model, but taking into account the renormalized energy ϵ_d. In this case the problem must be solved self-consistently: the average occupation

$\langle n_d \rangle$ of the level depends on the level position (see eqn (8.9)), but on the other hand the magnitude of ϵ_d depends on the average occupation.

Chapter 3 (see Sec. 3.4 and Exercise 3.5) presents a calculation of the transmission probability through a double barrier structure. There, the constructive interference of different paths through the structure involving different numbers of scatterings at the barriers and the coherent evolution between the scattering events leads to the Fabry–Perot resonant tunnelling. Whenever the energy of the incoming electrons equals the energy of the standing wave formed between the barriers, the transmission is greatly enhanced compared to the other energies. For energies close to ϵ_l, the transmission probability through such a structure can be expressed via the Breit–Wigner form (3.74). A similar expression can be found from the Anderson Hamiltonian in the non-interacting limit. An exact calculation based on the Heisenberg equation of motion involves the introduction of the (non-equilibrium) Green's function methods, which is outside the scope of this book. The full non-equilibrium calculation is based on the method established by Meir and Wingreen (1992), who showed that the energy-dependent transmission probability through the level with energy ϵ_d is

$$T_\sigma(\epsilon) = 2 \frac{\Gamma_L \Gamma_R}{\Gamma_L + \Gamma_R} A_{l\sigma}(\epsilon). \tag{8.13}$$

Assuming that the broadening of the level is entirely due to the coupling to the electrodes, $\Gamma = \Gamma_L + \Gamma_R$, we obtain the Breit–Wigner form for the transmission

$$T_\sigma(\epsilon) = \frac{\Gamma_L \Gamma_R}{(\epsilon - \epsilon_d)^2 + (\Gamma/2)^2}, \tag{8.14}$$

similar to that discussed in Sec. 3.4.

For a small bias voltage, the conductance is obtained from

$$G = \frac{e^2}{h} \sum_\sigma T_\sigma(\mu), \tag{8.15}$$

where $\mu = \mu_L \approx \mu_R$ is the chemical potential in the leads. Taking into account the interaction in the mean-field approximation, we can find the renormalized position of the energy level from eqn (8.12) using

$$\langle n_\sigma \rangle = \frac{1}{2} - \frac{1}{\pi} \arctan \left(\frac{2(\epsilon_l + \frac{U}{2} \langle n_{\bar{\sigma}} \rangle - \mu)}{\Gamma} \right), \tag{8.16}$$

where $\bar{\sigma}$ denotes the spin opposite to σ.

Such a mean-field model does not describe the transport process very well in the limit $U > \Gamma$, as it does not take into account the fact that the instantaneous position of the energy level depends on the instantaneous occupation of the quantum dot. While one electron occupies the dot level, a second electron trying to transmit into it is required to have a higher (by U) energy than in the case of an empty quantum dot. This effect can be taken into account by lifting the requirement $\langle n_\uparrow \rangle = \langle n_\downarrow \rangle$ and solving eqn (8.16) without this assumption. Within a certain

[11]See Sec. 10.4.2 of (Bruus and Flensberg, 2004).

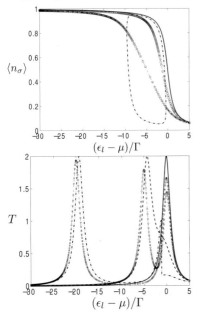

Fig. 8.4 Top: Occupation probability $\langle n_\sigma \rangle$ of a single-level quantum dot as a function of the position ϵ_l of the energy level w.r.t. the chemical potential μ of the leads. The curves are obtained for $U/\Gamma = 0$ (solid line), $U/\Gamma = 2.5$ (dashed lines and circles) and $U/\Gamma = 10$ (dotted lines and squares). The bare lines show the occupations in the case when $\langle n_\uparrow \rangle$ is allowed to differ from $\langle n_\downarrow \rangle$, and the lines with symbols for the case $\langle n_\uparrow \rangle = \langle n_\downarrow \rangle$. Bottom: the corresponding transmission probabilities summed over spin, in the self-consistent mean-field model (bare lines) and using eqn (8.17). In both cases $\Gamma_L = \Gamma_R = \Gamma/2$.

[12]A general description of the Coulomb blockade limit on QD transmission can also be found at (Beenakker, 1991a). This paper also takes into account the effects of the asymmetry of the setup.

[13]The voltage drop is included in the chemical potentials of the electrodes.

range of parameters (see an example in Fig. 8.4) this yields two possible solutions: one with $\langle n_\uparrow \rangle = \langle n_\downarrow \rangle$, and one where they differ. Including the latter solution in eqn (8.14) yields for large values of U/Γ two peaks for the transmission probability, roughly when $\epsilon_l = \mu$ and when $\epsilon_l = \mu - U$, as shown in Fig. 8.4.

This non-restricted self-consistent mean-field theory describes the transport through the quantum dot qualitatively correctly. A more refined approach based on the kinetic equation description of the quantum dot[11] yields a spectral function that is a sum of two Lorentzians,

$$A_{l\sigma}(\epsilon) = \frac{(1 - \langle n_\sigma \rangle)\Gamma}{(\epsilon - \epsilon_l)^2 + (\Gamma/2)^2} + \frac{\langle n_\sigma \rangle \Gamma}{(\epsilon - \epsilon_l - U)^2 + (\Gamma/2)^2} \qquad (8.17)$$

This function hence describes first adding an electron into the bare level ϵ_l when the quantum dot is empty ($\epsilon = \epsilon_l$) and then adding a second electron to the quantum dot when the first state is occupied ($\epsilon = \epsilon_l + U$). Experimentally, the position ϵ_l of the energy level relative to the Fermi level of the electrodes can be tuned with a gate voltage (as for example in Fig. 8.1). The corresponding transmission probability is compared to the non-restricted self-consistent mean field model in Fig. 8.4. Although quantitatively the two models do not yield the same results, qualitative aspects are captured by both of them.

8.3 Weakly transmitting limit: Coulomb blockade

When $\Gamma \ll U$, the quantum dot enters the state of Coulomb blockade. Then some of the features found in the transmission through the dot are similar to that discussed in Ch. 7, but the observably large level spacing $\Delta\epsilon$ adds another ingredient to the story. In the following discussion, I only consider the limit of sequential tunnelling, neglecting the higher-order effects such as cotunnelling. I take into account any number of energy levels inside the quantum dot; $\Delta\epsilon$ should be regarded as just an average level spacing. I also describe the interaction energy similarly to the previous chapter, i.e., the charging energy for n electrons on the quantum dot is given by eqn (7.5). For simplicity, I only consider the left-right symmetric setting in the linear response limit of small voltages.[12]

Consider the energy required for an electron tunnelling from the quantum dot into one of the electrodes. Now the energy change must include the energy ϵ_l of the level l from which tunnelling takes place,[13]

$$\delta E_L^-(n, l) = \epsilon_l + E_{\text{ch}}(n) - E_{\text{ch}}(n - 1). \qquad (8.18)$$

The energy required for the tunnelling of an electron into state l from the left lead is

$$\delta E_L^+(n, l) = \epsilon_l + E_{\text{ch}}(n + 1) - E_{\text{ch}}(n). \qquad (8.19)$$

These thus satisfy the condition $E_L^-(n + 1, l) = E_L^+(n, l)$.

The equation for the current has to take into account the fermionic character of the energy levels: at most only one electron may occupy level ϵ_l.[14] The current through the left barrier is given by

$$I_L = \tag{8.20}$$

$$-\frac{e}{\hbar} \sum_{\{n_i\}} \sum_{l=1}^{\infty} \Gamma_L^l P(\{n_i\}) \{\delta_{n_l,0} f_L(\delta E_L^+(n,l)) - \delta_{n_l,1}[1 - f_L(\delta E_L^-(n,l))]\},$$

[14]Note that this includes spin in label l and allows in principle for degenerate states.

where $f_L(E; \mu_L) = \{1 + \exp[(E - \mu_L)/(k_B T)]\}^{-1}$ is the Fermi distribution function of the left electrode with potential $\mu_L = E_F - eV/2$, $\Gamma_L^l = \pi N_L \Omega_L |t_L|^2$ is the coupling strength from state l to the left electrode and N_L is the density of states in the left electrode. The right electrode is at potential $\mu_R = E_F + eV/2$, and the tunnelling into it is described by $\Gamma_R = \pi N_R \Omega_R |t_R|^2$. The current I_L equals the current $-I_R$ through the right barrier. Here we consider only the left–right symmetric system where $\Gamma_L = \Gamma_R \equiv \Gamma/2$. The sum over $\{n_i\} = \{n_1, n_2, \dots\}$ is over all the realizations of occupation numbers n_i of the different levels. Naturally, each of these numbers can only be either zero or 1. The sum is weighted by the probability $P(\{n_i\})$ for the corresponding realization of occupation numbers. The latter can be obtained in the general case from a master equation. To make things easier, let us concentrate on the linear-response limit, where current (8.20) is linearized with respect to voltage V. In that case, for a left–right-symmetric system, the lowest-order corrections on the non-equilibrium probability $P(\{n_i\})$ from equilibrium are quadratic in V, and it is enough to use the equilibrium probability distribution.[15] This is given by

$$P_{\text{eq}}(\{n_i\}) = Z^{-1} \exp\left[-\frac{\sum_{i=1}^{\infty} \epsilon_i n_i + E_{\text{ch}}(n)}{k_B T} \right], \tag{8.21}$$

[15]This can be justified by symmetry: in a left–right symmetric system where voltage drops at both junctions are equal and both electrodes are at the same temperature we have to have the symmetry $P(\{n_i\}, V) = P(\{n_i\}, -V)$ as changing the sign of V corresponds to inverting the structure. Therefore, V-dependent corrections in P have to be at least quadratic.

where the energies ϵ_i are calculated with respect to the Fermi level E_F, $n = \sum_i n_i$ is the number of electrons on the quantum dot, and the partition function

$$Z = \sum_{\{n_i\}} \exp\left[-\frac{\sum_{i=1}^{\infty} \epsilon_i n_i + E_{\text{ch}}(n)}{k_B T} \right] \tag{8.22}$$

ensures the normalization of $P_{\text{eq}}(\{n_i\})$.

In the linear limit we can approximate $f_L(E; eV/2) \approx f^0(E; 0) + f'(E; 0) eV/2$. Defining $\tilde{n} = \sum_{i \neq l} n_i$, we find that the energy changes in the two terms in (8.20) are $\delta E_L^+(\tilde{n}, l)$ and $\delta E_L^-(\tilde{n}+1, l)$, respectively. As these two are equal (see eqns (8.18,8.19)), we find for the current

$$I = -\frac{e^2}{\hbar} V \sum_l \sum_{\{n_i\}} \Gamma^l P_{\text{eq}}(\{n_i\}) f'(\epsilon) \left(\delta_{n_l,0} + \delta_{n_l,1} \right), \tag{8.23}$$

where $\epsilon = \epsilon_l + E_{\text{ch}}(\tilde{n}+1) - E_{\text{ch}}(\tilde{n})$. Using the equalities

$$P_{\text{eq}}(n_1, \dots n_{p-1}, 1, n_{p+1}, \dots) = P_{\text{eq}}(n_1, \dots n_{p-1}, 0, n_{p+1}, \dots) e^{-\epsilon/k_B T} \tag{8.24}$$

and

$$k_B T f'(\epsilon)(1 + e^{-\epsilon/k_B T}) = -f(\epsilon) \tag{8.25}$$

we get

$$I = \frac{e^2 V}{\hbar k_B T} \sum_l \sum_{\{n_i\}} \Gamma^l P_{\text{eq}}(\{n_i\}) f(\epsilon_l + E_{\text{ch}}(n+1) - E_{\text{ch}}(n)) \delta_{n_l,0}. \tag{8.26}$$

In the non-symmetric case we should replace Γ^l by $\Gamma_L^l \Gamma_R^l / (\Gamma_L^l + \Gamma_R^l)$ (Beenakker, 1991a).

This formula can be simplified somewhat further by replacing the sums over the different configurations of n_i by sums over charge states labelled by n. This can be accomplished by defining the probability that there are exactly n electrons on the quantum dot,

$$P_{\text{eq}}(n) = \sum_{\{n_i\}} P_{\text{eq}}(\{n_i\}) \delta_{n, \sum_i n_i}$$

$$= \sum_{\{n_i\}} Z^{-1} \exp\left(-\frac{\sum_{i=1}^{\infty} \epsilon_i n_i + E_{\text{ch}}(n)}{k_B T}\right) \delta_{n, \sum_i n_i}$$

$$= \frac{e^{-E_{ch}(n)/k_B T} \sum_{\{n_i\}} \exp\left(-\frac{1}{k_B T} \sum_{i=1}^{\infty} \epsilon_i n_i\right) \delta_{n, \sum_i n_i}}{\sum_n e^{-E_{ch}(n)/k_B T} \sum_{\{n_i\}} \exp\left(-\frac{1}{k_B T} \sum_{i=1}^{\infty} \epsilon_i n_i\right) \delta_{n, \sum_i n_i}},$$

$$= \frac{e^{-\Omega(n)/k_B T}}{\sum_n e^{-\Omega(n)/k_B T}}.$$

Here $\Omega(n) = F(n) + E_{\text{ch}}(n)$ is the thermodynamic potential of the dot, and

$$F(n) = -k_B T \ln\left[\sum_{\{n_i\}} \exp\left(-\frac{1}{k_B T} \sum_i \epsilon_i n_i\right) \delta_{n, \sum_i n_i}\right] \tag{8.27}$$

is the free energy related to the kinetic energy inside the quantum dot. Moreover, we need the conditional probability that the level l is occupied provided there are n electrons on the quantum dot:

$$P_{\text{eq}}(\epsilon_l | n) = \frac{1}{P_{\text{eq}}(n)} \sum_{n_i} P_{\text{eq}}(\{n_i\}) \delta_{n_l,1} \delta_{n, \sum_i n_i}$$

$$= \frac{1}{P_{\text{eq}}(n)} \frac{e^{-E_{ch}(n)/k_B T}}{\sum_n e^{-\Omega(n)/k_B T}} \sum_{\{n_i\}} \exp\left(-\frac{1}{k_B T} \sum_{i=1}^{\infty} \epsilon_i n_i\right) \delta_{n_l,1} \delta_{n, \sum_i n_i}$$

$$= e^{\frac{F(n)}{k_B T}} \sum_{\{n_i\}} \exp\left(-\frac{1}{k_B T} \sum_i \epsilon_i n_i\right) \delta_{n_l,1} \delta_{n, \sum_i n_i}. \tag{8.28}$$

In terms of these two probabilities the linear conductance $G = I/V$ of the quantum dot can be written as

$$G = \frac{e^2}{\hbar k_B T} \sum_l \Gamma^l \sum_{n=0}^{\infty} P_{\text{eq}}(n) f(\delta E_L^+(n, l))[1 - P_{\text{eq}}(\epsilon_l | n)]. \tag{8.29}$$

Using eqn (8.24) and the fact that $f(\epsilon)e^{\epsilon/k_BT} = 1 - f(\epsilon)$ this can be written also in the form

$$G = \frac{e^2}{\hbar k_B T} \sum_l \Gamma^l \sum_{n=1}^{\infty} P_{\text{eq}}(n) P_{\text{eq}}(\epsilon_l|n)[1 - f(\delta E_L^-(n,l))]. \quad (8.30)$$

Equations (8.29) and (8.30) can now be used to obtain various limits for the conductance of the quantum dot.

8.3.1 Metallic limit

In the limit $k_B T \gg \Delta\epsilon$ the discrete spectrum of the quantum dot levels no longer plays a role. In this case the conditional probability can be approximated by the Fermi function, $P_{\text{eq}}(\epsilon_l|n) = f(\epsilon_l - \mu(n))$, where $\mu(n) \approx \mu_L$ is the chemical potential of the quantum dot. The distribution function $P_{\text{eq}}(n)$ takes the Maxwell–Boltzmann form (see Exercise 7.4),

$$P_{\text{SET}}(n) = \frac{\exp(-E_{\text{ch}}(n)/k_B T)}{\sum_n \exp(-E_{\text{ch}}(n)/k_B T)}. \quad (8.31)$$

We may proceed further by replacing the summation over l in eqn (8.29) by an integral over E and including the density of states N_D of the quantum dot. In the case when we can disregard the energy dependence of N_D, we can use eqns (A.64)a and (A.64)h to obtain

$$G = \frac{e^2 N_D \Omega_D \Gamma}{\hbar k_B T} \sum_n P_{\text{SET}}(n) g(E_{\text{ch}}(n) - E_{\text{ch}}(n-1)), \quad \text{for } \Delta\epsilon \ll k_B T \ll E_C. \quad (8.32)$$

Here $g(x) = x/(1 - e^{-x/k_B T})$. This is the same conductance as one would obtain for a single-electron transistor using the methods of the previous chapter (see Exercise 7.5).

8.3.2 Two-state limit

At low temperatures $\Gamma \ll k_B T \ll \Delta\epsilon, E_C$, the energy levels in the quantum dot are filled sequentially, starting from the lowest-energy state and adding electrons one by one to the higher-energy states. In this case it is enough to find the state with the addition energy closest to the Fermi levels of the leads. This level $n = n_{\text{min}}$ minimizes the absolute value of

$$\Delta(n) = \epsilon_n + E_{\text{ch}}(n) - E_{\text{ch}}(n - 1), \quad (8.33)$$

where ϵ_n is calculated from the lead chemical potential μ. Whenever n_{min} is not occupied, the energy levels are occupied up to $n = n_{\text{min}} - 1$, which is thus also included in the calculation (see Fig. 8.5). In this case $P_{\text{eq}}(n)$ is negligibly small for the other levels, so that

$$P_{\text{eq}}(n_{\text{min}}) = \frac{e^{-\Omega(n_{\text{min}})/(k_B T)}}{e^{-\Omega(n_{\text{min}})/(k_B T)} + e^{-\Omega(n_{\text{min}}-1)/(k_B T)}} = f(\Delta_{\text{min}}), \quad (8.34)$$

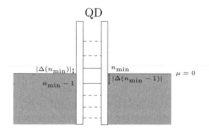

Fig. 8.5 At low temperatures it is enough to consider the two levels of the quantum dot closest to the chemical potential of the leads (solid lines) and neglect the other levels (dashed lines).

where $f(\epsilon)$ is a Fermi function and $\Delta_{\min} = \Omega(n_{\min}) - \Omega(n_{\min}-1)$. At low temperatures we can expand eqn (8.27) and obtain $\Omega(n) = \sum_{i=1}^{n} \epsilon_i + E_{ch}(n)$ and therefore $\Delta_{\min} = \Delta(n_{\min})$. Moreover, when $k_B T \ll \Delta\epsilon$ the extra electrons entering the quantum dot occupy the energy levels in the order of their energies, and therefore $F_{eq}(E_n|n) = 1$. Thus, the conductance (8.30) obtains the form

$$G = e\frac{\Gamma^{n_{\min}}}{\hbar k_B T} f(\Delta_{\min})(1 - f(\Delta_{\min})), \quad \text{for } \Gamma \ll k_B T \ll \Delta\epsilon, E_C. \quad (8.35)$$

This is the conductance for a thermally broadened resonance peak ($k_B T$ much larger than Γ), generalized to include the effects of a non-zero charging energy.

8.3.3   Addition spectrum

Let us consider the effect of a changing gate voltage (or alternatively, changing the Fermi level of the electrodes) on the conductance of the quantum dot. As in the case of a single-electron transistor, the conductance is peaked whenever the gate voltage satisfies a resonant condition $\Delta(n) = 0$ for some integer n, i.e.,

$$\epsilon_n + E_{ch}(n) - E_{ch}(n-1) = \epsilon_n + E_C \left(2n + 1 - 2Q_G/e\right) = 0. \quad (8.36)$$

Contrary to the single-electron transistor, the gate voltage dependence is not entirely periodic. Namely, the distance between subsequent peaks is

$$\Delta Q_G = \frac{e}{2}\left(\frac{\epsilon_{n+1}}{E_C} + 2(n+1) + 1\right) - \frac{e}{2}\left(\frac{\epsilon_n}{E_C} + 2n + 1\right) = \frac{e}{2}\frac{\Delta\epsilon}{E_C} + e. \quad (8.37)$$

Generally, the distance between the subsequent levels in the quantum dot is not constant, and therefore also the period of the gate voltage dependence is no longer constant when $\Delta\epsilon$ is of the same order or larger than E_C. For example, in the absence of a magnetic field the energy levels are spin degenerate. Then, $\epsilon_{2n+1} - \epsilon_{2n} = 0$, and plotting G vs. Q_G will lead to a doublet-type structure of the oscillations, with a period alternating between e and $e + e\Delta\epsilon/(2E_C)$. Figure 8.6 shows the $G(V_G)$ through a single-wall carbon nanotube weakly coupled to the electrodes. There the energy levels are fourfold degenerate due to the chirality of the carbon nanotubes.[16] This figure can be used to estimate the charging energy and the level spacing.

Another example on the addition spectrum comes from studies of circular quantum dots, which can be totally depleted from electrons. Then, adding electrons one by one allows the experimentalists to study the properties of the excitation spectrum, which is qualitatively similar to that in atoms: for example, Hund's rule is obeyed (shells are first filled with parallel spins until the shells are half full). Such quantum dots have therefore been called artificial atoms, and the researchers have been able to construct the first elements of the periodic table of semiconductor quantum dots; see (Kouwenhoven *et al.*, 2001).

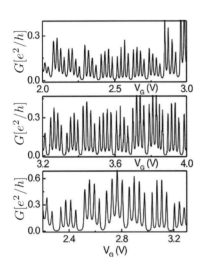

Fig. 8.6 Linear conductance of a metallic single-wall carbon nanotube weakly coupled to electrodes. The horizontal axis is the gate voltage on the quantum dot. The conductance peaks clearly show the fourfold degeneracy of the nanotube. (S. Sapmaz *et al.*, *Phys. Rev. B* **71**, 153402 (2005), Fig. 1.) © 2005 by the American Physical Society.

[16]This chirality corresponds to the *valley* symmetry in graphene, i.e., the presence of two non-equivalent reciprocal space vectors at the Fermi level. This is discussed in Ch. 10.

8.3.4 Charge sensing with quantum point contacts

Besides the addition spectrum probed by the direct gate-dependent linear conductance at low temperatures, the charge state of the quantum dot can be measured by utiling the quantum point contacts (QPCs) as briefly discussed in Sec. 3.1.4.[17] The QPCs are tuned close to the voltage values where the number of propagating modes changes, at which point the QPC current is a sensitive probe of the charge in a nearby quantum dot. Such a measurement can be used to determine the charge on the quantum dot or the charge configuration in a series of dots — provided the QPC tunnelling rate far exceeds the relaxation rate of the charge state on the dot.

Similar techniques can then also be used for the readout of the spin state of the quantum dot as discussed in (Hanson *et al.*, 2007) and in Exercise 8.8.

[17]For more details, see for example (Hanson *et al.*, 2007).

8.4 Kondo effect

Above sections discuss the conductance through a quantum dot in the weakly interacting and weakly transmitting limits, where the physics can be described in relatively simple terms, either with the self-consistent mean field description or with the master equation approach. In the case when both the transmission and the interactions are strong, the transport through quantum dots shows features related with the true many-body physics of an interacting electron system. A particular but representative example is the Kondo effect, whose theory was first constructed to explain an anomalous behaviour of the resistance of bulk metals at low temperatures.[18]

At low temperatures the resistivity of most conductors is governed by elastic scattering, whose strength is generally temperature independent (remember the Drude model explained briefly in Sec. 1.2.1 and derived in Ch. 2). Increasing the temperature to above some 10 K, other scattering mechanisms such as electron-phonon scattering start to influence the transport, decreasing the effective mean free path and thereby increasing the resistivity. Thus, resistivity $\rho(T)$ should be a monotonously increasing function of temperature. However, in many cases the experimentalists find a resistivity minimum at temperatures of the order of a few Kelvin, and rather than saturating, the resistivity of bulk metals starts to increase upon decreasing the temperature. This phenomenon was shown by Jun Kondo to be due to the presence of magnetic impurities which can flip their spin. Non-magnetic materials favor spin non-alignment, and the conduction electrons try to screen the spin of the magnetic impurity roughly in a similar way as an extra charge is screened by them. The screening takes place via a scattering mechanism that flips the spin of the magnetic impurity. Such a spin-flip scattering gives rise to an extra resonant level at the Fermi level around the magnetic impurity. Such a *Kondo resonance* enhances the scattering and thereby leads to an increased resistance in bulk metals.

[18]See (Kondo, 1964; Nozières, 1974). An introduction to the Kondo effect in quantum dots has been also given for example by Kouwenhoven and Glazman (2001).

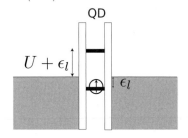

Fig. 8.7 When a quantum dot has an odd number of occupied levels, the spin of the electrons on the highest occupied level breaks the spin rotation symmetry of the system.

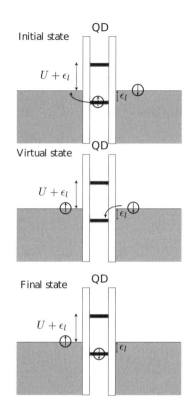

Fig. 8.8 Spin-flip process in a quantum dot occurs via a virtual state where the energy level ϵ_l is unoccupied.

[19]See Sec. 10.5 in (Bruus and Flensberg, 2004).

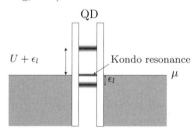

Fig. 8.9 Coupling to the electrodes leads to the formation of an additional energy level in the quantum dot, exactly at the position of the Fermi level of the leads. The nature of this Kondo resonance can be studied from the bias voltage or magnetic field dependence of transport through the quantum dot.

In quantum dots one can find a similar effect. An island with an odd number of electrons has one extra spin which in principle breaks the spin-rotation symmetry (see Fig. 8.7). But as the ensemble-averaged system has to be spin-rotation invariant, the conduction electrons from the leads try to screen the spin. In the lowest order in the coupling to the leads, the electron from level ϵ_l cannot escape to them nor can the electrons from the leads occupy the next level, lying way above the Fermi level. However, including higher orders in the coupling allows the electrons to tunnel off and on the island in a single coherent process, similarly to the process of cotunnelling described in Sec. 7.4. In such a process, the electron occupying level ϵ_l tunnels into one of the electrodes, and another electron tunnels back to level ϵ_l (see Fig. 8.8). This second electron does not have to have the same spin as the first electron. Thus including these higher-order processes leads to an effective spin-flip scattering for the quantum dot, and thereby the conduction electrons screen the spin of the island.

Contrary to the bulk metals, the Kondo effect in quantum dots leads to an increased conductance. The (logarithmic) onset of this increase can be found when calculating the second- and third-order corrections to the conductance in the case of an unpaired degenerate level.[19] However, the full effect cannot be captured by the perturbation theory. There are a few non-perturbative approaches which capture the Kondo physics with varying levels of accuracy.[20] These treatments are beyond the scope of this book, but in what follows I discuss some of their results.

One of the characteristic signatures of the Kondo effect is that the spectral function of the dot obtains an additional peak at the Fermi level of the leads (see Fig. 8.9). The width of this level is roughly given by the *Kondo temperature* (Haldane, 1978)

$$T_K = \frac{1}{2}\sqrt{\Gamma U} \exp\left(\frac{\pi(E_F - \epsilon_l)(U + E_F - \epsilon_l)}{\Gamma U}\right). \qquad (8.38)$$

As a result of this additional resonance, the conductance of the quantum dot is strongly enhanced, contrary to the bulk metals where the resistance is enhanced The difference between the two systems is clear: in quantum dots, the Kondo resonance facilitates transport which is otherwise blocked, whereas in bulk metals it leads to an increased scattering. In quantum dots, it turns out that the conductance is a universal function of the ratio T/T_K, i.e., the dependence on the exact position ϵ_l of the energy level, the interaction energy U and the level broadening Γ can be described with the single parameter T_K. When $T \ll T_K$, the conductance equals $2e^2/h$, i.e., the Kondo resonance leads to a unit transmission through the quantum dot even when the bare energy level

[20]For example, the renormalization group theory initiated by P. W. Anderson and K. Wilson in the 1960s, see for example (Wilson, 1975), where the renormalization group method is applied for the solution of the 'classical' Kondo problem. For other approaches to the classical Kondo problem, see the description in (Mahan, 2000). A detailed theoretical description of the Kondo effect in quantum dots is given, for example, in (Hershfield *et al.*, 1991).

is far off the Fermi level.

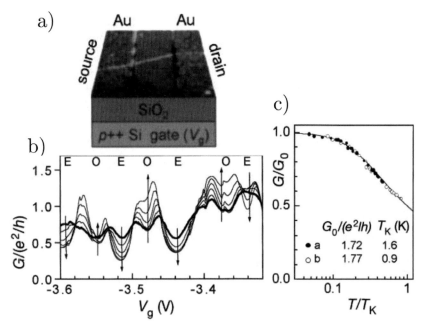

Fig. 8.10   Kondo effect in a single-walled carbon nanotube (SWNT): a) AFM image of the device. b) Linear conductance vs. gate voltage shows the even–odd (E–O) periodicity: at high temperatures (thicker lines), the resonances related to both cases with an even and odd number of electrons on the SWNT are roughly on the same level. Decreasing the temperature changes the conductance differently for the two cases: for the case with an even number of electrons the conductance decreases, and it increases for the case with an odd number of electrons. The different curves, in the direction of the arrows, are measured at $T = 780$, 560, 490, 320, 245, 180, 125 and 75 mK. c) Conductance vs. temperature for two cases with an odd number of electrons on the SWNT follows the universal scaling behaviour dependent only on the ratio T/T_K. (J. Nygård *et al.*, *Nature* **408**, 342 (2000), Figs. 1 and 2.) Used by permission from Macmillan Publishers Ltd., © 2000.

In general, the Kondo effect forms when there is a partially occupied N-fold degenerate level in the quantum dot. In most quantum dots, the Kondo effect thus leads to an even–odd asymmetry (due to spin) of the conductance peaks whenever the transmission is strong enough. An example of the measurement of the Kondo effect in a single-walled carbon nanotube is shown in Fig. 8.10.

8.5   Double quantum dots

Besides the single quantum dots described above, the transport properties of systems with two quantum dots in series have been studied in great detail. The interest towards these systems is driven by the idea to use the spin states of the double dots as quantum bits, but also other types of interesting qualitative effects related especially to their spin structure have been found; for example, the Pauli spin blockade. These phenomena are introduced in this section.[21]

The charging energy of coupled quantum dots can be described similarly to that of arrays of metallic single-electron transistor islands (see Sec. 7.6.3 in the previous chapter). In particular, the electrostatic interaction between the QDs 'tilts' the charging energy diagram for different stable charge states of the dot as shown in Fig. 7.23. Another reason for the tilting of this diagram is due to the usually occurring cross-coupling between the gate voltages, i.e., gate voltage V_{G1} coupled mostly to the

[21]This description follows closely that given in (Hanson *et al.*, 2007), where more details and references can be found.

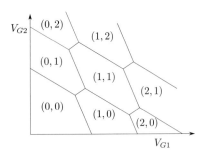

Fig. 8.11 Typical charge state diagram of a double quantum dot. The lowest-energy charge states of the two dots are indicated with the notation (N_1, N_2), where N_i refers to the number of conduction electrons on the ith dot.

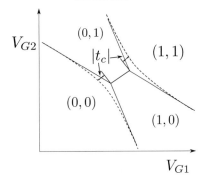

Fig. 8.12 Coupling t_c between the single dot levels smoothens the triple points in the double-dot charge state level diagrams.

---

[22] Note that by including spin it would seem that both electrons of the $(1,1)$ state could occupy the bonding orbital, and therefore the energy of the $(1,1)$ should be lowered compared to that seen in Fig. 8.12. However, in practice this is not the case, as typically the Coulomb interaction is stronger than the coupling energy, and the two electrons of the $(1,1)$ state mostly occupy the single-dot levels. Therefore Fig. 8.12 is valid even in this case.

[23] See Appendix A.4.

[24] This actually allows for the determination of g: for free electrons $g \approx 2$, but its particular value in the quantum dots depends on the material.

---

first dot has a non-zero capacitance also on the second dot, and vice versa for the gate voltage V_{G2}. Contrary to the metallic single-electron transistors, quantum dots can be entirely depleted from conduction electrons. Therefore, the typical charge state diagram for a double quantum dot shown in Fig. 8.11 starts from strictly zero conduction electrons on the dots. This diagram shows the emergence of the triple points—crossings between (n, m), $(n+1, m)$ and $(n, m+1)$ charge state configurations. As a non-zero current requires that the charge state of the dots can be tuned, at low temperatures current through the QD takes place only around such triple points.

8.5.1   Artificial molecules

There are also other qualitative differences of double QDs to multi-island metallic single-electron transistors than the possibility to empty the dots. If single quantum dots resemble atoms, double dots can be used to create artificial molecules. This can be observed in the addition energy as the coupling t_c between the states on the two different dots shows up in the level diagrams. The (non-interacting) double-dot eigenstates $\psi_{B,A}$ are thus the bonding and antibonding superpositions of the single dot eigenstates $\phi_{1,2}$ (see Exercise 8.6)

$$\psi_{B/A} = a\phi_1 \pm b\phi_2. \tag{8.39}$$

When the single-dot states have the same addition energy, the bonding orbital has an energy that is lower than the sum of the single-dot energies by the coupling energy $|t_c|$. This smoothens the triple points.

As a result, the energy level structure accessed in transport or charge sensing measurements is of the form shown in Fig. 8.12.[22]

8.5.2   Spin states in double quantum dots

Besides the direct coupling between the dots, an important property that can be accessed in quantum dot experiments is related to their spin states. Let us consider the triple point around the charge states $(0, 1)$, $(1, 1)$ and $(0, 2)$. The first of these states contains only one electron on the right dot, and it has only the two spin states $|\uparrow_2\rangle$ and $|\downarrow_2\rangle$. This notation $|\sigma_i\rangle$ refers to the spin state σ in the dot i. At a vanishing magnetic field these two spin states have the same energy. Applying a field lifts this degeneracy and introduces the Zeeman energy $\Delta\epsilon_Z = g\mu_B B$ between the two spin states,[23]

$$\epsilon_{l\sigma}(B) = \epsilon_{l\sigma}(0) + \sigma g\mu_B B/2, \tag{8.40}$$

where $\sigma = \pm 1$ for the spin parallel/antiparallel with the magnetic field. Here $\mu_B = e\hbar/(2m^*)$ is the Bohr magneton and g is the g-factor for the electrons in the quantum dot.[24]

In the case of two electrons in the double dot system, the spin part of the electron wave function is either of the singlet or the triplet form.

For example, the spin states in the charge state $(0, 2)$ are

$$
\begin{aligned}
S(0, 2) &= (|\uparrow_2\downarrow_2\rangle - |\downarrow_2\uparrow_2\rangle)/2 \\
T_+(0, 2) &= |\uparrow_2\uparrow_2\rangle \\
T_0(0, 2) &= (|\uparrow_2\downarrow_2\rangle + |\downarrow_2\uparrow_2\rangle)/2 \\
T_-(0, 2) &= |\downarrow_2\downarrow_2\rangle
\end{aligned}
\tag{8.41}
$$

and in the charge state $(1, 1)$ they are

$$
\begin{aligned}
S(1, 1) &= (|\uparrow_1\downarrow_2\rangle - |\downarrow_1\uparrow_2\rangle)/2 \\
T_+(1, 1) &= |\uparrow_1\uparrow_2\rangle \\
T_0(1, 1) &= (|\uparrow_1\downarrow_2\rangle + |\downarrow_1\uparrow_2\rangle)/2 \\
T_-(1, 1) &= |\downarrow_1\downarrow_2\rangle.
\end{aligned}
\tag{8.42}
$$

For a vanishing magnetic field, all the triplet states are degenerate. However, they typically have a higher energy than the singlet states. In the constant interaction model the energy difference between the $T(0, 2)$ and $S(0, 2)$ states is the energy level spacing $\Delta\epsilon = \epsilon_2 - \epsilon_1$ as the electrons on the single state occupy the same spin-degenerate single-electron state whereas the triplet state involves also the next excited single-electron state. On the other hand, the energy difference between $T(1, 1)$ and $S(1, 1)$ depends on the inter-dot tunnelling energy t_c, and for aligned single-electron states of the two dots, it is $J = 4t_c^2/E_C$ (Loss and DiVincenzo, 1998).

In the presence of the tunnel coupling between the two dots, the charge states $(1, 1)$ and $(0, 2)$ become hybridized (again, see Exercise 8.6). Due to the fact that typically the tunnelling amplitude is spin independent, only the singlets couple to each other and triplets to one another.

8.5.3   Pauli spin blockade

Pauli spin blockade is a manifestation of the spin states of a double quantum dot.[25] Consider a situation where the quantum dot is tuned to the region around the charge states $(0,1)$, $(0,2)$ and $(1,1)$. Applying a negative bias allows to move one electron from the right to the left electrode, such that the charge state transitions are $(0, 1) \rightarrow (0, 2) \rightarrow (1, 1) \rightarrow (0, 1)$. Consider the case when the right dot initially contains a spin up electron. Because of the Pauli principle, the electron transferred from the right lead must hence have spin down. Either of these electrons can then tunnel to the left island and the left lead.

A positive bias would lead to a sequence $(0, 1) \rightarrow (1, 1) \rightarrow (0, 2) \rightarrow (0, 1)$, where an electron from the left reservoir tunnels through the double quantum dot. Consider the case when the initial state again contains a spin up electron on the right dot. If the electron tunnelling from the left has spin down, it may tunnel into the right lead and from there to the right electrode. However, if the second electron also has spin up, it cannot enter the second dot because it would have to occupy a higher energy level; the triplet state $T_+(0, 2)$ would be too high in energy. Therefore, no current flows until either of the spins flip. As a

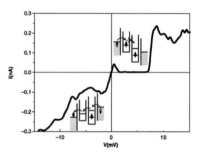

Fig. 8.13 Measured current–voltage curve of a double quantum dot in the Pauli spin blockade regime. The current at forward bias $V > 0$ is suppressed compared to that at backward bias $V < 0$ due to the Pauli blockade, as explained in the text. The insets show the schematics of the spin blockade. (K. Ono *et al.*, *Science* **297**, 1313 (2002), Fig. 2.) Reprinted with permission from AAAS.

[25]Note that similar considerations hold for systems with other conserved degrees of freedom than spin. For example, in graphene and carbon nanotubes, the electrons have an additional valley degeneracy (see Ch. 10), which also affects the Pauli spin blockade transport properties, see (Pályi and Burkard, 2009).

[26]The Pauli spin blockade is lifted in situations when the nuclear spins of the quantum dot host atoms are coupled with the spin state of the electrons (Koppens *et al.*, 2005). This interaction mixes the singlet $S(1,1)$ and the triplet $T(1,1)$ states, leading to a transition from the latter to the former, and therefore allowing the current to flow. Understanding this interaction mechanism is important for spin qubits, as fluctuations in the nuclear spin state lead to dephasing of the spin states.

[27]Besides double quantum dots, spin qubits have been fabricated also for single quantum dots. Eventually, a quantum computer would involve many quantum dot islands.

result, the current for the positive bias is suppressed compared to that of the negative bias. This *Pauli spin blockade* (Ono *et al.*, 2002) hence leads to current rectification (see Fig. 8.13).[26]

8.5.4   Spin qubits in quantum dots

The spin states of quantum dots can be used as elementary bits of a quantum computer: qubits. The original idea for this came from Loss and DiVincenzo (1998), who proposed using the spin of the excess electron on a single-electron quantum bit as the qubit. The 'up' spin refers to qubit state 0 and the 'down' spin to state 1. In many-island quantum dots the qubits can be controllably coupled by using external gate voltages to tune the heights of tunnel barriers separating the islands from each other. This would thus allow to realize *quantum gates*, operations on the qubits. Alternatively, the spin states of double quantum dots, eqn (8.42), could be used as the basis states.[27]

The main motivation behind the use of spins as the basis states, instead of the charge states that could in principle equally be used, is to avoid decoherence of the qubit from the fluctuations in the environment (see Sec. 6.6.2). In solid-state setups, the fluctuations of charge are quite generally much stronger than those of the spin. However, the problem is that the spin states cannot be read off directly, but typically need to be converted into a charge signal; for example, by the use of the Pauli spin blockade (see Exercise 8.8).

An example spin qubit system is shown in Fig. 8.14.

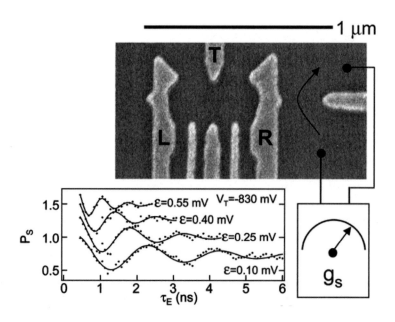

Fig. 8.14 Top: Electron micrograph of a double quantum dot working as a spin qubit. Voltages applied on gates denoted L and R were used to control the charge states of the islands, whereas the gate voltage T was used to control the tunnelling amplitude between them. The quantum point contact on the right was used to measure the charge state of the right dot. Bottom left: The measured coherent oscillations between the singlet state S and the triplet state T_0. The other triplet states were split from them with a magnetic field. (J.R. Petta *et al.*, *Science* **309**, 2180 (2005), Fig. 1), a qubit measured in the Marcus Lab, Harvard University. Reprinted with permission from AAAS.

Overview

As shown in this chapter, quantum dots are very versatile objects with a rich physics. Although there are a number of limits where the transport properties are quite poorly known, the theories for the better understood limits apply for many types of systems: molecules, carbon nanotubes, graphene quantum dots and 'artificial atoms' made by semiconductor technology. The physics of these systems is still under intense investigation: on the one hand, the better understood effects are searched for in ever new types of materials posing different types of relations between the fundamental scales, such as $\Gamma, k_B T, \Delta\epsilon$ and U, or the transport can be used for the level spectroscopy of the quantum dots, and on the other hand the fundamental phenomena are still researched actively: an example of such a novel phenomenon is the two-channel Kondo effect (Potok *et al.*, 2007).

The quest for realizing quantum computing in quantum-dot structures has uncovered also other types of physics relevant for these structures. For example, studying one of the most relevant decoherence mechanisms—that due to the coupling of the electrons to the fluctuating nuclear spins—has revealed the possibility of mediating magnetic interactions between the nuclei, and for example controlling their transition into a ferromagnetic state (Żak *et al.*, 2012).

Further reading

- A detailed theory of transport through quantum dots in various different regimes, employing non-equilibrium Green's function techniques, is discussed in (Bruus and Flensberg, 2004).
- Many of the specific properties of quantum dots are discussed in reviews, such as (Kouwenhoven *et al.*, 2001; van der Wiel *et al.*, 2002; Hanson *et al.*, 2007).

- Molecular electronics systems can be often regarded as specific realizations of quantum dots. These are described in (Cuevas and Scheer, 2010).

Exercises

(8.1) Estimate the magnitude of the energy level spacing in a) a metallic Cu island of length 1 μm, width 200 nm and thickness 50 nm, b) single-wall carbon nanotube (SWNT) of length 1 μm, and c) a fullerene C_{60} molecule. The density of states in Cu at the Fermi level is of the order of $N_0 \approx 10^{47}$ J^{-1} m^{-3}.

A metallic SWNT has four degenerate propagating modes at the Fermi level. The energy level spacing can be obtained from the longitudinal modes. The Fermi velocity of SWNTs is of the order of 10^6 m/s. (See for example (Lemay *et al.*, 2001).) To produce a rough idea for the level spacing in C_{60}, approximate the molecule by a rectangular box with size $d \times d \times d$, $d \approx 1$ nm being the approximate diameter of C_{60}. Use the fact that a neutral C_{60} has sixty conduction electrons. To get an idea of when the level spacing is relevant, compare it to the 'thermal' energy scale $K \cdot k_B$.

(8.2) Prove the normalization of the spectral function, eqn (8.5).

(8.3) In the mean field model, the average charge on the quantum dot increases with an increasing potential μ according to an effective capacitance

$$C^* = e^2 \frac{d \langle n \rangle}{d \mu}, \qquad (8.43)$$

where $\langle n \rangle$ is the average number of electrons on the quantum dot. a) Find C^* for the case of a quantum dot with a single interacting spin-degenerate level. b) Find C^* in the limit $U \gg \Gamma, \epsilon_l - \mu$. Denote this by C_g. c) Find C_d such that the general result C^* can be written as a series capacitance of C_g and C_d.

(8.4) Using the limits eqn (8.32) and (8.35), show that for a metallic dot the conductance peak has the form

$$G(\Delta_{\min}) = G_{\text{met}} \frac{\Delta_{\min}/k_B T}{\sinh\left(\frac{\Delta_{\min}}{k_B T}\right)}. \qquad (8.44)$$

and for the two-level dot it is

$$G(\Delta_{\min}) = \frac{G_{2l}}{\cosh^2\left(\frac{\Delta_{\min}}{2k_B T}\right)} \qquad (8.45)$$

What are G_{met} and G_{2l}?

(8.5) Compare eqn (8.45) to the the self-consistent mean-field conductance

$$G_{\text{mf}} = -\frac{2e^2}{h} \int_{-\infty}^{\infty} dE\, T(E) f'(E - \mu), \qquad (8.46)$$

in the limit $\Gamma \ll k_B T$. Here $f(E) = 1/(\exp(E/k_B T) + 1)$ is the Fermi function and $T(E)$ can be obtained using eqns (8.13) and (8.17). Note that in the specified limit you can take $\Gamma \to 0$ inside the spectral function. Remember to include the spin degeneracy in eqn (8.45).

(8.6) **Avoided level crossing.** Consider a system of two coupled energy levels, ϵ_1 and ϵ_2, with a coupling amplitude t. Such a system is thus described by the Hamiltonian

$$H = \epsilon_1 \hat{d}_1^\dagger \hat{d}_1 + \epsilon_2 \hat{d}_2^\dagger \hat{d}_2 + t \hat{d}_1^\dagger \hat{d}_2 + \text{h.c.} \qquad (8.47)$$

Diagonalize this Hamiltonian by introducing $\hat{d}_{b/a} = a\hat{d}_1 \pm b\hat{d}_2$ and finding the scalars a, b that make the Hamiltonian diagonal. Show that the eigenenergies $\epsilon_b < \epsilon_a$ satisfy

$$\begin{aligned} \epsilon_b(\epsilon_1 - \epsilon_2 \ll |t|) &= \epsilon_a(\epsilon_1 - \epsilon_2 \gg |t|) = \epsilon_1 \\ \epsilon_a(\epsilon_1 - \epsilon_2 \ll |t|) &= \epsilon_b(\epsilon_1 - \epsilon_2 \gg |t|) = \epsilon_2. \end{aligned} \qquad (8.48)$$

What is the energy splitting $\epsilon_a - \epsilon_b$ when $\epsilon_1 = \epsilon_2$? Using this information, sketch the behaviour of the eigenenergies when ϵ_1 and ϵ_2 cross as a function of some control parameter (say, gate voltage or a magnetic field in a (double) quantum dot).

(8.7) Find the eigenenergies of the spin states in eqns (8.41) and (8.42) in the presence of a magnetic field B. Include the single-electron level spacing $\Delta\epsilon$, the inter-dot coupling energy J and the Zeeman energy $E_Z = \pm g\mu_B B/2$, which is positive for spin down and negative for spin up. Plot these energies and find the condition when one of the triplet states becomes degenerate with the singlet state.

(8.8) **Question on a scientific paper.** Take the paper Elzerman *et al.*, *Nature* **430**, 431 (2004), which shows how a spin-state of a quantum dot can be mapped into a charge signal. Consider first Fig. 1. a) Explain how the spin-selective tunnelling can be realized by the use of a magnetic field. b) How does the quantum point contact have to be gated to obtain maximal resolution to the charge state of the quantum dot? What determines this resolution? c) Figure 2 shows the spin-read-out scheme and Fig. 3a the measured QPC response. Why does the lower curve in Fig. 3a constitute a readout of spin down? d) Find out how this measurement can be used to find the spin lifetime (Fig. 3c) in the quantum dot.

Tunnel junctions with superconductors

9

The simplest mesoscopic systems with superconducting elements are the tunnel junctions between two superconductors (SIS system) or a superconductor and a normal metal (SIN or NIS). In the former there are essentially two regimes of interest: at a voltage-biased junction the main measured effect is the quasiparticle tunnelling between the superconductors. This process is governed mainly by the superconducting density of states. At a vanishing bias voltage across the junction, or at low voltages in some cases, the Josephson effect is the dominating phenomenon. In the latter system (NIS) only the quasiparticle current can be observed. In this chapter we focus on these processes and find that they lead to a multitude of effects that can be used, e.g., for thermometry, cooling, accurate charge pumping, magnetic field sensing and quantum computing.

This chapter separates into two more or less independent parts: Secs. 9.1 and 9.2 discuss effects related to quasiparticle current, and Secs. 9.3 and 9.4 concentrate on Josephson junctions and the quantum effects in them.

9.1 Tunnel contacts without Josephson coupling

9.1.1 NIS contact

The current through a NIS contact follows eqn (1.12) with one of the densities of states assumed constant, $N_R(E) = N_F^R$, and the other one assumed of the form of eqn (5.18), i.e.,

$$I = \frac{1}{eR_T N_F^S} \int_{-\infty}^{\infty} dE N_S(E)[f(E, T_R) - f(E + eV, T_L)]. \quad (9.1)$$

This cannot be analytically integrated for an arbitrary voltage and temperature, but its numerical evaluation is straightforward; see Fig. 9.1. We can see that in the limit $T \to 0$ the form of the differential conductance dI/dV turns towards the superconducting density of states. Indeed, it is obtained by

$$\frac{dI}{dV} = \frac{1}{R_T} \int_{-\infty}^{\infty} \frac{N_S(E)}{N_F^S} \left[-\frac{\partial f(E + eV)}{\partial(eV)} \right] dE. \quad (9.2)$$

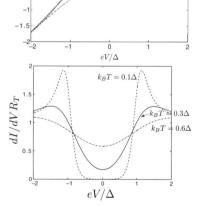

Fig. 9.1 Top: Current–voltage curve of a NIS junction at three different temperatures. Bottom: Corresponding differential conductance.

The function $-\partial f(E+eV)/\partial(eV) = [4k_BT\cosh^2((E+V)/(2k_BT))]^{-1}$ is a Bell-shaped curve peaked at eV and with a width $\sim 4k_BT$. Therefore with $T \to 0$, we get

$$\frac{dI}{dV} \overset{T \to 0}{\to} \frac{N_S(e|V|)}{N_F^S R_T}. \tag{9.3}$$

This allows for a measurement of the superconducting density of states via the differential conductance.

As discussed in Sec. 9.2 below, NIS junctions are versatile objects, allowing, for example, to perform local electron thermometry and cooling.

9.1.2   SIS contact

In the absence of Coulomb blockade effects and supercurrent, the current through a SIS contact is obtained from

$$I = \frac{1}{eR_T N_F^L N_F^R} \int_{-\infty}^{\infty} dE N_S(E) N_S(E+eV)[f(E,T_L) - f(E+eV,T_R)]. \tag{9.4}$$

Now a current can flow if the voltage exceeds twice the magnitude of the gap. This can be explained using the *semiconductor model* depicted in Fig. 9.2. In this picture the current flows horizontally from left to right, and in order for a finite current to flow, the electron has to leave from an occupied state and enter an unoccupied one. At a low temperature this is possible only if the difference in the Fermi levels of the two superconductors exceeds 2Δ. The SIS current–voltage curves are shown in Fig. 9.3.

9.1.3   Superconducting SET

Superconductivity modifies the behaviour of single-electron transistors. The lowest-order effects are twofold. Firstly, the presence of the gap in the energy spectrum gives rise to an additional energy scale for the transition rates and for the charging energy corresponding to single-electron (single-quasiparticle) excitations. Secondly, the presence of Josephson

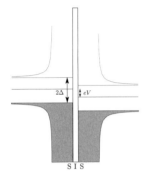

Fig. 9.2 Semiconductor model for the SIS junction. At low temperatures current flow is allowed only if $eV > 2\Delta$.

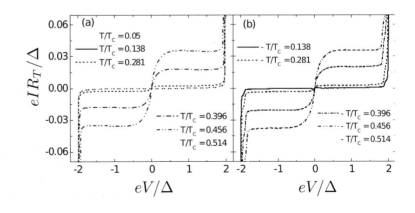

Fig. 9.3 (a) Calculated and (b) measured current–voltage curves of a SIS tunnel junction at a few temperatures. Supercurrent has been suppressed. Courtesy of Alexander Savin and Jukka Pekola.

coupling induces a contribution to the current, as the Cooper pairs tunnel through the structure. The latter effect is described in Sec. 9.4.

The modification of the quasiparticle tunnelling rates is fairly straightforward. The tunnelling rate for a junction between normal metals, the second line in eqn (7.15), is derived assuming an energy-independent density of states. The presence of superconductivity changes the relation to

$$\Gamma_{LI}(n) = \frac{1}{e^2 R_L}$$
$$\times \int_{-\infty}^{\infty} d\epsilon_k \int_{-\infty}^{\infty} d\epsilon_q f_L(\epsilon_k) \frac{N_L(E) N_I(E)}{N_L^0 N_I^0} [1 - f_I(\epsilon_q)] \delta(\delta E_{ch} + \epsilon_q - \epsilon_k),$$
$$(9.5)$$

where $N_{L/I}(E)$ should be replaced by eqn (5.18) for the superconducting parts of the transistor. This integral can no longer be performed in closed form, but it can be represented using the current $I_t(V)$ of a non-interacting case, eqn (9.1) and Fig. 9.1 for the NIS, and eqn (9.4) and Fig. 9.3 for the SIS case:

$$\Gamma_{LI}(n) = \frac{1}{e} I_t \left(\frac{\delta E_{ch}}{e} \right) \frac{1}{\exp[\delta E_{ch}/(k_B T)] - 1}. \qquad (9.6)$$

The presence of the gap gives rise to an additional suppression of the quasiparticle tunnelling rates compared to the NIN case. In Exercise 9.1 you will find the threshold voltage for the quasiparticle current in different types of superconducting SETs.

For a single junction (NIS or SIS), the environmental Coulomb blockade on the quasiparticle current can be described similarly to the NIN case; i.e., by replacing the $\delta(E)$-function in eqn (9.5) by the function $P(E)$ characterizing the environment. As explained in Complement 9.7, the dynamical Coulomb blockade in Josephson junctions is qualitatively different.

Complement 9.1 Higher order effects

In the second order in the tunnel coupling, superconductivity gives rise to additional effects to cotunnelling. These are Andreev reflection in NIS junctions,[1] and Josephson-quasiparticle processes in SISIS-SETs. The latter combine quasiparticle and Cooper pair tunnelling and take place in two different types of cycles: the Josephson-quasiparticle (JQP) and double Josephson-quasiparticle (DJQP) cycles (see Figs. 9.4 and 9.5).[2] In these cycles one or more Cooper pairs tunnel through one of the junctions, and two quasiparticles tunnel through the other junction. These processes are resonant (i.e., happen only for a restricted region of bias voltages), and thus they give rise to peaks in the current–voltage characteristics. The Josephson-quasiparticle processes in superconducting SETs have been employed, for example, in the measurement of charge qubits discussed in Sec. 9.4.3 and for cooling nanoelectromechanical resonators (Naik *et al.*, 2006).

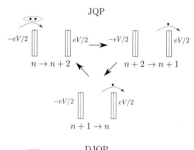

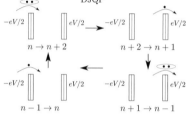

Fig. 9.4 Schematic illustrations of Josephson-quasiparticle (JQP) and double Josephson-quasiparticle (DJQP) cycles. The numbers indicate the changes in the charge state in the processes.

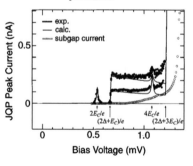

Fig. 9.5 Measured current–voltage curve of a superconducting SET showing the JQP peaks at $V = 2E_C/e$ and $V = 4E_C/e$. (Y. Nakamura *et al.*, *Phys. Rev. B* **53**, 8234 (1996), Fig. 4.) © 1996 by the American Physical Society.

[1]For a description of this process, see Ch. 3 in (Dittrich *et al.*, 1998).

[2]For details, see for example (Nakamura *et al.*, 1996) and (Clerk *et al.*, 2002).

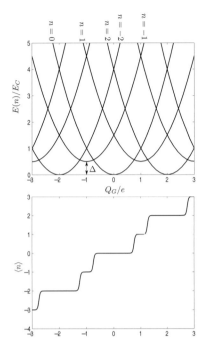

Fig. 9.6 Parity effect in the charging energy (top) and the average charge (bottom) as a function of the gate charge. Here $\Delta = E_C/2$.

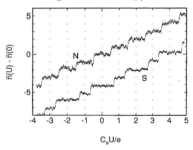

Fig. 9.7 Parity effect measured in the charge of a superconducting single-electron transistor island. N denotes the case of an island in the normal state and S in the superconducting state. (P. Lafarge *et al.*, *Phys. Rev. Lett.* **70**, 994 (1993), Fig. 4.) © 1993 by the American Physical Society.

[3] See (Schön and Zaikin, 1994) or Ch. 3 in (Dittrich *et al.*, 1998).

Complement 9.2 Parity effect

If there is an odd number of conduction electrons in a superconductor, not all of them can be paired to Cooper pairs. The unpaired electron necessarily has an excess kinetic energy of at least Δ (and because of the BCS divergence, most of the excess electron states are near this). Therefore, in effect the charging energy obtains an extra contribution depending on the parity of the electron number,

$$E_{\mathrm{ch}}(n, Q_G) = \frac{(ne - Q_G)^2}{2C} + \Delta_n, \qquad (9.7)$$

where $\Delta_n = 0$ for even n and $\Delta_n = \Delta$ for an odd n. Here Q_G is the gate charge and C is the sum capacitance as discussed in Ch. 7. This function and the corresponding ground-state charge are plotted in Fig. 9.6. As a result, the charge, and correspondingly the current, are no longer a e-periodic but $2e$-periodic functions.

The parity effect in the current was first measured by Tuominen *et al.* (1992) and in the charge by Lafarge *et al.* (1993) (see Fig. 9.7). This effect can be observed at low temperatures, $T < T_{cr}$, with $T_{cr} \ll \Delta/k_B$. Above T_{cr}, the usual e-periodic conductance is obtained. The crossover temperature can be roughly estimated by following what happens to the odd, unpaired, quasiparticle. Let us denote the rate of tunnelling out of the island for the unpaired quasiparticle by γ, and that for the other, paired quasiparticles which can tunnel out by Γ. The latter consists of an effective number $N_{\mathrm{eff}} \gg 1$ of electrons defined in eqn (9.8) below—for typical mesoscopic islands $N_{\mathrm{eff}} \approx 10^4$. Within the Coulomb blockade, Γ is exponentially suppressed but has the large prefactor N_{eff}. On the contrary, γ has a small prefactor but may not contain the exponential prefactor due to the excitation energy $\approx \Delta$ released in the tunnelling process. A more precise analysis based on analyzing the regime of temperatures where the potential shift caused by the extra quasiparticle exceeds the temperature[3] shows that the number of other electrons is temperature dependent,

$$N_{\mathrm{eff}} = N_I^F \Omega_I \sqrt{2\pi \Delta k_B T} = \frac{\sqrt{2\pi \Delta k_B T}}{\delta_I}, \qquad (9.8)$$

where N_I^F is the normal-metal density of states of the island at the Fermi energy, Ω_I is the volume, and δ_I is the energy level spacing of the island.

The parity effects survive as long as this single-electron rate is observable, i.e., $\gamma \geq \Gamma$. The crossover temperature can thus be found from $N_{\mathrm{eff}}(T_{cr}) = e^{\Delta/(k_B T_{cr})}$. As T_{cr} depends on the volume of the island only logarithmically, for typical Al islands ($N_F = 1.45 \times 10^{47}$ m^{-3}J^{-1}, $\Omega_I = 10^{-22} \ldots 10^{-20}$ m^3) $N_{\mathrm{eff}} = 200 \ldots 20000$ and this crossover temperature is between 200 and 300 mK.

The suppression of the parity effect and other effects related with non-paired quasiparticles on superconductor islands are often called *quasiparticle poisoning*.

9.2   SINIS heat transport and pumping

The presence of the superconducting gap makes the tunnel junctions with superconducting parts non-linear: electronic conductance of the junction depends on voltage and on temperature. This mere fact allows the use of NIS junctions for thermometry. However, not only measuring the temperature, biasing these systems with a constant dc voltage also allows manipulation of the electron heat flow and thereby the temperature, either cooling or heating the electrons on the normal side of the junction. And, as I show towards the end of this section, the charging effects in small SINIS junctions may be used for an efficient pumping of electrons via application of an alternating gate voltage.

9.2.1   Thermometry with (SI)NIS junctions

Consider the current–voltage (I–V) curve of the NIS junction presented in Fig. 9.1 with a fixed current instead of a fixed voltage. The measured voltage turns out to be a monotonous function of the temperature (see Fig. 9.8), and can hence be regarded as a thermometric quantity. Contrary to the Coulomb blockade thermometer introduced in Ch. 7, NIS junction is a secondary thermometer: the measured voltage for a given current depends on the resistance R_T of the tunnel junction and the gap Δ of the superconductor. However, it is quite versatile and is easily manufactured, and the back-action on the electron system from the measurement can be made quite small, since heat flow into the superconductors is blocked at sub-gap energies. In Exercise 9.3 you will show that the NIS current–voltage curve is independent of the temperature of the superconductor—a clear benefit for the temperature measurement. For example, the temperatures shown in Fig. 9.11 were measured with a NIS thermometer.

Besides working as a thermometer, NIS junctions have also been used to measure the full non-equilibrium distribution function of electrons, as illustrated in Example 2.4.

As indicated by their quasiparticle I–V curve, Fig. 9.3, also SIS junctions can be used for thermometry. However, in this case the supercurrent through the junction, in small junctions often showing up also at finite voltages (see Complement 9.7 below), has to be suppressed by making two adjacent SIS contacts, joining the superconductors into a loop, and tuning the magnetic flux close to half a flux quantum (see the description of SQUIDs in Sec. 9.3.1).

9.2.2   Electron cooling and refrigeration

As indicated in Fig. 9.1, the charge current through a NIS junction is at voltages smaller than the gap Δ greatly reduced compared to the corresponding NIN tunnel junction. A similar reduction can be found also for the heat current. However, the behaviour of the heat current

$$\dot{Q}_{\mathrm{NIS}} = \frac{1}{e^2 R_T} \int dE (E - eV) \frac{N_S(E)}{N_F} (f_N(E - eV) - f_S(E)) \quad (9.9)$$

through a NIS junction as a function of the bias voltage between the two is quite peculiar: at low bias voltages V, the heat current from the normal metal is positive, indicating cooling of the normal metal. Only for $e|V| \gtrsim |\Delta|$ the sign of $\dot{Q}_{\mathrm{NIS}}$ changes to negative, as one would get by including only the Joule heating effect. This behaviour is shown in Fig. 9.9.

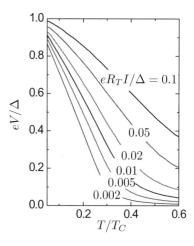

Fig. 9.8 NIS voltage for a given current I at different temperatures of the electron system.

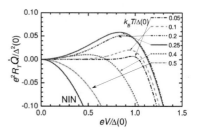

Fig. 9.9 Heat current through a NIS junction as a function of the bias voltage V for different temperatures. Also shown is the strictly negative heat current for a NIN junction.

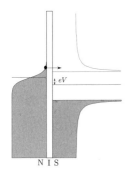

Fig. 9.10 Semiconductor model for a NIS junction shows the principle of the cooling effect: only high-energy ('hot') excitations can escape from the normal metal into the superconductor.

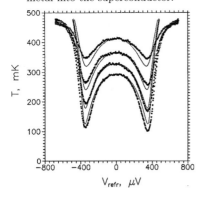

Fig. 9.11 Example of a measured $T_N(V)$ in a SINIS system and the electron temperature on the island as a function of the bias voltage. The curves are for different temperatures of the refrigerator (essentially equal to T_S and T_{ph}). Dots are experimental data and lines are theoretical fits to eqn (9.10). (M.M. Leivo *et al.*, *Appl. Phys. Lett.*, **68**, 1996 (1996), Fig. 2.) © 1996, American Institute of Physics.

[4]Typical nomenclature is to refer to cooling when the aim is to fix the operating temperature by tackling unwanted heating effects, and refrigeration when trying to lower the temperature. However, this is not strictly followed even in this book.

The principle of NIS cooling can be explained by following the semiconductor model as illustrated in Fig. 9.10. At low bias voltages the electrons with low energies $E < |\Delta| - eV$ cannot tunnel into the superconductor. However, at finite temperatures there are electrons on the high-energy states which can tunnel. The result of this energy-selective tunnelling is cooling of the normal metal.

As $\dot{Q}_{NIS}$ is quadratic in the bias voltage, this cooling is not similar to the usual Peltier effect (see Sec. 2.8), for which the sign of the heat current depends on the sign of the bias voltage. Rather, because $\dot{Q}$ is insensitive to the sign of the voltage, the cooling effect can be enhanced by connecting two NIS junctions in series, to realize a SINIS (superconductor–insulator–normal-metal–insulator–superconductor) system. In this case the normal metal forms an 'island' whose temperature can be lowered by a proper biasing of the structure. The resultant temperature T_N of the electrons on this island is obtained by a heat-balance equation

$$2\dot{Q}_{NIS}(V; T_N, T_S) + \dot{Q}_{e-ph}(T_N, T_{ph}) = 0, \qquad (9.10)$$

where $\dot{Q}_{e-ph}$ describes the heat transport to the phonon system at temperature T_{ph} (see eqn (2.39)) and the prefactor 2 in the first term is due to the presence of two junctions. This heat balance thus depends on two fixed temperatures T_S and T_{ph} of the electrons in the superconductor and of the phonons, and of the bias voltage V. In some cases more elements need to be included into the heat-balance model, such as the coupling between the phonons in the film and those in the substrate, but most often eqn (9.10) is sufficient. Figure 9.11 shows an example of the measured T_N vs. V in a SINIS junction. Up to some constraints the measured $T_N(V)$ curves correspond rather well to the temperature T_N obtained from eqn (9.10).

SINIS refrigerators are especially efficient for lowering the electron temperature on the normal-metal island; the lattice temperature does not typically change in these experiments because of their stronger coupling to the substrate. One of the present goals is to use the SINIS as a targeted microrefrigerator which could also refrigerate[4] the lattice. This requires a strong isolation of the phonons from the substrate and can be accomplished by building suspended membranes. An example of such a device is presented in Fig. 9.12. There, a 'macroscopic' cube with size of a few hundred μm was refrigerated from the starting temperature of 320 mK to 240 mK.

Complement 9.3 Brownian refrigeration
Biasing the SINIS with a constant dc bias voltage is not the only way to find electron cooling (although probably it is the easiest method). It was recently shown theoretically that even a heated resistor in the environment of the SINIS can lead to a cooling of the normal island (Pekola and Hekking, 2007). In this case, the NIS heat current from the normal island to the superconductor is obtained from

$$\dot{Q}_{NIS} = \frac{2}{e^2 R_T} \int dE' dE \, E' \frac{N_S(E)}{N_F} f_N(E')[1 - f_S(E)] P(E' - E) \qquad (9.11)$$

a)

b)

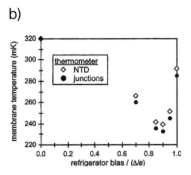

Fig. 9.12 a) Scanning electron micrograph of a NIS refrigerator and the suspended membrane which is refrigerated. The 'large' cube with a side of $250\ \mu m$ is a germanium resistance thermometer. The circle shows the location of one of the NIS junctions. b) Temperatures of the lattice and the electrons vs. the bias voltage over the NIS junctions. (A.M. Clark *et al.*, *Appl. Phys. Lett.* **86**, 173508 (2005), Figs. 2 and 3.) © 2005, American Institute of Physics.

and the heat current from the superconductor into the normal metal is

$$\dot{Q}_{SIN} = \frac{2}{e^2 R_T} \int dE' dE\, E \frac{N_S(E)}{N_F} f_S(E)[1 - f_N(E')]P(E - E'). \quad (9.12)$$

Here the function $P(E)$ describes the emission or absorption of the energy to/from the electromagnetic environment of the SINIS device as discussed in Sec. 7.5. An illustration of this effect as a function of the temperature T_R of the environment is shown in Fig. 9.13. Note that in this case the bias voltage across the SINIS vanishes.

This type of a Brownian refrigerator is unlikely to become the method of choice for electron refrigeration. However, its significance lies in its relation to basic thermodynamics: it is a realization of the famous Maxwell's demon. In Maxwell's original suggestion, this demon acts between two subsystems, A and B, allowing only the hot particles to pass from container A to B and the cold particles to pass from container B to A. In this process, container B heats up and container A cools down. As the excess heat of container B can be used for work, this seems to be inconsistent with the second law of thermodynamics. However, taking into account the work done by the demon on the system shows that the total entropy of the whole system has to increase and therefore this process does not violate the second law. A SINIS Brownian refrigerator operates exactly like a Maxwell's demon: it takes energy from the environment to separate the highly energetic electrons (going to the superconductors) from the less energetic ones (staying in the normal metal). However, the total entropy change is also here positive due to the work done by the electromagnetic environment.

Fig. 9.13 Cooling power calculated from eqn (9.11) as a function of the relative temperatures T_R/T_N of the environment and of the normal island for different ratios E_C/Δ. Here the environment is described by a large resistance, producing a Gaussian $P(E)$ as in eqn (7.48). The figure has been calculated for $T_N = T_S = 0.03\Delta/k_B$. Note that the cooling power has a peak at around $T_R/T_N \approx \Delta/E_C$.

Complement 9.4 Electronic refrigerator with charging effects
Apart from the dynamical Coulomb blockade effects, effects related to the direct charge quantization on the SINIS island can also be found if the junctions have a sufficiently low capacitance. Generally a finite charging energy E_C leads to a decreasing cooling power, and hence for the direct biased refrigeration small junctions are not optimal. However, the charging effects lead to the possibility of operating the cooler with other types of control variables; for example, the gate voltage.[5] Figure 9.14 shows an example of a heat transistor where the heat flow can be controlled with the gate voltage. The theoretical modelling of such a device can be carried out by solving the master equations for the SINIS charge state by using the tunnelling rates given by eqn (9.6) and then calculating the heat flow separately for each charge state. The latter are

[5]The electronic refrigeration with a radio-frequency gate voltage (a single-electron refrigerator) was discussed theoretically in (Pekola *et al.*, 2007) and later demonstrated in (Saira *et al.*, 2007) and (Kafanov *et al.*, 2009). This device was called a heat transistor in analogy to the conventional field-effect transistor.

Fig. 9.14 a) Scanning electron micrograph of the SINIS heat transistor and its measurement setup. In the middle is the copper island, and the four superconducting (S) contacts are made of aluminium. Two of the contacts can be used for controlling the heat current, and two of them for thermometry. b) Electron temperature vs. bias voltage V_{DS} extracted from the measurements of the heat transistor. The two curves are for two values of the gate voltage, the squares are for an 'open gate' where the gate charge is at the degeneracy point, $Q_g = e/2$, and the triangles denote the 'closed gate' $Q_g = 0$. (O.-P. Saira *et al.*, *Phys. Rev. Lett.* **99**, 027203 (2007), Figs. 1 and 3.) © 2007 by the American Physical Society.

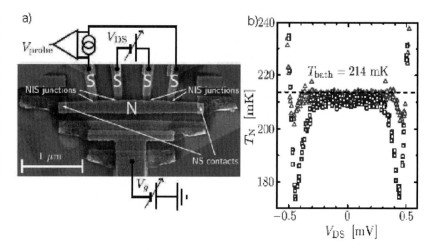

given by

$$\dot{Q}^{+} = \frac{1}{e^2 R_T} \int dE (E - \delta E_{\mathrm{ch}}^{+}) \frac{N_S(E)}{N_F} f(E; T_S)[1 - f(E - \delta E_{\mathrm{ch}}^{+}; T_N)]$$
$$\dot{Q}^{-} = \frac{1}{e^2 R_T} \int dE (E + \delta E_{\mathrm{ch}}^{-}) \frac{N_S(E)}{N_F} f(E; T_S)[1 - f(E + \delta E_{\mathrm{ch}}^{-}; T_N)]. \tag{9.13}$$

Here the first line refers to the heat flux for the incoming electrons and the latter for the outgoing ones. The energy changes for these processes in a symmetric SINIS are $\delta E_{\mathrm{ch}}^{+/-} = \pm 2E_C(n + Q_g/e \pm 1/2) \pm eV/2$. This system of equations (solving the master equation and the associated heat flows) has to be then coupled to a heat balance equation of the type eqn (9.10) to find the electronic temperature T_N.

Complement 9.5 Accurate charge pumping with SINIS devices
Apart from controlling heat flow, a SINIS device in the charging regime can be used to create very accurate electron pumping (Pekola *et al.*, 2008). This device is operated with a constant bias voltage V satisfying $k_B T < e|V| < |\Delta|$ and an alternating gate voltage $V_g = V_o + V_{g0}\cos(2\pi f t)$. The direct current is strongly suppressed below the gap voltage (the $I - V$ curves are similar to those of a SIS junction, see Fig. 9.3), but the bias voltage breaks the left–right symmetry, allowing for a finite average current in the direction of the voltage. The current is found to be very close to nef, where $n \sim C_g V_{g0}/e$ depends on the gate amplitude (see Fig. 9.15, which shows the characteristics of such a hybrid turnstile). The main advantages of this device over the normal-metal counterpart[6] are the simpler construction (only two junctions are required compared to the several junctions in the normal-metal case) and the enhanced operation frequency range, allowing for higher currents. The latter is due to the presence of the gap in the superconducting density of states; it reduces the probability for unwanted tunnelling events. Moreover, a single-island device is also less vulnerable to background charge fluctuations or, more precisely, their effect can to some extent be corrected. This is one of the most serious problems limiting the operation of charge-based devices.

[6] See the introduction in Sec. 7.6.3.

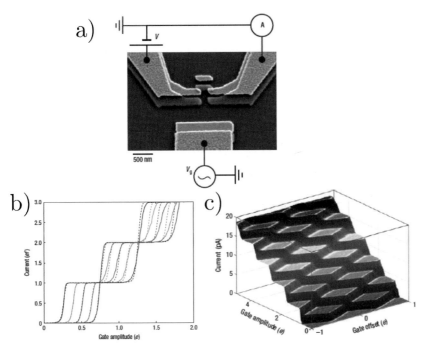

Fig. 9.15 a) Electron micrograph of a superconductor (aluminium)–normal-metal (copper) hybrid turnstile. b) Pumped current as a function of gate charge amplitude for five different gate offsets V_o. The plateaus correspond to the current $I = nef$. c) Similar to b), but with a varying gate offset. This figure shows the diamond-like structure of the pumped current. (J.P. Pekola *et al.*, *Nature Phys.* **4**, 120 (2008), Figs. 1 and 2.) Used by permission from Macmillan Publishers Ltd., © 2008.

9.3 Josephson junctions

Josephson junctions are versatile objects. They can be used to realize highly accurate voltmeters, magnetometers, thermometers or radiation detectors, to mention just a few of their applications. Combining the Josephson and charging effects, they can be used as realizations of quantum bits: qubits. The versatility of the Josephson junctions lies in the fact that they are non-linear circuit elements, but often in a very controlled manner (the non-linearity stems from the dc Josephson relation, eqn (5.28)).

In what follows I detail some of the basic properties of Josephson junctions and give examples of their use in magnetometry and quantum computing.

9.3.1 SQUIDs

The discussion on the Josephson effect in the previous chapter is based on the concept of the phase difference φ across the junction, without paying attention to how such a phase difference could be created. The microscopic theory of superconductivity tells that the phase is connected to the momentum of the superconducting condensate, i.e., of the Cooper pairs. Hence, a gauge-invariant way to treat the phase difference is to include the effect of the vector potential, such that the total gauge-

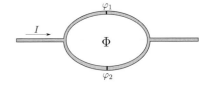

Fig. 9.16 Dc SQUID.

invariant phase difference between some points A and B is

$$\phi = \varphi_{AB} - \frac{2\pi}{\Phi_S} \int_A^B \mathbf{A} \cdot \vec{dl}. \tag{9.14}$$

Here it is also customary to use the flux quantum for Cooper pairs,

$$\Phi_S \equiv \frac{h}{2e} = \Phi_0/2. \tag{9.15}$$

If the magnetic field is sufficiently high to give rise to a flux exceeding a flux quantum through the junction area, this relation leads to the *Fraunhofer pattern* in the current–field relation[7]. However, in the following we assume that the fields are much lower.

[7]See Ch. 6 in (Tinkham, 1996).

Consider a superconducting loop containing two Josephson junctions and connected to an external circuit as shown in Fig. 9.16. This system is called a dc-SQUID (Superconducting QUantum Interference Device). The single-junction version is called an rf-SQUID, but I refer to books on superconductivity for the description of its behaviour. In a SQUID, the presence of the magnetic flux gives rise to a phase gradient within the ring. This in turn induces a circulating supercurrent, which screens the flux. The total flux Φ across the ring is a sum of the applied flux and that produced by the screening current. Now, starting from the upper junction, the total phase difference gathered within one cycle around the ring is

$$\phi_{\text{tot}} = -\varphi_1 + \frac{2\pi}{\Phi_S} \int_1^2 \mathbf{A} \cdot \vec{dl} + \varphi_2 + \frac{2\pi}{\Phi_S} \int_2^1 \mathbf{A} \cdot \vec{dl}. \tag{9.16}$$

Here the first integral is taken around the right side of the ring, and the second around the left. Including the small effect of the vector potential on the junctions, we obtain

$$\phi_{\text{tot}} = \varphi_2 - \varphi_1 + \frac{2\pi}{\Phi_S} \oint \mathbf{A} \cdot \vec{dl} = \varphi_2 - \varphi_1 + \frac{2\pi\Phi}{\Phi_S}. \tag{9.17}$$

The phase at a given position should be single-valued (up to 2π), i.e., the total phase difference around the ring should be $\phi_{\text{tot}} = 2n\pi$. This means that

$$\varphi_1 - \varphi_2 = \frac{2\pi\Phi}{\Phi_S} \quad (\text{mod } 2\pi). \tag{9.18}$$

Now the total current through this ring is given by

$$I = I_{c1}\sin(\varphi_1) + I_{c2}\sin\left(\varphi_1 - \frac{2\pi\Phi}{\Phi_S}\right), \tag{9.19}$$

where I_{c1} and I_{c2} are the critical currents of the two junctions. In the symmetric case $I_{c1} = I_{c2} \equiv I_c$, for a given flux Φ, the maximum external current that can flow as a supercurrent through this system is

$$I = 2I_c \left| \cos\left(\frac{\pi\Phi}{\Phi_S}\right) \right|. \tag{9.20}$$

Hence, the critical current of the ring can be modulated with a flux. Because of this property, the SQUIDs are very sensitive magnetometers.

9.3.2 Resistively and capacitively shunted junction model

If a current larger than the critical current I_C is applied across a Josephson junction, a voltage starts to develop across it. This voltage can be described with the resistively and capacitively shunted junction (RCSJ) model. Consider the circuit of Fig. 9.17. The Josephson junction is denoted with the cross inside a box.[8] It is in parallel with a shunt resistance R and its own capacitance C. Resistance R may be due to the quasiparticle current (in which case it would be dependent on the driving current), but it may also be just some additional shunt resistance placed in parallel with the junction.

The driving current I divides into three parts, $I = I_c \sin(\varphi) + I_C + I_R$. In the capacitive path the current is given by $I_C = \dot{q} = C\dot{V}$, and in the resistive path it is $I_R = V/R$, where V is the voltage over the circuit elements. Using the ac Josephson relation, eqn (5.29), this equality can also be written as

$$I = I_c \sin(\varphi) + \frac{\hbar}{2e}\dot{\varphi}/R + C\frac{\hbar}{2e}\ddot{\varphi}. \tag{9.21}$$

Let us make this equation dimensionless, dividing both sides by I_C and introducing a dimensionless time variable $\tau = \omega_p t$, the plasma frequency

$$\omega_p = \sqrt{2eI_c/\hbar C} \tag{9.22}$$

and the quality factor

$$Q = \omega_p RC. \tag{9.23}$$

With these definitions, eqn (9.21) turns into

$$\ddot{\varphi} + Q^{-1}\dot{\varphi} + \sin(\varphi) = I/I_c. \tag{9.24}$$

There is a mechanical analogue for this model. This equation can be seen as the equation of motion for a particle with mass $(\hbar/(2e))^2 C$ moving along the φ-axis in an effective potential

$$U(\varphi) = -E_J \cos(\varphi) - (\hbar I/(2e))\varphi, \tag{9.25}$$

where $E_J = \hbar I_c/(2e)$ is the *Josephson energy*. Because of its shape, this potential is called the tilted washboard potential (see Fig. 9.18). The second term in eqn (9.24) describes friction, which is consistent with an idea of the resistor as a dissipative element.[9]

Generally, the RCSJ model can be separated into overdamped ($Q < 1/2$) and underdamped ($Q > 1/2$) regimes, based on the behaviour of the phase within a single well of the potential (9.25). Assume we drive the phase to some initial value φ_0 with the external current, and then turn the external current suddenly off. If φ_0 is not $2n\pi$, corresponding to a minimum of $-E_J \cos(\varphi)$, the phase starts a damped oscillation in this potential, approaching the potential minimum (see Fig. 9.19 for an example). In the underdamped regime it oscillates many times before decaying into the minimum, and in the overdamped regime the decay time is less than the oscillation period.

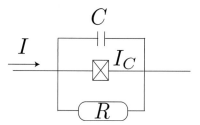

Fig. 9.17 Equivalent circuit for the RCSJ model with a frequency-independent shunt resistor.

[8]This is a symbol for a Josephson junction where the capacitance can be neglected. Including the capacitance would give rise to an additional line in the middle.

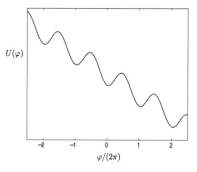

Fig. 9.18 Tilted washboard potential.

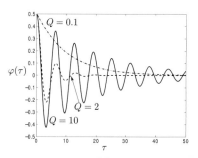

Fig. 9.19 Oscillations and decay of the phase within the RCSJ model, eqn (9.24), for different values of Q and with $\varphi(0) = 0.5$ and $\varphi'(0) = 0$. The phase does on average Q oscillations before the amplitude decays by a factor e^{-1}.

[9]Note that friction is the smaller the higher the parallel resistance is.

9.3.3   Overdamped regime

In the overdamped regime, the behaviour of the Josephson junction is regular, and an equation for the average voltage across the junction may be derived (see Exercise 9.5):

$$\langle V \rangle = R\sqrt{I^2 - I_c^2}\,\theta(|I| - I_c)\,\mathrm{sgn}(I). \qquad (9.26)$$

This current–voltage curve is plotted in Fig. 9.20 and is valid at a vanishing temperature.[10] At a non-zero temperature, there is a finite probability for the junction to 'escape' into the finite-voltage state at some lower current than I_c (see Sec. 9.3.5 below). This process makes the resultant I–V curve smoother, and the measured critical current (i.e., maximum current with which no voltage appears) lower than I_c.

[10]Compare this to the $I - V$ curve of a Coulomb blockaded junction, Fig. 7.13: the form of the curve is similar, but the roles of the current and voltage are interchanged. This is a signature of a duality between the two effects, as discussed, for example, in (Weiss, 1999).

9.3.4   Underdamped regime

When the quality factor is much larger than $1/2$, the $I - V$ curves become hysteretic. Upon increasing the current I from zero, the junction remains in the supercurrent-carrying state up to a current close to I_c, at which point it jumps discontinuously (at $T = 0$) to some finite value close to $2\Delta/e$. If I is now reduced, the junction does not enter the zero-voltage state before the current reaches a *retrapping current* $I_r \approx 4I_c/(\pi Q)$.[11] This hysteretic behaviour can be roughly explained with the mechanical analogue. Assume a particle is located in one of the potential wells. Tilting the potential, the particle escapes from the well only once the potential minima have almost vanished. The 'running state' of the particle ($\dot{\phi} = \frac{2eV}{\hbar} \neq 0$) corresponds to the finite voltage. Now, levelling the potential back towards the stable position does not immediately retrap the particle in one of the wells. The physical reason for the dependence $\propto 1/Q$ is that for retrapping, the energy fed by the driving current as the phase advances by 2π has to be dissipated at the same time.

[11]For a circuit model more complicated than that applied here, the switching behaviour probes Q at the plasma frequency and the retrapping at low frequencies corresponding to the measurement time scales.

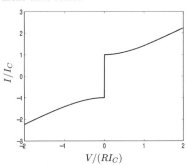

Fig. 9.20 I–V curve of an overdamped Josephson junction.

9.3.5   Escape process

The above description of the Josephson junction is effectively a zero-temperature description. At a finite temperature, the thermal fluctuations of the current cause an extra driving term in eqn (9.24),

$$\ddot{\varphi} + Q^{-1}\dot{\varphi} + \sin(\varphi) = I/I_c + \delta I(t)/I_c. \qquad (9.27)$$

This is a Langevin equation for the phase. The fluctuations are treated with the quantity $\delta I(t)$, which satisfies

$$\langle \delta I(t) \rangle = 0 \qquad (9.28)$$

and

$$\langle \delta I(t)\delta I(0) \rangle = \tilde{S}_I(t). \qquad (9.29)$$

For macroscopic shunt resistors, the Fourier transform of $S_I(t)$ satisfies the fluctuation–dissipation theorem, eqn (6.7),

$$S_I(\omega) \equiv \int dt\, e^{i\omega t} \tilde{S}_I(t) = \frac{2k_B T}{R}. \tag{9.30}$$

This is valid at low (plasma) frequencies $\hbar\omega_p \ll k_B T$.

The thermal fluctuations make it possible for the phase to thermally escape from the potential well, as schematically illustrated in Fig. 9.21. The rate for this escape process is obtained from the Arrhenius law,

$$\Gamma(I) = \frac{\omega_A(I)}{2\pi} \exp\left(-\frac{\Delta U(I)}{k_B T}\right). \tag{9.31}$$

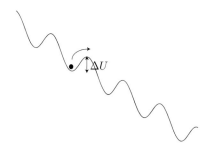

Fig. 9.21 Thermal escape of the 'phase particle' from a potential well.

Here, $\omega_A \approx \omega_p$ is the frequency that characterizes the dynamics of the phase particle inside a single potential well. The frequency $\omega_A/(2\pi)$ is hence an attempt frequency for the escape. On each attempt there is a probability $\exp(-\Delta U/(k_B T))$ of success, i.e., this is the probability that the instantaneous value of the current (including the fluctuations) exceeds the critical current. The attempt frequency has to be taken for the tilted potential, and it reads (see Exercise 9.6)

$$\omega_A = \omega_p[1 - (I/I_c)^2]^{1/4}. \tag{9.32}$$

The height of the potential well depends on the driving current, vanishing at I_c. In Exercise 9.6 you will show that this is

$$\Delta U = \frac{8\sqrt{2}}{3} E_J \left(1 - \frac{I}{I_c}\right)^{3/2}. \tag{9.33}$$

This escape probability can be measured directly, as each escape of the junction produces an observable voltage pulse, of the order of $2\Delta/e$. Measuring the state of the junction for time τ, the probability for escape is

$$P_e(\tau; I) = \exp(-\Gamma(I)\tau). \tag{9.34}$$

A typical escape probability as a function of the bias current I_b constitutes an S-shaped curve, where P increases from zero to unity within a width $\sim (k_B T)^{2/3}$ somewhat below I_c (see Fig. 9.22).

When the temperature is lowered below $\hbar\omega_p/k_B$, quantum fluctuations should be included in the description of the escape process. In this regime the escape rate is no longer governed by eqn (9.31), but it becomes independent of temperature: rather than hopping over the potential barrier, the phase particle tunnels through it. This is the regime of *macroscopic quantum tunnelling*, as it governs the behaviour of a collective quantum variable ϕ, which describes a quantity containing a macroscopic amount of particles. Macroscopic quantum tunnelling in Josephson junctions was first observed in the early 1980s (Voss and Webb, 1981; Devoret *et al.*, 1985).

As the escape probability is strongly dependent on the strength of current fluctuations, it can be also used to measure them.[12] The escape

[12]This measurement is an example of an 'on-chip' noise measurement, which has some advantages compared to the traditional measurement schemes utilizing macroscopic measuring setups since the frequency bandwidth of on-chip measurement is not limited by the measuring wires and can hence be very large. This scheme has been used to measure both the second (Pekola *et al.*, 2005) and third moments (Timofeev *et al.*, 2007; Huard *et al.*, 2007) of driven fluctuations.

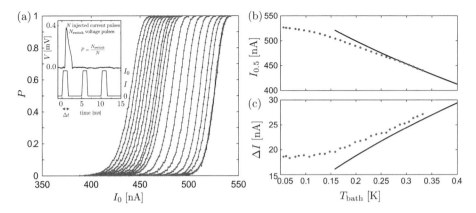

Fig. 9.22 Measured escape probabilities in a Josephson junction crossing over from the thermal activation regime to macroscopic quantum tunnelling. (a) Escape histograms, showing the probability P of escape vs. bias current I_0. Inset shows how the histogram is measured: the junction is biased for time τ with the current I_0, and the voltage over the junction, resulting from the escape, is measured. (b) Temperature dependence of the current that yields $P = 0.5$. (c) Width of the histogram vs. temperature. Both (b) and (c) show the crossover from thermal activation at $T \gtrsim 0.2$ K to macroscopic quantum tunnelling at $T \lesssim 0.2$ K. This data was measured in the Low Temperature Laboratory, Aalto University. Courtesy of Joonas Peltonen (Peltonen, 2011).

[13]In the overdamped limit, a closed-form equation for the thermally induced average voltage can be found; see (Ivanchenko and Zilberman, 1969).

process has also been used to measure the state of quantum bits, where the two states of the qubit can be translated into two different bias currents over the Josephson junction (Vion *et al.*, 2002).

As a result of the finite escape probability below the critical current, the average current–voltage curves are smeared. The actual details of the smearing depend on the value of the Q-factor, and we do not detail them here.[13]

9.4   Quantum effects in small Josephson junctions

In small Josephson junctions, the charging energy for the Cooper pair tunnelling through the junctions becomes essential. For single junctions, this shows up in the dynamical Coulomb blockade regime (see Complement 9.7 below), and for double junction systems or Cooper pair boxes comprising superconducting islands the orthodox theory of Coulomb blockade (Ch. 7) should be generalized to include the Josephson effect. In this case, a fully quantum-mechanical description of the junction dynamics is necessary. The first question in this case is on the form of the Hamiltonian. The potential energy of the Josephson junction is above shown to be of the form of eqn (9.25). The kinetic energy requires the identification of a canonical conjugate variable for the phase. This is the polarization charge Q on the junction(s), and the corresponding kinetic energy is the charging energy. Thus, the total Hamiltonian is[14]

[14]Note that replacing a single junction by a SQUID allows control of E_J separately with a magnetic field. In such setups there are thus two external control fields: gate charge Q_g and the magnetic field, allowing in principle for a full external control of the states. This is important for quantum computing.

$$H = \frac{(Q - Q_g)^2}{2C} - E_J \cos(\varphi) - \frac{\hbar}{2e} I_b \varphi, \qquad (9.35)$$

where I_b is the bias current. This conjugate pair of variables can be justified through the Hamilton equations for the Josephson junction (see Exercise 9.8), which produce the ac and dc Josephson relations. The theory for the quantized Josephson junction is defined by assuming that φ and Q are operators that satisfy the commutation relation,

$$[\varphi, Q] = 2ei. \tag{9.36}$$

The charge on this equation is naturally the charge of a Cooper pair.

The mechanical analogue of the tilted washboard potential turns into an analogue of a quantum-mechanical description of a particle in a periodic potential, following the correspondence described in Table 9.1.

Particle	Josephson junction
$H = \frac{p^2}{2m} - U\cos\left(\frac{x}{a}\right) - Fx$	$H = \frac{(Q-Q_g)^2}{2C} - E_J\cos(\varphi) - \frac{\hbar}{2e}I_b\varphi$
coordinate x	phase φ
momentum $p = -\frac{\hbar}{i}\partial_x$	$\propto$ charge $\frac{\hbar Q}{2e} = -2ei\partial_\varphi$
velocity $v = \frac{dx}{dt} = \frac{p}{m}$	$\propto$ voltage $\frac{2eV}{\hbar} = \frac{\partial\varphi}{\partial t} = \left(\frac{2e}{\hbar}\right)^2\frac{1}{C}\frac{\hbar Q}{2e}$
mass m	$\propto$ capacitance $\left(\frac{\hbar}{2e}\right)^2 C$
force F	$\propto$ bias current $\frac{\hbar}{2e}I_b$.

Table 9.1 Translation between the quantum theory of a particle in a periodic potential and the quantum theory of a Josephson junction.

The state of the junction is characterized by a macroscopic wave function Ψ. In the time-independent case it satisfies the Schrödinger equation

$$4E_C(-i\partial_\varphi - Q_g/e)^2\Psi_n - E_J\cos(\varphi)\Psi_n = E_n\Psi_n. \tag{9.37}$$

This is called the Mathieu differential equation, and its eigenfunctions Ψ_n are Mathieu functions.[15] In the next two subsections, I consider its eigenenergies in two opposite limits $E_J \gg E_C$ and $E_J \ll E_C$.

9.4.1   'Tight-binding limit'

In the tight-binding limit $E_J \gg E_C$ the lowest eigenstates are peaked around the minima of the potential well, i.e., $\varphi \approx 2n\pi$. In this case we can find the eigenenergies by expanding the potential around one of the minima, $\cos(\varphi) \approx 1 - \varphi^2/2$. As a result, we get the Hamiltonian of a Harmonic oscillator. For example, for $Q_g = 2ne$ it is

$$H\Psi_n = -4E_C\partial_\varphi^2\Psi_n + \frac{1}{2}E_J\varphi^2\Psi_n = E_n\Psi_n. \tag{9.38}$$

---

[15]See   http://mathworld.wolfram.com/MathieuDifferentialEquation.html   and http://mathworld.wolfram.com/MathieuFunction.html. The easiest way to compute its eigenvalues in the context of Josephson junctions is using the charge basis in eqn (9.43).

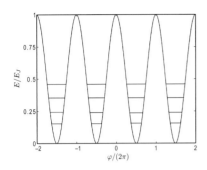

Fig. 9.23 Energy levels in the tight-binding limit $E_J \gg E_C$.

[16] Note that here we totally disregard the state with an odd number of quasiparticles, and n refers directly to the number of Cooper pairs on the island. This limit is justified for $\Delta > E_C$ and for temperatures below the crossover temperature T_{cr} for parity effects.

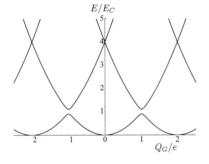

Fig. 9.24 Coulomb parabola for Cooper pairs in the case $E_J = 0.2E_C$.

[17] See Exercise 8.6.

The solutions can be written in terms of the Hermite polynomials as explained in the basic courses on quantum mechanics. The eigenenergies are

$$E_n = (n + 1/2)\hbar\omega_p, \tag{9.39}$$

where ω_p is the plasma frequency (see Exercise 9.7 and Fig. 9.23).

9.4.2 'Nearly free-electron limit'

In the weak-coupling case $E_J \ll E_C$ it is useful to start the description from the case of a completely isolated charge, $E_J = 0$. In this case the solutions to eqn (9.37) are plane waves corresponding to different fixed charges Q,

$$\Psi_Q(\varphi) = A_Q e^{i\frac{Q}{2e}\varphi}. \tag{9.40}$$

This is the analogue of the electron plane wave $\sim e^{i\vec{p}\cdot\vec{x}}$. The charge Q is generally of the form $Q = 2en + q$, where q denotes the polarization charge. For the different n we get the Coulomb parabola, depicted in Fig. 9.24.[16]

The polarization charge can be controlled in a geometry of a Cooper pair box, i.e., the single-electron box (see Fig. 7.4) with superconducting electrodes, with a finite Josephson coupling between the island and the left lead. Therefore, we concentrate on this case in the following.

It turns out that the introduction of a small $E_J \ll E_C$ gives rise to an avoided crossing[17] of the energy levels corresponding to the different n. In this limit, it is useful to write the Hamiltonian in terms of the basis functions corresponding to the different charge states $|n\rangle$. This Hamiltonian is obtained by calculating the matrix elements of the original Hamiltonian, eqn (9.35), in the basis defined by the charge states. With this scheme we obtain

$$\langle n|\frac{Q^2}{2C}|m\rangle = 4E_C(n - n_g)^2\delta_{nm}, \tag{9.41}$$

directly from the definition of Q. Here $n_g = Q_g/(2e)$ is the gate charge. The Josephson term can be written as $E_J(e^{i\varphi} + e^{-i\varphi})/2$. The matrix element of these operators can be found in the 'real-space' representation of the wave functions,

$$\langle n|e^{\pm i\varphi}|m\rangle = A_n A_m \int e^{-i(n-n_g)\varphi} e^{\pm i\varphi} e^{i(m-n_g)\varphi} d\varphi$$
$$= A_n A_m \int e^{i(m\pm 1-n)\varphi} = 2\pi A_n A_m \delta_{n,m\pm 1}. \tag{9.42}$$

As plane waves are not normalizable, the correct normalization would require the introduction of some distant boundary conditions. This technique is discussed in the basic quantum-mechanics courses, and we do not dwell on it here. From such an analysis we determine that the prefactor $2\pi A_n A_{n\pm 1} = 1$. With these matrix elements we can insert the resolution of unity, $1 = \sum_n |n\rangle\langle n|$, and get the desired form for the

Hamiltonian,

$$H = \sum_{n,m} |n\rangle\langle n|H|m\rangle\langle m|$$

$$= \sum_{n} 4E_C(n - n_g)^2|n\rangle\langle n| - \frac{1}{2}E_J(|n\rangle\langle n+1| + |n+1\rangle\langle n|). \tag{9.43}$$

This may now be reduced further to analyze the spectrum at $E_J \ll E_C$.

Near the crossing point of the energies of two charge states, $n_g \approx 1/2$, the Hamiltonian can be described by the two-state model, involving only the two crossing charge states, say $|0\rangle$ and $|1\rangle$. In this basis, the Hamiltonian is a 2×2 matrix, specified by the Pauli matrices,

$$H = -\frac{1}{2}B_z\sigma_z - \frac{1}{2}B_x\sigma_x. \tag{9.44}$$

Here the 'magnetic fields' are $B_z = 4E_C(1 - 2n_g)$ and $B_x = E_J$. The eigenfunctions of this Hamiltonian are

$$|e1\rangle = \cos(\eta/2)|0\rangle + \sin(\eta/2)|1\rangle$$
$$|e2\rangle = -\sin(\eta/2)(\eta/2)|0\rangle + \cos(\eta/2)|1\rangle, \tag{9.45}$$

where $\eta = \arctan(B_x/B_z)$. The corresponding eigenenergies are $E_\pm = \pm\sqrt{B_x^2 + B_z^2}/2 = E_J/(2\sin(\eta))$.

It is straightforward to see that in the limit $n_g \ll 1/2$, the ground state goes towards the charge state $|0\rangle$ and the excited state towards $|1\rangle$, and vice versa for $n_g \gg 1/2$. At the degeneracy point $n_g = 1/2$, the difference between the eigenenergies is E_J, which is then also the splitting of the two energy bands (see Fig. 9.25).

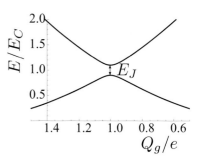

Fig. 9.25 Avoided crossing in the eigenenergies of a superconducting SET in the nearly free-electron limit $E_J \ll E_C$. Here $E_J = 0.2E_C$.

Complement 9.6 Josephson inductance and dynamic capacitance

Depending on the ratio between the two characteristic energies, Josephson energy E_J and the charging energy E_C, the Josephson junction or a superconducting SET can act either as a tunable inductance (single junctions and SSETs for $E_J \gg E_C$) or a tunable capacitance (SSETs with $E_J \ll E_C$). Moreover, in both cases the inductance or capacitance is a non-linear function of the driving, such that the reactance seen by the rest of the circuit depends on how heavily the junction(s) is (are) driven.

For a single Josephson junction, the inductance can be readily obtained from the Josephson relation. Consider the time derivative of the supercurrent,

$$\dot{I}_S = \frac{\partial I_S}{\partial \phi}\dot{\phi} = \frac{2e}{\hbar}\frac{\partial I_S}{\partial \phi}V \equiv L_K^{-1}V. \tag{9.46}$$

This defines the *kinetic* or *Josephson inductance* L_K, which in the case of a tunnel junction ($I_S = I_C\sin(\phi)$) is

$$L_K = \frac{\hbar}{2eI_C\cos(\phi)}. \tag{9.47}$$

The inductance can hence be tuned in a SQUID geometry via a magnetic flux or via tuning the critical current. This tunability of the inductance is used in superconducting kinetic inductance radiation detectors (KIDs) (Day *et al.*,

2003) and, for example, in circuit quantum electrodynamics (see Sec. 9.4.3 below), which couples quantum bits with resonances in transmission lines.

The Josephson inductance of superconducting SETs has been studied recently, for example, by Paila *et al.* (2009).

The band model of a Josephson junction can be also used to define a 'dynamic capacitance', analogous to the effective mass m_e in the theory of electrons in a lattice. This dynamic capacitance is defined as

$$\frac{1}{C^*} = \frac{\partial^2 E_0}{\partial q^2}, \tag{9.48}$$

where q is the polarization charge, and E_0 is the ground-state energy. As this dynamic capacitance depends on the gate charge, it can be used as a tunable capacitance, for example in LC-circuits.[18]

[18]This was first measured by Sillanpää *et al.* (2005).

Complement 9.7 Dynamical Coulomb blockade on Josephson junctions

When placed in a dissipative environment, single Josephson junctions show features of dynamical Coulomb blockade. This effect can be studied by using the same $P(E)$ theory as for the normal-metal junctions in Sec. 7.5. Why this is so can be argued in a straightforward manner by considering the supercurrent term in the Hamiltonian of a Josephson junction

$$H_J = \frac{E_J}{2} \left(e^{i\phi} + e^{-i\phi} \right). \tag{9.49}$$

This term carries electrons from the right junction to the left junction (term $e^{i\phi}$) and from the left to the right ($e^{-i\phi}$). Let us consider what happens if the phase contains a fluctuating part. The rate of tunnelling from left to right can be calculated from

$$\Gamma_{LR}(V) = \frac{\pi}{2\hbar} E_J^2 \sum_{i,f} P_i^{\text{env}} \langle i|_{\text{env}} e^{-i\phi} |f\rangle_{\text{env}} \delta(E_i - E_f). \tag{9.50}$$

Proceeding similarly as in Sec. 7.5, we can include the effect of phase fluctuations on the supercurrent and obtain

$$I_S = 2e(\Gamma_{LR}(V) - \Gamma_{LR}(-V)) = \frac{\pi e E_J^2}{\hbar} \left[\tilde{P}(2eV) - \tilde{P}(-2eV) \right]. \tag{9.51}$$

Here the probability $\tilde{P}(E)$ of absorbing the energy E from the environment is otherwise the same as that defined in Sec. 7.5, but the definition of the phase differs by a factor of 2 (compare the ac Josephson relation (5.29) with eqn (7.34)). Because of this, the resistance quantum R_K in eqn (7.46) has to be replaced by $R_S = R_K/4 = h/(4e^2)$.

Because of the phase fluctuations, the time-averaged Josephson current may become finite also at non-zero voltages. This is relevant, for example, in the studies of SIS current–voltage characteristics at low but finite voltages: part of the signal is due to the quasiparticles, part of it due to the supercurrent. A more interesting effect happens when the resistance of the environment crosses R_S. As shown in the following paragraph, at low voltages the time-averaged supercurrent is proportional to V^{2R/R_S-1} and the differential conductance to V^{2R/R_S-2}. For $R < R_S$, this differential conductance diverges as $V \to 0$, indicating the onset of a coherent supercurrent. However, for $R > R_S$, the differential conductance becomes zero as $V \to 0$, and hence there is a transition to the insulating state. This is a quantum phase transition, originally discussed by Albert Schmidt (1983).

Supercurrent at low voltages in the presence of phase fluctuations
Let us calculate the supercurrent (9.51) at zero temperature using the model described in Fig. 9.17. The phase correlation function and the associated $P(E)$ functions are given in eqns (7.50)–(7.52). To obtain $I_S(V)$ at low voltages $V \ll \hbar/(2eRC)$, we note that it is sufficient to describe $J(t)$ at large values of $|t| \gg \tau$. In this limit

$$J(t) \approx -\frac{R}{R_S}\left(2\ln|t/\tau| + 2\gamma - i\pi\right). \tag{9.52}$$

With this approximation, we get[19]

$$P(E \ll \hbar/\tau) = 2\Gamma(1 - 2R/R_S)\sin(2\pi R/R_S)\tau^{2R/R_S}e^{-2\gamma R/R_S}(\hbar E)^{2R/R_S-1}, \tag{9.53}$$

where $\Gamma(x)$ is the Γ-function. Using this $P(E)$, we get for the current

$$I_S = \frac{2\pi e E_J}{\hbar}\Gamma(1 - 2R/R_S)\sin(2\pi R/R_S)\tau^{2R/R_S}e^{-2\gamma R/R_S}(\hbar 2eV)^{2R/R_S-1}. \tag{9.54}$$

As a function of voltage, this shows a supercurrent-insulator transition, as discussed above.

[19]Note that for $R > R_S/2$, the integral needs to be cut off at low $|t|$. In this case, we would need to include more terms from eqn (7.50), but I do not go into detail of this here.

9.4.3 Superconducting qubits

The presence of the macroscopic superconducting coherence allows the use of superconducting circuits as realizations of quantum bits, qubits. Such quantum bits are the elementary elements of a quantum computer, which allows exploitation of the quantum-mechanical superposition principle for efficient computing.[20] These fulfil the five criteria which have been commonly accepted for realistic qubit realizations (DiVincenzo, 2000): (i) Each of the qubit realizations introduced in this section comprise well-characterized quantum two-level systems where the third and higher states can be made to lie much above the ground and the first excited state, and each system is at least in principle scalable to many qubits; (ii) Each qubit can be initialized to the ground state; (iii) All qubits have a long coherence time compared to the time it takes to apply quantum operations changing the state of the qubits; (iv) All types of one-qubit operations can be realized, i.e., arbitrary tuning of B_z and B_x in the generic qubit Hamiltonian of the type (9.44), and one non-trivial two-qubit operation: unitary transformation involving two qubits that cannot be reduced to simple one-qubit transformations. This requires controllable qubit–qubit coupling. Finally, the fifth requirement, (v) the capability of measuring each qubit separately, is also possible with each of the superconducting qubits.

In the following I outline the main types of superconducting qubits and mention the later developments. On introducing the qubits, the emphasis is on defining what the qubit states are and how the single-qubit Hamiltonian is constructed. I do not dwell on the details of coupling and measuring the different types of qubits, which is widely discussed in recent literature.

[20]Details of quantum computing are discussed in many textbooks, for example (Nielsen and Chuang, 2000) and (Nakahara and Ohmi, 2008). I will not dwell on them here.

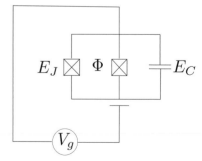

Fig. 9.26 Cooper pair box which can be utilized as a charge qubit: in the charge state basis, the energy level separation can be controlled with the gate voltage V_g and the coupling between the levels by the flux Φ, which tunes the effective Josephson energy.

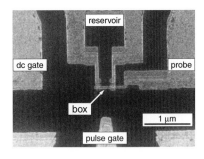

Fig. 9.27 The first superconducting qubit, a Cooper pair box, measured in NEC Tsukuba, Japan. The two gates were used for manipulating the charge state, and the probe for measuring the state and the loop on the top allowed for manipulation of the Josephson coupling. For more details, see (Nakamura *et al.*, 1999). Figure courtesy of Yasunobu Nakamura, University of Tokyo.

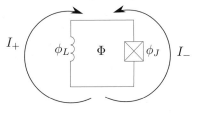

Fig. 9.28 Simplest type of a flux qubit is a radio frequency SQUID containing a Josephson junction and a superconducting loop with inductance L. The qubit states are the persistent currents rotating in clockwise or anticlockwise directions, and their occupations can be controlled with the flux Φ.

Charge qubit

Charge qubit is a superconducting version of the single-electron box discussed in Sec. 7.1.1. There, the tunnelling element is replaced by a pair of Josephson junctions as depicted in Fig. 9.26. In the limit $\Delta \gg E_C$ at low temperatures the states with an odd number of electrons on the island can be disregarded, and it is enough to study the dynamics of Cooper pairs. The full Hamiltonian for this Cooper pair box is given by eqn (9.35) and in the 'charging' limit $E_J \ll E_C$ its two-state version is described in eqn (9.44).

Charge qubit was the first realization of superconducting qubits (Nakamura *et al.*, 1999) (see Fig. 9.27). However, as it is based on the direct manipulation of charge, it is prone to low-frequency offset charge fluctuations which are an outstanding problem in the field. Due to these, the observed phase coherence times are typically smaller in charge qubits compared to other types of superconducting qubits.

Flux or persistent current qubit

Flux qubits rely on the opposite parameter regime of Josephson junctions, $E_J \gg E_C$. Instead of the Cooper pair box system, the flux qubits are formed in a radio frequency SQUID setting; see Fig. 9.28 for the simplest version of them. To avoid the harmonic-oscillator-type excitation spectrum discussed in Sec. 9.4.1, which would require describing higher-order states besides the two lowest ones, these qubits make use of the inductance of the superconducting loops, whose potential energy is a quadratic function of the phase drop ϕ_L across the inductor. From similar considerations as for the dc SQUID (Sec. 9.3.1), in the presence of a finite flux through the loop the phase across the Josephson junction is

$$\phi_J = \phi_L - \frac{2\pi\Phi}{\Phi_S}. \tag{9.55}$$

The total Hamiltonian of the flux qubit can hence be written as

$$H_{\text{fluxqubit}} = \frac{Q^2}{2C} + \frac{(2\pi\Phi_S)^2}{2L}\phi_L^2 - E_J \cos\left(\phi_L - \frac{2\pi\Phi}{\Phi_S}\right). \tag{9.56}$$

The system becomes interesting for qubit operations when the applied flux is close to half a flux quantum, $\Phi \approx \Phi_S/2 + f\Phi_S/(2\pi)$, in which case the potential energy of the system can be written as

$$U(\phi_L) = E_J \left[\beta\phi_L^2 + \cos\left(\phi_L - f\right)\right], \tag{9.57}$$

where $\beta = 2\pi^2\Phi_S^2/(LE_J)$ is a dimensionless parameter characterizing the inductor. This potential energy is plotted in Fig. 9.29. The two minima which at low values of β occur close to $\phi_L \approx \pi$ describe a SQUID with screening currents moving in the clockwise or anticlockwise directions. The flux qubit eigenstates are the superpositions of these states, due to the finite tunnel coupling across the shallow potential barrier with height approximatively E_J. The energies of the minima can be controlled by varying the flux f: for a small but finite f one of the minima is favored. The coupling energy E_S between the two states with opposite screening currents depends on the plasma frequency of the junctions and exponentially on the barrier height. Control of E_S and f therefore allows controlling the σ_z and σ_x fields describing the states with opposite screening currents, and therefore all single-qubit operations can be performed.

Phase qubit

The third traditional version of superconducting qubits is based on the tilted washboard potential depicted in Figs. 9.18 and 9.30. The potential wells in the case $I_b < I_C$ support quasi-eigenstates[21] which in the limit $I_b \ll I_C$ are those of a harmonic oscillator. With an increasing I_b, the potential becomes anharmonic—especially the cubic term in ϕ becomes relevant—and as a result, the spacing between the energy levels is no longer constant. By choosing the bias current properly, the ground and the lowest excited state can then be used as the two states of a qubit (Martinis *et al.*, 2002).

The energy spacing ΔE between the two levels (σ_z-term in the qubit Hamiltonian (9.44)) equals the plasma frequency normalized by the dc current, eqn (9.32), and it can be controlled by tuning the dc bias current. The coupling between the levels (σ_x- and σ_y-terms) is typically induced by applying, on top of the dc bias current, microwave pulses with angular frequencies ω at resonance with the energy spacing, i.e., $\hbar\omega \approx \Delta E$. The qubit measurement employs the measurement of the escape process (see Sec. 9.3.5): In the macroscopic tunnelling regime, the probability of escape depends exponentially on which energy level in the washboard potential the process takes place.[22] The measurement is therefore done by applying a rapid dc pulse to tilt the potential closer to the escape threshold and then waiting for the voltage pulse. If the waiting time is chosen suitably a measured voltage pulse indicates that the qubit occupied the excited state, whereas the lack of voltage detection indicates that the qubit was in the ground state.

Other types of superconducting qubits

Recent interest in superconducting qubits has given rise to other variations of the basic qubit types (see Fig. 9.31 for a sample of recent qubit designs). The main motivation for introducing these new types has been to eliminate or at least weaken the harmful effect of charge fluctuations on the qubit dephasing: *quantronium* (Vion *et al.*, 2002) uses a charge qubit concept, but with an increased E_J to increase the E_J/E_C; *transmon* (Koch *et al.*, 2007) does the same, but by decreasing E_C and furthermore coupling the charge qubit to a transmission line resonator; and *fluxonium* (Manucharyan *et al.*, 2009) is a hybrid between charge and flux qubits, where a small Josephson junction is shunted with a large (kinetic) inductance due to a series of larger Josephson junctions (see Complement 9.6).

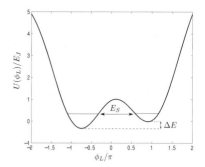

Fig. 9.29 Potential $U(\phi_L)$ employed in the flux qubit, plotted with $\beta = 0.1$ and $f = 0.3$. The two minima refer to screening currents rotating in opposite directions. For $f \neq 0$, one of the directions has a lower energy. The states with the different screening currents are coupled by a coupling energy E_S, related to the tunnelling through the shallow potential well.

[21]Quasi-eigenstates because the lifetime of these states is finite for any non-zero bias current I_b. However, this lifetime is extremely long when the bias current is not close to I_C.

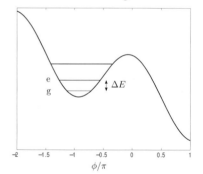

Fig. 9.30 The two lowest-energy quasi-eigenstates of the wells of the tilted washboard potential are used as the states of phase qubits.

[22]Typically it differs by a factor $10^2 \ldots 10^3$ between the ground and excited states.

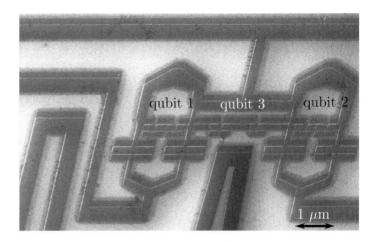

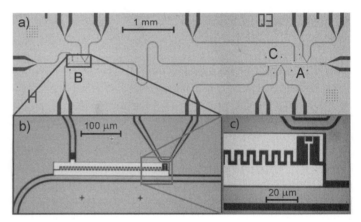

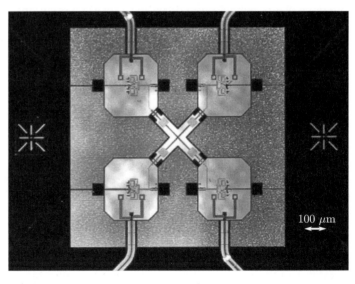

Fig. 9.31 Zoo of superconducting qubits (scale bars to (a) and (c) added by the author): a) Three coupled flux qubits, courtesy of Fumiki Yoshihara and Jaw-Shen Tsai, Riken Advanced Science Institute and NEC Tsukuba, Japan (Harrabi *et al.*, 2009); b) three coupled transmon qubits, courtesy of Johannes Fink, Quantum Device Lab, ETH Zürich (Baur *et al.*, 2012); and c) four coupled phase qubits, courtesy of Erik Lucero, Martinis group, University of California Santa Barbara (Neeley *et al.*, 2010).

Besides the work on the qubits themselves, the research on superconducting qubits has in the past few years started a totally new research field: studies of artificial atom–photon interactions between superconducting qubits and superconducting transmission line resonators. This field of research has been dubbed *circuit quantum electrodynamics*, pioneered by the work of Wallraff *et al.* (2004).

Further reading

- Many basic and not so basic properties of Josephson junctions are described in books on superconductivity: for example (Tinkham, 1996) and (Barone and Paterno, 1982).
- For a review of the heat transport effects in hybrid normal-metal–superconductor junctions, see (Giazotto *et al.*, 2006).
- Dissipative effects and the charge-phase duality are discussed in (Weiss, 1999).

- Superconducting qubits are discussed in many reviews, for example (Makhlin *et al.*, 2001) and (Devoret *et al.*, 2004), and at least shortly in a couple of books on quantum computing, for example (Nielsen and Chuang, 2000) and (Nakahara and Ohmi, 2008).

Exercises

(9.1) At $T = 0$, find the threshold voltage where quasiparticle current starts to flow (in the sequential tunnelling regime) for a) NSN, b) SNN, c) SSN, d) SNS, and e) SSS superconducting SETs (for example, NSN SET refers to a SET composed of normal-metal electrodes and a superconducting island).

(9.2) Assume $T_N \leq T_S \ll \Delta/k_B$. Show that the NIS heat current for bias voltage close to Δ is

$$\dot{Q}_{\text{opt}} \propto \frac{\Delta^2}{e^2 R_T} \left(\frac{k_B T_N}{\Delta} \right)^{3/2}. \qquad (9.58)$$

Hint: when $T_N, T_S \ll \Delta/k_B$, you can use asymptotic forms for the Fermi functions, $f(E \gg k_B T) \approx \exp(-E/k_B T)$ and $f(E \ll k_B T) \approx 1 - \exp(E/k_B T)$.

(9.3) Show that the $I-V$ curve of a NIS junction depends only on the temperature T_N of the normal metal, not on the temperature T_S of the superconductor.

(9.4) Consider the Brownian refrigerator described in Complement 9.3. Assume $T_N = T_S \ll \Delta/k_B$. Find approximative forms for the heat currents $\dot{Q}_{NIS}$ and $\dot{Q}_{SIN}$ (eqns (9.11) and (9.12)) in the case when the environment is at a vanishing temperature and described by a large resistance $R \gg R_K$. From these results, find the heat current flowing to/from the environment. Which parts of the total system heat up and which cool down? Hint: use again the high-energy approximations for the Fermi functions as explained in Exercise 9.2.

(9.5) Consider the RCSJ model in the strongly overdamped limit, $Q \ll 1$. In this case you can neglect the first term in eqn (9.24). From the resultant equation, find the average voltage across the junction as a function of the driving current, both for $I < I_c$ and for $I > I_c$. Hint: Separate the coordinates φ and τ and integrate over one period $[0, 2\pi]$ in the phase. Use this for the averaged phase.

(9.6) Derive eqns (9.32) and (9.33) by a) finding the phase difference $\phi_{\min}$ that gives a local minimum in the tilted washboard potential $U(\phi)$, b) expanding $U(\phi)$ to the second order in $\phi - \phi_{\min}$ and c) finding the phase $\phi_{\max}$ that gives the local maximum of $U(\phi)$. The attempt frequency ω_A is proportional to the prefactor of the second-order term derived in b) and the potential height is $\Delta U = U(\phi_{\max}) - U(\phi_{\min})$.

(9.7) a) Combine the Josephson inductance, eqn (9.47), with the capacitance C of the junction, and find the resonant frequency ω_0 of the resultant LC-oscillator. b) Express ω_0 using the Josephson energy E_J and the charging energy E_C. Use this to justify eqn (9.39).

(9.8) Show that the Hamilton equations (the equations describing the dynamics of the canonical coordinates),

$$\frac{\hbar}{2e}\dot{Q} = -\frac{\partial H}{\partial \phi} \tag{9.59a}$$

$$\dot{\phi} = \frac{2e}{\hbar}\frac{\partial H}{\partial Q} \tag{9.59b}$$

for the Josephson junction in the non-driven case

(eqn (9.35)) produce the dc and ac Josephson relations.

(9.9) **Question on a scientific paper.** Take the paper Vion *et al.*, *Nature* **296**, 886 (2002), presenting the 'quantronium' qubit and its characterization. The measurement circuit is in Fig. 1. a) Identify the 'qubit' from the circuit. b) Explain how the Josephson escape mechanism in the junction denoted E_{J0} can be used for the measurement of the qubit state. What determines the relative error (or *fidelity*) of this measurement? The calculated transition frequency ν_{01} vs. phase ϕ across the quantronium and the gate charge N_g are shown in Fig. 2a and the corresponding measurements in Fig. 2b. c) How is ν_{01} related to the quantronium energy level diagram, and how can it in principle be calculated? d) Bonus question: the real 'quantum' characteristics of the device is obtained by measuring the Rabi oscillations (Fig. 3a) or the Ramsey fringes (Fig. 3b). Find out how this was done.

Graphene

10

Graphene—a single layer of graphite—is the basis of many different-dimensional allotropes of carbon (see Fig. 10.1): roll it into a ball shape and you get fullerene (found experimentally in 1985), wind it into a cylinder-shaped wire and you get a carbon nanotube (discovered in 1991), and stack it into layers to form graphite (discovered in ancient times and named in the 1700s). It has been known since the mid-1900s (Wallace, 1947) that a single graphene layer would have an exotic linear dispersion relation which bears close resemblance to that found for massless relativistic particles. Yet for a long time this model was considered to be merely a mathematical nuisance, not connected to experimental reality. All that changed in 2004 when the group led by Andre Geim and Konstantin Novoselov of the University of Manchester found a relatively simple technique of fabricating graphitic samples containing fairly large regions with only a single graphene layer.[1] In fact, these graphene layers are quite easy to make, the harder part was to find them after fabrication as tools to image single-atom thick samples were not very commonplace. The major finding of the Manchester group was a subtle optical effect a single-layer graphene produces when placed on top of a SiO_2 substrate of suitable width (Novoselov *et al.*, 2004). Later it was found by the same group that as a result of the peculiar dispersion relation (see Sec. 10.1.1), the quantum Hall effect, describing electron transport in high magnetic fields, is essentially different in single-layer than in multi-layer (including bilayer) graphene samples or, for example, in regular semiconductor systems (Novoselov *et al.*, 2005). This showed that Wallace's mathematical nuisance is in fact a proper model for experiments in graphene.

These experimental findings spurred tremendous activity among other research groups. By today there have been numerous predictions about different types of effects related to electron transport and other properties in graphene. These include the nature of different types of scattering, interaction effects, magnetic field effects, localization, graphene placed between different types of reservoirs, high-frequency effects, and many others.[2] Experimental activity is also booming, and nowadays hundreds or thousands of research groups around the world routinely fabricate graphene samples. Researchers describe graphene with many superlatives: it is the thinnest imaginable material (one atomic layer thick), it has the highest thermal conductivity, supports the highest current density, has the highest intrinsic mobility, and the longest mean free path at room temperature. Also, its mechanical properties are remarkable (see graphene NEMS in Ch. 11. It is the most stretchable crystal (allowing up to 20% deformation), and is the strongest material ever measured.

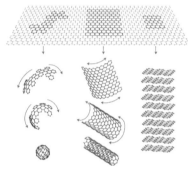

Fig. 10.1 Graphene is a mother of other carbon allotropes. (A. Geim and Novoselov, *Nature Materials* **6**, 183 (2007), Fig. 1.) Reprinted by permission from Macmillan Publishers Ltd., © 2007.

[1]In 2010 Geim and Novoselov were awarded the Nobel prize in physics for this discovery.

[2]For a thorough review of the properties of graphene, see (Castro Neto *et al.*, 2009), (Das Sarma *et al.*, 2011).

In this chapter I describe some of the exotic phenomena found in graphene, but to produce a comprehensive picture of all its properties would require a separate book. The main point here is to show what is fundamentally different in graphene compared to other conductors. After derivation of the peculiar dispersion relations for monolayer, bilayer and multilayer graphene, the exotic properties of the monolayers are illustrated by discussing three phenomena related to the Dirac-type dispersion relation and the presence of additional degrees of freedom (valley and pseudospin). The first two examples are quite counterintuitive. First I discuss the phenomenon of Klein tunnelling, which predicts that electrons encountering a high potential barrier are not reflected but transmit through it as if this potential did not exist at all. Second, I show that ballistic scattering-free graphene may show features similar to those found in ordinary diffusive electron systems. Both of these seemingly contradictory effects can be explained by utilizing the low-energy Hamiltonian (10.10) and rather straigthforward wave function matching. The third phenomenon I describe is related to nanoribbons made of graphene. I show how the transverse states depend on the Dirac Hamiltonian and exact types of the boundaries of these nanoribbons.

10.1 Electron dispersion relation in monolayer graphene

In this section I derive the electron dispersion relation and the effective low-energy Hamiltonian describing electrons in single-layer graphene, and in the following section for bilayer and a specific type of multilayer graphene. All these systems are qualitatively different from ordinary electron systems, as the elementary excitations can be described by *spinors*. In the case of single-layer graphene, these spinors are massless, and their low-energy Hamiltonian is similar to the Dirac Hamiltonian for massless particles. In the bilayer case the coupling between the layers induces a finite mass proportional to this coupling, yet the spinor wave function still makes bilayer graphene different from ordinary metals. Stacking more layers on top of each other, the resultant low-energy dispersion turns out to depend sensitively on the nature of stacking. As shown in Sec. 10.2, for one type of stacking even more exotic dispersion of the form $\epsilon_k \propto k^N$, where N is the number of layers, can be found.

The derivation of the dispersion relations is done with the use of the tight-binding approach. The specific type of a dispersion can been seen in this approach to be connected to the symmetry of the carbon lattice.

10.1.1 Massless Dirac fermions in graphene

The low-energy Hamiltonian of the electrons in graphene can be derived from the tight-binding description, which starts from identifying the graphene lattice structure. Graphene is formed from carbon atoms that are arranged in a honeycomb lattice, shown in Fig. 10.2. The lattice spacing is $a = 1.42$ Å. The lattice is described by two lattice vectors,

$$\vec{a}_{1,2} = \mp \frac{\sqrt{3}a}{2}\hat{e}_x + \frac{3a}{2}\hat{e}_y. \tag{10.1}$$

Each unit cell contains two atoms, denoted by labels A and B, a *sublattice*. Each A atom is connected to three B atoms and each B atom to three A atoms. The nearest-neighbour vectors in real space are

$$\vec{\delta}_1 = -\frac{\sqrt{3}a}{2}\hat{e}_x - \frac{a}{2}\hat{e}_y, \quad \vec{\delta}_2 = \frac{\sqrt{3}a}{2}\hat{e}_x - \frac{a}{2}\hat{e}_y, \quad \vec{\delta}_3 = a\hat{e}_y. \tag{10.2}$$

The strength of this coupling can be described with the hopping energy γ_0. Therefore, including only the nearest-neighbour hopping, the tight-binding Hamiltonian for this lattice reads

$$H_g = -\gamma_0 \sum_{\vec{R}} \psi(\vec{R})^\dagger \left[\psi(\vec{R} + \vec{\delta}_1) + \psi(\vec{R} + \vec{\delta}_2) + \psi(\vec{R} + \vec{\delta}_3) \right] + \text{h.c.}, \tag{10.3}$$

where $\psi(\vec{R})^\dagger$ creates a particle in position $\vec{R}$ and $\psi(\vec{R})$ annihilates it. The sum goes over all the A atoms in the lattice.

We can define the annihilation operators (Fourier transforms) for the A and B electrons in the momentum state $\vec{k}$ through

$$c_{A,\vec{k}} \equiv \frac{1}{\sqrt{N}} \sum_{\vec{R}} e^{i\vec{k}\cdot\vec{R}} \psi(\vec{R}), \quad c_{B,\vec{k}} \equiv \frac{1}{\sqrt{N}} \sum_{\vec{R}} e^{i\vec{k}\cdot\vec{R}} \psi(\vec{R} + \vec{\delta}_3) \tag{10.4}$$

and likewise for the creation operators. Here N is the number of lattice points in the graphene sheet. The position-dependent annihilation operators can now be written in terms of the momentum-state operators as

$$\psi(\vec{R}) = \frac{1}{\sqrt{N}} \sum_{\vec{k}} e^{-i\vec{k}\cdot\vec{R}} c_{A,\vec{k}}, \quad \psi(\vec{R} + \hat{\delta}_3) = \frac{1}{\sqrt{N}} \sum_{\vec{k}} e^{-i\vec{k}\cdot\vec{R}} c_{B,\vec{k}}. \tag{10.5}$$

With these definitions we can write the Hamiltonian as

$$H_g = -\gamma_0 \sum_{\vec{k}} \left(1 + e^{i\vec{k}\cdot\vec{a}_1} + e^{i\vec{k}\cdot\vec{a}_2} \right) c_{A,\vec{k}}^\dagger c_{B,\vec{k}} + \text{h.c.}$$

$$\equiv -\gamma_0 \sum_{\vec{k}} \gamma_{\vec{k}} c_{A,\vec{k}}^\dagger c_{B,\vec{k}} + \text{h.c.} \tag{10.6}$$

Denoting the sublattice indices via 1×2 vectors, the Hamiltonian can be expressed as a matrix

$$H_g = -\gamma_0 \sum_{\vec{k}} \begin{pmatrix} 0 & \gamma_{\vec{k}} \\ \gamma_{\vec{k}}^* & 0 \end{pmatrix}. \tag{10.7}$$

The summand has the eigenenergies

$$\epsilon_{\vec{k}} = \pm\gamma_0|\gamma_{\vec{k}}| = \pm\gamma_0\sqrt{3 + 2\cos(\sqrt{3}k_x a) + 4\cos\left(\frac{3k_y a}{2}\right)\cos\left(\frac{\sqrt{3}k_x a}{2}\right)}. \tag{10.8}$$

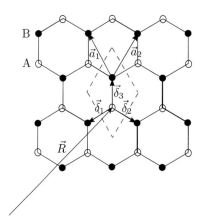

Fig. 10.2 In graphene, carbon atoms form a single layer honeycomb lattice. The carbon atoms reside at the edges of the honeycomb; the lines denote the nearest-neighbour coupling between the atoms. The area enclosed by the dashed line denotes one unit cell with two atoms, labelled A and B and denoted with open and closed circles. The vectors connecting atom B to the three neighbouring atoms A are denoted with δ_i.

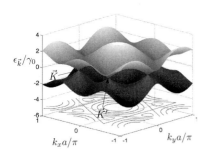

Fig. 10.3 Graphene dispersion relation $\epsilon_{\vec{k}}$, eqn (10.8). The two non-equivalent valleys are marked with $\vec{K}$ and $\vec{K}'$.

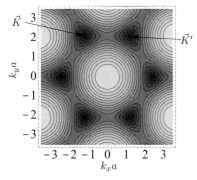

Fig. 10.4 Contour plot of graphene dispersion relation $\epsilon_{\vec{k}}$, eqn (10.8). Note that only two of the valleys, marked with $\vec{K}$ and $\vec{K}'$, are non-equivalent, and the rest are connected to these two via integer multiples of reciprocal lattice vectors.

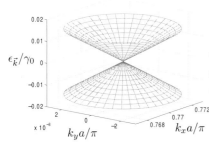

Fig. 10.5 Low-energy part of the graphene dispersion relation around one of the valleys illustrates the 'Dirac cones'.

[3]See, for example, Fig. 1 in (Boukhvalov *et al.*, 2008).

This dispersion relation is plotted in Figs. 10.3 and 10.4. As the reciprocal lattice vectors are

$$\vec{b}_{1/2} = \frac{2\pi}{3a}\left(\mp\sqrt{3}\vec{e}_x + \vec{e}_y\right), \qquad (10.9)$$

only two of the energy minima — or valleys — for each honeycomb cell are non-equivalent in the reciprocal lattice. The points where $\epsilon_{\vec{k}} = 0$ are typically called *Dirac points*. The low-energy dispersion relation is plotted in Fig. 10.5. Linearizing around the Dirac point (see Exercise 10.2), we get a low-energy Hamiltonian

$$H_g \approx \hbar v_F \begin{pmatrix} 0 & \pm k_x - ik_y \\ \pm k_x + ik_y & 0 \end{pmatrix} = \hbar v_F \left(\pm k_x \sigma_x + k_y \sigma_y\right). \quad (10.10)$$

The two signs are for the two valleys. This Hamiltonian has the form of the Dirac Hamiltonian for massless particles, but the speed of light has been replaced by $v_F = 3\gamma_0 a/(2\hbar) \approx 10^6$ m/s. Although this result is here derived based on the tight-binding approximation, the low-energy Hamiltonian (10.10) has the same form regardless of this approximation.[3].

The linear dispersion relation and the presence of the *pseudospin*, corresponding to the sublattice index A or B, results in many exotic electronic properties only seen in graphene. This is one reason why the successful fabrication and measurements of single graphene sheets have resulted in a massive activity in the detailed study of its properties. At the time of writing this book, graphene is probably the most popular research topic in condensed-matter physics.

10.1.2 Eigensolutions in monolayer graphene

The low-energy Schrödinger (Dirac) equation of ballistic graphene can obtained from the real-space version of eqn (10.10),

$$\begin{pmatrix} \mu_g & \hbar v_F\left(\pm i\frac{\partial}{\partial x} + \frac{\partial}{\partial y}\right) \\ \hbar v_F\left(\pm i\frac{\partial}{\partial x} - \frac{\partial}{\partial y}\right) & \mu_g \end{pmatrix}\begin{pmatrix} \psi_A(x,y) \\ \psi_B(x,y) \end{pmatrix} = E\begin{pmatrix} \psi_A(x,y) \\ \psi_B(x,y) \end{pmatrix}.$$
$$(10.11)$$

The possible non-zero chemical potential is denoted μ_g. It can be tuned externally either by doping or by applying a gate voltage. Equation (10.11) has the normalized eigensolutions of the form

$$\Psi(x,y) = \frac{1}{\sqrt{2}}\begin{pmatrix} e^{\mp i\phi/2} \\ se^{\pm i\phi/2} \end{pmatrix}e^{i(\pm k_x x + k_y y)}, \qquad (10.12)$$

where $\phi = \arctan(k_y/k_x)$ denotes the direction of propagation, and $s = \text{sgn}(E)$, i.e., electrons (states in the conduction band, i.e., $E > 0$), have an opposite sign to holes (states in the valence band, $E < 0$). The energy corresponding to this solution is $E = \mu_g + \hbar v_F\sqrt{k_x^2 + k_y^2}$, where μ_g is the chemical potential. The two valleys (upper and lower signs) are related by the time-reversal transformation $\vec{k} \leftrightarrow -\vec{k}$. Another interesting property of these wave functions can be seen by adding 2π to the phase

ϕ: the wave function changes its sign, i.e., obtains an added phase of only π. This is typically called the Berry phase, and is characteristic of spinors.[4].

An important property of the graphene states is their helicity,[5] which is positive for particles with spin in the same direction as their direction of motion, and negative otherwise. Helicity is defined by

$$\hat{h} = \frac{1}{2}\sigma \cdot \frac{\vec{k}}{k}. \qquad (10.13)$$

Using the eigensolutions $\Psi(x,y)$ from eqn (10.12) it is straightforward to show that $\hat{h}\Psi(x,y) = \frac{s}{2}\Psi(x,y)$, i.e., electrons have a positive and holes a negative helicity.[6] In graphene helicity is conserved at low energies for a potential that is symmetric in the sublattice space.

From the low-energy dispersion relation it is straightforward to derive the density of states close to the Fermi energy $E = 0$ corresponding to neutral graphene (see Exercise 10.4),

$$N_g(E) = \frac{2A_c}{\pi} \frac{|E|}{v_F^2}, \qquad (10.14)$$

where $A_c = 3\sqrt{3}a^2/2$ is the unit cell area. Hence, the density of states vanishes at the Dirac point.

10.2 Bilayer and more

The interlayer coupling of multilayer graphene is much weaker than the intralayer hopping. Because of this, stacking graphene layers amounts to coupling Dirac-type particles. For bilayer graphene, this coupling results in the formation of a quadratic dispersion relation around the Dirac point, but preserving the pseudospin structure of the model. Moreover, breaking the symmetry between the two layers allows the creation of a tunable band gap (Ohta *et al.*, 2006). This gap allows the fabrication of field-effect transistors, where the current in the 'on' and 'off' states may differ by orders of magnitude (Zhang *et al.*, 2009).

The basic tight-binding description of bilayer graphene is shown in Fig. 10.6. It is customary to assume that the bilayer follows the same stacking as graphite, and also use the tight-binding parameters known from graphite.[7] In this stacking the layers are first placed on top of each other, and one of the lattices is shifted such that the A atoms (open circles in Fig. 10.6) of the first layer are on top of the B atoms (filled circles) of the second layer, whereas the B atoms of the first layer are on top of the centre of the hexagon of the second layer, and A atoms of the second layer below the centre of the hexagon of the first layer.[8]

[4]For the relation of the Berry phase to the quantum Hall effect in graphene, see (Zhang *et al.*, 2005).

[5]Or chirality, which for massless particles equals helicity.

[6]This applies for valley $\vec{K}$ — for the other valley $\vec{K}'$ the helicity for electrons and holes is inverted.

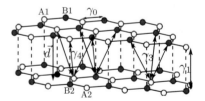

Fig. 10.6 Tight-binding hoppings in bilayer graphene.

[7]The exact values of these parameters may depend on the number of layers, but their orders of magnitude remains the same.

[8]Note that the labelling of the atoms in separate layers is arbitrary. Another possible notation is to describe A atoms on top of each other and B atoms at the centre of the other hexagon, i.e., with reversed labels A and B in the second layer.

The tight-binding Hamiltonian for bilayer graphene has the form

$$
\begin{aligned}
H_b = & -\gamma_0 \sum_{\vec{R}_1} \sum_{j=1,2,3} \left[\psi_1^A(\vec{R}_1)^\dagger \psi_1^B(\vec{R}_1 + \vec{\delta}_j) + \text{h.c.} \right] \\
& -\gamma_0 \sum_{\vec{R}_2} \sum_{j=1,2,3} \left[\psi_2^A(\vec{R}_2)^\dagger \psi_2^B(\vec{R}_2 + \vec{\delta}_j) + \text{h.c.} \right] \\
& -\gamma_1 \sum_{\vec{R}_1} \left[\psi_1^A(\vec{R}_1)^\dagger \psi_2^B(\vec{R}_2 + \vec{\delta}_2) + \text{h.c.} \right] \\
& -\gamma_3 \sum_{\vec{R}_1} \sum_{j=1,2,3} \left[\psi_1^B(\vec{R}_1 + \vec{\delta}_1)^\dagger \psi_2^A(\vec{R}_1 + \vec{d} - \vec{\delta}_j) + \text{h.c.} \right].
\end{aligned}
\tag{10.15}
$$

Here $\vec{R}_i$ ($i = 1, 2$) denotes the locations of the A atoms in layer i. Moreover, $\vec{\delta}_j$ are the vectors pointing from the A atom to the neighbouring B atoms, defined in eqn (10.2), and $\vec{d} = d\hat{u}_z$ with $d \approx 3\text{Å}$ is the vector perpendicular to the graphene planes, connecting the nearest-neigbour atoms on different planes. For graphite the couplings have the strengths (Dresselhaus and Dresselhaus, 2002) $\gamma_0 = 3.2$ eV, $\gamma_1 = 0.39$ eV and $\gamma_3 = 0.32$ eV.[9] We disregard the coupling γ_4 marked in Fig. 10.6. In graphite, it is $\gamma_4 = 0.04$ eV.

Defining creation and annihilation operators as above, we can write the bilayer Hamiltonian as

$$
H_b = -\sum_k \begin{pmatrix} c_{A1}^\dagger \\ c_{B1}^\dagger \\ c_{B2}^\dagger \\ c_{A2}^\dagger \end{pmatrix} \begin{pmatrix} \delta\mu & \gamma_{\vec{k}}\gamma_0 & 0 & \gamma_1 \\ \gamma_{\vec{k}}^*\gamma_0 & \delta\mu & \gamma_3\gamma_{\vec{k}}^* & 0 \\ 0 & \gamma_1 & -\delta\mu & \gamma_0\gamma_{\vec{k}}^* \\ \gamma_3\gamma_{\vec{k}} & 0 & \gamma_0\gamma_{\vec{k}} & -\delta\mu \end{pmatrix} \begin{pmatrix} c_{A1} \\ c_{B1} \\ c_{B2} \\ c_{A2} \end{pmatrix}.
\tag{10.16}
$$

Here $\delta\mu$ denotes a possible potential difference between the layers.

Let us analyze this Hamiltonian close to the single-layer valley $\vec{K}$ (the other valley $\vec{K}'$ goes analogously), i.e., we describe the wave vectors $\vec{k}$ around $\vec{K}$. As there $\gamma_{\vec{k}} \propto |k|$ and $\gamma_3 \ll \gamma_0$, we can as a first approximation disregard γ_3 and approximate $\gamma_0\gamma_{\vec{k}} \approx v_F(k_x - ik_y)$. The resultant Hamiltonian has four branches of solutions (see Fig. 10.7), two residing around $\epsilon = \pm\sqrt{\gamma_1^2 + \delta\mu^2}$ and other two around $\epsilon = \pm\delta\mu$. For low potentials $\delta\mu$ the latter are close to the Fermi energy $E = 0$ and therefore more relevant.

For low values of k the eigenvectors of this bilayer Hamiltonian corresponding to the low-energy solutions are

$$
\psi_- = \begin{pmatrix} -\dfrac{k_x+ik_y}{\gamma_1} \\ \dfrac{(k_x+ik_y)^2}{2\gamma_1\delta\mu} \\ \dfrac{2\delta\mu(k_x+ik_y)}{\gamma_1^2} \\ 1 \end{pmatrix}, \quad \psi_+ = \begin{pmatrix} -\dfrac{2\delta\mu(k_x-ik_y)}{\gamma_1^2} \\ 1 \\ -\dfrac{k_x-ik_y}{\gamma_1} \\ -\dfrac{(k_x-ik_y)^2}{2\gamma_1\delta\mu} \end{pmatrix}.
\tag{10.17}
$$

The eigenvalues corresponding to these eigenvectors at $|k \approx 0|$, $H\psi_\pm = \pm\epsilon_{\vec{k}}\psi_\pm$ are

$$
\epsilon_{\vec{k}} \approx \delta\mu - v_F^2 \frac{2(k_x^2 + k_y^2)\delta\mu}{\gamma_1^2} + v_F^2 \frac{k_x^2 + k_y^2}{\gamma_1}.
\tag{10.18}
$$

[9] Note that the exact magnitudes of these coupling strengths depend on which property of graphite they have been fitted and how many couplings are included in the fit. Moreover, the values are slightly different for a few-layer graphene compared to bulk graphite. For this discussion it is the relative order of these couplings that is more important than the exact numerical values.

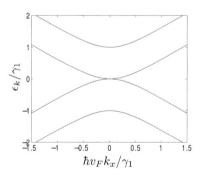

Fig. 10.7 Electron dispersion relation of bilayer graphene in the nearest-neighbour approximation with $\delta\mu = 0$ close to the Dirac points at $\vec{K}$ and $\vec{K}'$. The low-energy Hamiltonian (10.19) describes the two low-energy states residing around $\epsilon_k \approx 0$. Note that the gap to the higher-order states is of the order of 0.4 eV, which in temperature units would be 4600 K.

This form applies for low momenta satisfying $v_F|k| \ll \gamma_1$. For most relevant parameter values, when $\delta\mu \ll \gamma_1$, the second term can rather safely be disregarded.

The bilayer dispersion relation around the Dirac point for two values of $\delta\mu$ is plotted in Fig. 10.8.

In the absence of the potential difference between the two layers, $\delta\mu = 0$, the effective low-energy Hamiltonian for the bilayer becomes

$$H_b = \begin{pmatrix} 0 & \frac{v_F^2 k^2}{\gamma_1} \\ \frac{v_F^2 (k^*)^2}{\gamma_1} & 0 \end{pmatrix}, \tag{10.19}$$

where $k = k_x + ik_y$. This is a 2×2 matrix as it is projected into the eigenspace spanned by the two eigenvectors with lowest energies. Hence, due to the coupling between the layers the charge carriers obtain an effective mass proportional to γ_1. This changes the properties of bilayer graphene compared to the single layer quite dramatically. However, the bilayer still does not behave quite like an ordinary metal, as it possesses the pseudospin index (chirality) absent in ordinary metals.

The possibility of opening up a gap by applying a potential difference between the layers is also an extremely relevant property of bilayer graphene. This is due to the fact that it may allow realizing field-effect transistors with a large difference in the current in the 'on' or 'off' state (Zhang *et al.*, 2009).

10.2.1   Multilayer graphene

Multilayer graphene with more than two layers comes in two possible forms: one with Bernal (1924) stacking, and another one with rhombohedral (McClure, 1969) stacking. The previous is the commonly found stacking of graphite, but the latter also exists in a metastable form. As shown in this section, the rhombohedral stacking leads to quite an exotic electronic dispersion for the lowest-energy states.

Both sequences are extensions of the bilayer stacking: one of the unit cell atoms (say, B) of the second graphene layer lies on top of one of the unit cell atoms (say, A) of the first graphene layer, and the second atom lies on top of the centre of the honeycomb of the other layer. The previous atoms are then coupled stronger than the latter. The difference in the stacking sequences is in the way of choosing how the line of strongest interlayer coupling (denoted γ_1) continues. In the Bernal stacking (also called ABA stacking, see Fig. 10.9) the line of strongest couplings is straight, whereas in the rhombohedral or ABC stacking shown in Fig. 10.10 it follows an 'armchair'-like pattern.

In the following, I describe the multilayers with the same hopping parameters γ_0, γ_1 and γ_3 as above. Their exact values depend on the stacking, but orders of magnitudes do not. Since the intralayer coupling γ_0 is much stronger than that between the layers, we can perform the approximations leading to the Hamiltonian in eqn (10.10), and concentrate on the dispersion around one of the valleys, say K. For simplicity,

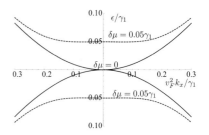

Fig. 10.8 Dispersion relation of the lowest-energy eigenstates in bilayer graphene for two values of $\delta\mu$.

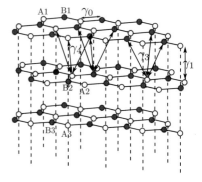

Fig. 10.9 Bernal graphene stacking: the nearest-neighbour interlayer coupling γ_1 (dashed) goes along a single line (from A1 to B2, then from B2 to A3, then from A3 to B4, and so forth).

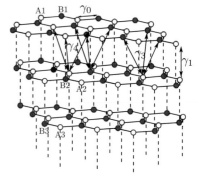

Fig. 10.10 Rhombohedral graphene stacking: the nearest-neighbour interlayer coupling γ_1 (dashed) is 'up' from every A atom and 'down' from every B atom.

[10]The couplings with strength γ_3 and γ_4 couple three atoms from one layer to one atom in the other layer. Therefore, these couplings obtain a similar phase factor γ_k as in the single-layer case, and lead to a coupling strength proportional to the momentum around the valleys K and K'. This is why they can be disregarded close to the Dirac point.

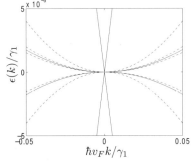

Fig. 10.11 Low-energy spectrum $\epsilon(k)$ for bernally stacked multilayer graphene with $N = 3$ (solid lines) and $N = 4$ (dashed lines) layers, within the approximative model of eqn (10.20).

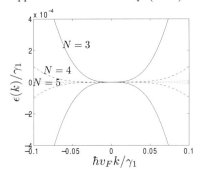

Fig. 10.12 Low-energy spectrum $\epsilon(k)$ for rhombohedrally stacked multilayer graphene with $N = 3$ (solid lines) and $N = 4$ (dashed lines) and $N = 5$ (dotted lines) layers, within the approximative model of eqn (10.21).

we can describe the electronic properties of the multilayers by including only the strongest interlayer coupling γ_1.[10] For bernal stacking, this leads to the Hamiltonian of the form (for the definition of the ladder operators $\sigma_{\uparrow/\downarrow}$, see Appendix A.6)

$$H^{\text{bernal}} = \begin{pmatrix} \hbar v_F \vec{\sigma} \cdot \vec{k} & \gamma_1 \sigma_\uparrow & 0 & 0 & 0 & \cdots \\ \gamma_1 \sigma_\downarrow & \hbar v_F \vec{\sigma} \cdot \vec{k} & \gamma_1 \sigma_\downarrow & 0 & 0 & \cdots \\ 0 & \gamma_1 \sigma_\uparrow & \hbar v_F \vec{\sigma} \cdot \vec{k} & \gamma_1 \sigma_\uparrow & 0 & \cdots \\ 0 & 0 & \gamma_1 \sigma_\downarrow & \hbar v_F \vec{\sigma} \cdot \vec{k} & \gamma_1 \sigma_\downarrow & \cdots \\ \vdots & \vdots & \vdots & \vdots & \vdots & \vdots \end{pmatrix},$$

(10.20)

where the rows and columns refer to the different layers. The Hamiltonian contains N rows and columns for N layers, and the interlayer couplings alternate between $\sigma_\uparrow$ and $\sigma_\downarrow$.

The energy spectrum splits into $2N$ branches, so that four branches lie around $\epsilon = 0$ and the other branches are gapped (see Fig. 10.11). For an odd N there are two branches with linear dispersion and two with a quadratic dispersion, whereas for even N all branches come with quadratic dispersion. For more details of the spectrum of few-layer graphene, and the effect of the higher-order couplings, see (Partoens and Peeters, 2006) or (Latil and Henrard, 2006).

For rhombohedral multilayers, the Hamiltonian is transformed to

$$H^{\text{rhg}} = \begin{pmatrix} \hbar v_F \vec{\sigma} \cdot \vec{k} & \gamma_1 \sigma_\uparrow & 0 & 0 & 0 & \cdots \\ t\sigma_\downarrow & \hbar v_F \vec{\sigma} \cdot \vec{k} & \gamma_1 \sigma_\uparrow & 0 & 0 & \cdots \\ 0 & \gamma_1 \sigma_\downarrow & \hbar v_F \vec{\sigma} \cdot \vec{k} & \gamma_1 \sigma_\uparrow & 0 & \cdots \\ 0 & 0 & \gamma_1 \sigma_\downarrow & \hbar v_F \vec{\sigma} \cdot \vec{k} & \gamma_1 \sigma_\uparrow & \cdots \\ \vdots & \vdots & \vdots & \vdots & \vdots & \vdots \end{pmatrix}.$$

(10.21)

Now the low-energy electronic dispersion is qualitatively different from the single- and bilayer cases. As discussed in more detail in Complement 10.1, there are now only two low-energy branches, with dispersion $\epsilon_k \propto \pm k^N$, where N is the number of layers. These low-energy states are localized mostly on the surfaces of multilayer graphene. Projected on these states, the effective low-energy Hamiltonian of multilayer rhombohedral graphene is

$$H_{\text{rhg}} = \gamma_1 \left(\frac{\hbar v_F}{\gamma_1} \right)^N \begin{pmatrix} 0 & (i\partial_x + \partial_y)^N \\ (i\partial_x - \partial_y)^N & 0 \end{pmatrix}.$$

(10.22)

This Hamiltonian describes exotic Dirac-type particles with dispersion $\epsilon_k \propto k^N$, where N may be very large. The low-energy dispersion for few-layer rhombohedral graphene within the nearest-neighbour coupling approximation is plotted in Fig. 10.12. In the limit $N \gg 1$, the surface states form a flat band, i.e., are essentially dispersionless for a large range of momenta. Such a flat band dispersion is typically susceptible to interaction effects. For example, it has been suggested that rhombohedral graphene multilayers could support surface superconductivity with a high transition temperature (Kopnin *et al.*, 2011).

At the time of writing, the first experiments on three-layer rhombohedral graphene have shown the consequences of this unusual dispersion; see for example (Bao *et al.*, 2011).

Majority of graphene theory and experiments have concentrated on monolayer graphene. In the following sections the discussion is also limited to monolayers. This does not mean that multilayers would be any less interesting. Therefore, I invite the reader to try to investigate similar phenomena under the general Hamiltonian (10.22). In many cases this will probably lead to a lot of fascinating physics.

Complement 10.1 Low-energy states of rhombohedral graphene

Let us consider the nearest-neighbour Hamiltonian of N-layer graphene with rhombohedral coupling, eqn (10.21). In the following, we find the two low-energy eigenstates of this system, and construct an effective Hamiltonian describing the dynamics within the space spanned by these eigenstates. As it turns out, these eigenstates are surface states, mostly localized on the upper- and lowermost surfaces of the system.[11]

[11] Within the nearest-neighbour model, the existence of these surface states can be argued from the topological properties of bulk rhombohedral graphite, see (Heikkilä and Volovik, 2011).

The low-energy eigenstates can be found by concentrating around the low-momentum region $k = \sqrt{k_x^2 + k_y^2} \approx 0$.[12] At $k = 0$, the Hamiltonian has two zero-energy eigenstates of the form

$$\psi_A(k=0) = |A\rangle_N, \quad \psi_B(k=0)|B\rangle_1, \tag{10.23}$$

[12] Note that electron momentum $\vec{p} = \hbar\vec{k}$ is directly related to the wave vector $\vec{k}$ and hence these terms are used interchangably.

where the notation $|\sigma\rangle_i$ refers to an electron with pseudospin (sublattice index) $\sigma \in \{A, B\}$ on layer i. For a non-zero but small momentum $k \ll \gamma_1/(\hbar v_F)$, the eigenstates go to the form

$$\psi_\pm(k) = \sum_{n=1}^{N} \alpha_{nA}(\vec{k})|A\rangle_n \pm \alpha_{nB}(\vec{k})|B\rangle_n, \tag{10.24}$$

where the coefficients $\alpha_{nA/B}$ of the eigenstates on different layers and sublattices should be determined from the eigenvalue equation

$$\begin{aligned} H^{rhg}(k)\psi_\pm(k) &= \epsilon(k)\psi(k) \\ &\approx \frac{\epsilon(k)}{\sqrt{2}}[e^{i\theta(\vec{k})}\psi_A(k=0) + \psi_B(k=0)] + \epsilon(k=0)\psi(k) \\ &= \frac{\epsilon(k)}{\sqrt{2}}[e^{i\theta(\vec{k})}\psi_A(k=0) + \psi_B(k=0)]. \end{aligned} \tag{10.25}$$

In the second, approximate, equality we expand the equation in momentum p, assuming that $\epsilon(k) \propto k^M$ with $M > 0$, and in the third we use the fact that $\epsilon(k=0) = 0$, discussed above. In this procedure, the phase $\theta(\vec{k})$ between the zero-momentum eigenstates has to be chosen so that the resultant energy $\epsilon(k)$ is real.

Writing eqn (10.25) for the coefficients $\alpha_{nA/B}$ yields

$$\sum_{n=1}^{N} [\alpha_{nA}\hbar v_F(k_x + ik_y)|B\rangle_n + \alpha_{nB}\hbar v_F(k_x - ik_y)|A\rangle_n$$

$$+\gamma_1\alpha_{nA}|B\rangle_{n+1}(1 - \delta_{nN}) + \gamma_1\alpha_{nB}|A\rangle_{n-1}(1 - \delta_{n1})] = \frac{\epsilon(k)}{\sqrt{2}}(|B\rangle_1 + e^{i\theta(\vec{k})}|A\rangle_N). \tag{10.26}$$

Here the Kronecker δ-functions take care of the lack of hopping away from the surface layers. To determine the coefficients $\alpha_{nA/B}$, we denote $k_x \pm ik_y =$

$ke^{\pm i\phi}$, where $\phi = \arctan(k_y/k_x)$, and multiply eqn (10.26) first by $\langle A|_m$ and then by $\langle B|_m$, where $m = 1, 2, \ldots, N$. As the states on different layers are orthogonal to each other, this yields

$$\hbar v_F k e^{i\phi} \alpha_{1A} = \frac{\epsilon(k)}{\sqrt{2}} \tag{10.27a}$$

$$\hbar v_F k e^{-i\phi} \alpha_{NB} = \frac{\epsilon(k) e^{i\theta(\vec{k})}}{\sqrt{2}} \tag{10.27b}$$

$$\hbar v_F k e^{i\phi} \alpha_{mA} = -\gamma_1 \alpha_{m-1A}, \quad m \neq 1 \tag{10.27c}$$

$$\hbar v_F k e^{-i\phi} \alpha_{mB} = -\gamma_1 \alpha_{m+1B}, \quad m \neq N. \tag{10.27d}$$

Solving these equations yields[13]

$$\alpha_{mA} = -\left(\frac{-\gamma_1}{\hbar v_F k e^{i\phi}}\right)^m \frac{\epsilon(k)}{\sqrt{2}\gamma_1}, \quad \alpha_{mB} = -\left(\frac{-\gamma_1}{\hbar v_F k e^{-i\phi}}\right)^{N-m+1} \frac{\epsilon(k) e^{i\theta(\vec{k})}}{\sqrt{2}\gamma_1}. \tag{10.28}$$

This solution should be consistent with a combination of the zero-momentum eigenstates, eqn (10.23). This consistency can be checked by requiring

$$\lim_{p\to 0} \alpha_{NA} = \lim_{p\to 0} -\left(\frac{-\gamma_1}{\hbar v_F k e^{i\phi}}\right)^N \frac{\epsilon(k)}{\sqrt{2}\gamma_1} = \frac{e^{i\theta(\phi)}}{\sqrt{2}} \quad \text{and}$$

$$\lim_{p\to 0} \alpha_{1B} = \lim_{p\to 0} -\left(\frac{-\gamma_1}{\hbar v_F k e^{-i\phi}}\right)^N \frac{\epsilon(k) e^{i\theta(\vec{k})}}{\sqrt{2}\gamma_1} = \frac{1}{\sqrt{2}}. \tag{10.29}$$

These requirements can be satisfied with[14] $\theta(\phi) = -N(\phi + \pi) + \pi$ or $\theta(\phi) = -N(\phi + \pi)$ and

$$\epsilon(k) = \pm\gamma_1 \left(\frac{\hbar v_F k}{\gamma_1}\right)^N, \tag{10.30}$$

where the first choice of the $\theta(\phi)$ corresponds to the positive energy and the second choice to the negative energy. The corrections to this energy are at lowest of the order k^{N+1}, and they become small in the limit of small momenta or, equivalently, for energies small compared to γ_1.[15] The result in eqn (10.30) is consistent also for the single-layer ($N = 1$) linear dispersion and bilayer ($N = 2$) quadratic dispersion. In the limit $N \gg 1$, the energy for the surface states is essentially zero for all $k < \gamma_1/(\hbar v_F)$, and the states form a flat energy band.

For low momenta, $k \ll \gamma_1/(\hbar v_F)$, the components of the low-energy eigenstates decay into the bulk of the system: the A component residing preferentially on the lower surface ($m = N$) and the B component on the upper surface ($m = 1$).[16] The decay length depends logarithmically on the size of the momentum vector: for example, for the B component $\psi(z = ma) \propto (\hbar v_F k/\gamma_1)^m \equiv \tilde{k}^m \sim e^{m \ln \tilde{k}} = e^{-z|\ln \tilde{k}|/a}$. The low-energy eigenstates are thus *surface states*.

Let us now define the two mutually orthogonal pseudospin vectors, $|\psi_A\rangle$ and $|\psi_B\rangle$, choosing their phase and normalization so that the low-energy eigenvectors corresponding to the two choices of the sign of the energy are[17]

$$|\psi_\pm(k)\rangle = \frac{1}{\sqrt{2}}(|\psi_A\rangle \pm e^{iN\phi}|\psi_B\rangle) \tag{10.31}$$

or

$$|\psi_A(k)\rangle = \frac{1}{\sqrt{2}}(|\psi_+(k)\rangle + |\psi_-(k)\rangle), \quad |\psi_B(k)\rangle = \frac{1}{\sqrt{2}}(|\psi_+(k)\rangle + \psi_-(k)\rangle)e^{-iN\phi}. \tag{10.32}$$

[13]The eigenfunctions defined in eqn (10.24) with the coefficients defined in eqn (10.28) are normalized at $p \to 0$. At a non-zero momentum they are normalized to the same accuracy at which we calculate the energy, i.e., up to the order $(\hbar v_F k/t)^N$.

[14]The πs are there to take care of the signs.

[15]It turns out that there is indeed a correction of the order k^{N+2}, which becomes important in the presence of interactions (Kopnin *et al.*, 2011).

[16]This symmetry-breaking is due to the chosen type of spin matrix in eqn (10.21) and is dependent on the valley index: for the other valley, $\sigma_\uparrow$ should be replaced by $\sigma_\downarrow$, and the role of the pseudospins are reversed.

[17]The two eigenstates satisfy $H|\psi_+\rangle = \hbar v_F k^N/\gamma_1^{N-1}|\psi_+\rangle$ and $H|\psi_-\rangle = -\hbar v_F k^N/\gamma_1^{N-1}|\psi_-\rangle$.

We may now project the Hamiltonian on the pseudospin states, noting that (see Exercise 10.6)

$$\langle\psi_A|H|\psi_A\rangle = \langle\psi_B|H|\psi_B\rangle = 0 \tag{10.33a}$$

$$\langle\psi_A|H|\psi_B\rangle = \langle\psi_B|H|\psi_A\rangle^* = \epsilon_k e^{-iN\phi} = \gamma_1 \left(\frac{\hbar v_F}{\gamma_1}\right)^N (k_x - ik_y)^N, \tag{10.33b}$$

Fourier transforming into the real space, we get the effective Hamiltonian of eqn (10.22).

10.3 Ray optics with electrons: np and npn junctions

In ballistic conductors, when interference effects may be disregarded, propagating electron states can often be described within the semiclassical picture, similar to the ray optics picture of electromagnetic waves. A given electronic state is described by its angle ϕ of propagation, and the group velocity corresponding to its state. The latter is defined as $\vec{v} = dE/d(\hbar k)$, and in graphene close to the Dirac point this group velocity is $+v_F$ for electrons ($E > 0$) and $-v_F$ for holes ($E < 0$). This difference, and the ability to make sharp boundaries between positively (p, $\mu_g < 0$) and negatively (n, $\mu_g > 0$) charged regions, allows graphene to be used for guiding electron beams. This is demonstrated in Secs. 10.3.1 and 10.3.2 below, which first discuss the angular dependence of the transmission through a simple np junction and then a pnp junction, which is a realization of a tunnelling thought experiment discussed by Klein.

Strictly speaking, both of these effects require ballistic graphene. While in typical experiments the graphene itself is extremely defect-free, the problem usually arises from charged defects on the substrate. These lead to local variations in the gate potential, creating electron and hole 'puddles' where the potential is slightly shifted in either direction from the Dirac point even in an otherwise charge neutral situation (Martin *et al.*, 2008).[18] Because of this, to reach as ballistic graphene as possible, experimentalists suspend the graphene to hang freely between the electrodes. However, such a scheme makes device fabrication more difficult and cannot be used for all applications.

10.3.1 Graphene np junction

Consider a graphene sample with an electrochemical potential profile $\mu_g(x)$ as in Fig. 10.13.[19] For $x < 0$, $\mu_g(x) = \mu_L > 0$ and $\mu_g(x > 0) = \mu_R < 0$. On the left-hand side of the junction the charge carriers are hence electrons (Fermi level is in the conduction band) and on the right they are holes (Fermi level is in the valence band). An electron wave adjacent from the left with an incoming angle ϕ either reflects back or transmits into a hole state with a group velocity pointing away from the junction. Our aim is to solve the Schrödinger equation (10.11) in this system by matching the wave functions at the junction. Among other

[18]At the time of writing, puddles are still a problem in non-suspended graphene. However, using a different substrate, based on boron nitride instead of the usual SiO_2, has allowed a decrease in these potential variations (Dean *et al.*, 2010).

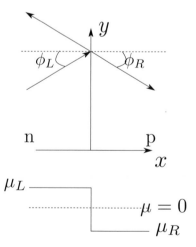

Fig. 10.13 Graphene np junction.

[19]Here I follow the ideas presented in (Cheianov *et al.*, 2007).

things, such a procedure gives us the transmission probability amplitude t through the system. The total wave function of the system is given by

$$\psi(x,y) = \psi_L(x,y) + \psi_R(x,y), \qquad (10.34)$$

[20]Note that here I have multiplied the eigenfunction of eqn (10.12) with a trivial phase factor $e^{i\phi/2}$, which moves all the phase factors to the B component.

where[20]

$$\psi_L(x,y) = \frac{1}{\sqrt{2}}\begin{pmatrix} 1 \\ -e^{i\phi_L} \end{pmatrix} e^{i(k_x^L x + k_y y)} + \frac{r}{\sqrt{2}}\begin{pmatrix} 1 \\ -e^{i(\pi-\phi_L)} \end{pmatrix} e^{i(-k_x^L x + k_y y)}$$

$$\psi_R(x,y) = \frac{t}{\sqrt{2}}\begin{pmatrix} 1 \\ e^{i(\pi-\phi_R)} \end{pmatrix} e^{i(-k_x^R x + k_y y)}. \qquad (10.35)$$

Here, at $E = 0$, $k_x^{L/R} = k_{L/R}\cos(\phi_{L/R})$ with $k_{L/R} = |\mu_{L/R}|/\hbar v_F$ is the part of momentum that is not conserved due to the presence of the potential step. The outgoing angle ϕ_R is determined by requiring the conservation of momentum in the y-direction, i.e.,

$$\frac{\sin(\phi_L)}{\sin(\phi_R)} = -\frac{k_R}{k_L} = \frac{\mu_R}{\mu_L}, \qquad (10.36)$$

which is nothing but the Snell's law for the transmitted electrons, but with a *negative* refractive index.

As the Dirac equation is linear it is sufficient to require continuity of the wave functions. For the Schrödinger or Bogoliubov–de Gennes equation (see Sec. 5.4.1), one would have to require also the continuity of the derivative. Continuity at $x = 0$ yields the transmission probability as a function of the incoming and outgoing angles ϕ_L and ϕ_R,

$$|t|^2 = \frac{\cos^2(\phi_L)}{\cos^2(\phi_L/2 - \phi_R/2)}. \qquad (10.37)$$

In particular, when $\mu_L = -\mu_R$ we get $\phi_R = \phi_L$ and $|t|^2 = \cos^2(\phi_L)$.

The negative refractive index allows to focus an electron beam. A system with such a property has been shown by Veselago (1968) to create a perfect lens, which is not limited by diffraction but only by the roughness of the np boundary.

The first experiments on this type of optical phenomena have appeared recently. For example, Williams *et al.* (2011) showed that a graphene sample containing n and p regions can be used to guide electrons similarly to what is done in fibre optics.

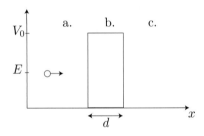

Fig. 10.14 Electron with energy E approaching a potential barrier with height V_0. Classically the electron would entirely reflect from the barrier, quantum-mechanically a non-relativistic electron has a small probability of tunnelling through it, and a relativistic electron may tunnel through it with a high probability. Here all the energies are expressed as a distance to the Dirac point $E = 0$. The system can be considered as an npn junction, as in semiconductor structures.

[21]For a more thorough treatment of the Klein tunnelling in graphene, see (Katsnelson *et al.*, 2006) or (Allain and Fuchs, 2011).

10.3.2 Klein tunnelling

Klein tunnelling is a phenomenon predicted by Klein (1929) in the context of relativistic particles. It is often also referred to as the Klein paradox, but it is not really a paradox as it can be well understood with the use of the Dirac equation and the coupling between electrons and holes that it implies.[21]

Consider a ballistic graphene layer with translational symmetry in the y-direction and the potential profile $U(x)$ shown in Fig. 10.14. Assume that both E and V_0 are low enough, so that we can employ the low-energy Hamiltonian (10.10) and the resultant eigensolutions, eqn (10.12).

In a ballistic graphene system depicted in Fig. 10.14, we can use the ansatz wave function

$$\psi(x,y) = \psi_a(x,y) + \psi_b(x,y) + \psi_c(x,y), \qquad (10.38)$$

where ψ_a, ψ_b and ψ_c are defined on the x-intervals specified in the figure and they are of the form

$$\psi_a(x,y) = \frac{1}{\sqrt{2}}\begin{pmatrix} 1 \\ se^{i\phi} \end{pmatrix} e^{i(k_x x + k_y y)} + \frac{r}{\sqrt{2}}\begin{pmatrix} 1 \\ se^{i(\pi-\phi)} \end{pmatrix} e^{i(-k_x x + k_y y)}$$

$$\psi_b(x,y) = \frac{a}{\sqrt{2}}\begin{pmatrix} 1 \\ s'e^{i\theta} \end{pmatrix} e^{i(q_x x + k_y y)} + \frac{b}{\sqrt{2}}\begin{pmatrix} 1 \\ s'e^{i(\pi-\theta)} \end{pmatrix} e^{i(-q_x x + k_y y)},$$

$$\psi_c(x,y) = \frac{t}{\sqrt{2}}\begin{pmatrix} 1 \\ se^{i\phi} \end{pmatrix} e^{i(k_x x + k_y y)}. \qquad (10.39)$$

Here $q_x = \sqrt{(E-V_0)^2/(\hbar v_F)^2 - k_y^2}$, $\theta = \arctan(k_y/q_x)$ (see Fig. 10.15) and $s' = \mathrm{sgn}(E - V_0)$. The different coefficients can be found by requiring the wave function to be continuous at the interfaces, i.e., $x = 0$ and $x = d$:

$$\psi_a(0,y) = \psi_b(0,y) \quad \text{and} \quad \psi_b(d,y) = \psi_c(d,y). \qquad (10.40)$$

Matching the wave function coefficients is lengthy but straightforward.[22] This procedure yields, for example, for the transmission coefficient of eqn (10.39)

$$t = \frac{e^{-idk_x} s \cos(\theta)\cos(\phi)s'}{s\cos(\theta)\cos(\phi)\cos(dq_x)\, s' + i\sin(dq_x)\left(s\sin(\theta)\sin(\phi)s' - 1\right)}. \qquad (10.41)$$

The transmission probability $T = tt^*$ is

$$T = \frac{4\cos^2(\theta)\cos^2(\phi)}{D_1 + D_2}. \qquad (10.42)$$

with

$$D_1 = \left(\cos(2\theta) + \cos(2\phi) - 2\right)\cos(2dq_x) \quad \text{and}$$
$$D_2 = \left(\cos(2\theta)\cos(2\phi) - ss'8\sin(\theta)\sin(\phi)\sin^2(dq_x)\right) + 3.$$

This function is plotted in Fig. 10.16 as a function of ϕ for a few values of V_0 for particles at the Dirac point ($E = 0$). For a large range of angles the transmission is unity even for a high and wide potential barrier. Especially for normal incidence ($k_y = 0$, i.e., $\phi = \theta = 0$) and for values of q_x satisfying $q_x d = n\pi$ the transmission is unity, irrespective of the other details of the junction.

Klein tunnelling can be roughly understood by noting that at the edges of the potential barrier the positive-energy excitations (electrons) on regions a and c couple to negative-energy excitations (holes) of region

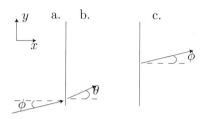

Fig. 10.15 Electron propagation angles ϕ and θ in the Klein tunnelling experiment.

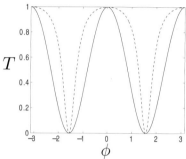

Fig. 10.16 Transmission probability through a graphene barrier at the Dirac point ($E = 0$) as a function of the angle of incidence ϕ of the electron for two values of the barrier height $V_0 = 20\hbar v_F/d$ (solid line) and $V_0 = 60\hbar v_F/d$ (dashed line). For example, for $d = 100$ nm these would be 0.13 eV and 0.39 eV, respectively.

[22] At least with mathematical software.

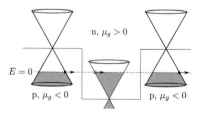

Fig. 10.17 Klein tunnelling is due to electron–hole conversion at the pn/np junctions.

b as in Fig. 10.17. The transmission is then finite due to the fact that holes can propagate inside the barrier. In normal electron systems the branch crossing between electron and hole states is not possible, and therefore Klein tunnelling is typically not encountered. Alternatively, the perfect transmission at normal incidence can be understood via the conservation of helicity of the charge carriers in the presence of scattering that does not break the sublattice symmetry.

Note that the above derivation assumes that the edges of the potential barriers are sharp compared to the Fermi wavelength within region *b*. In practice, such sharp contacts are difficult to realize. It has been shown (Cheianov and Fal'ko, 2006) that for a smooth contact (potential changing from 0 to V_0 on a scale longer than Fermi wavelength) the transmission probability is still unity for normal incidence but decays exponentially as $T(\phi) \sim e^{-\pi(k_F d)\sin^2(\phi)}$ away from it. However, especially in small systems the large transmission for $k_y \approx 0$ should show up in the experimentally measured transmission properties.

Observation of Klein tunnelling has been difficult due to the fact that typical mean free paths in graphene are of the order of 100 nm, and sharp potential variations within scales below that are difficult to achieve. However, there are recent reports of the evidence of Klein tunnelling in mildly disordered systems; see for example (Young and Kim, 2009).

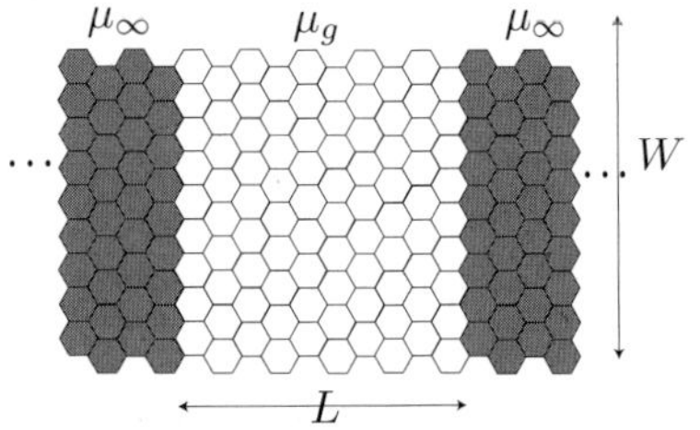

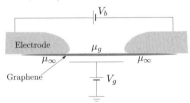

Fig. 10.18 Graphene wire of width W and length L between two electrodes (shaded regions). The Fermi level μ_g of the wire can be varied by tuning a gate voltage capacitively connected to the graphene strip. In the model applied here, we assume the leads to be composed of doped graphene regions with potential $|\mu_\infty| \gg |\mu_g|$.

[23] For further details, see (Tworzydło *et al.*, 2006).

Fig. 10.19 Graphene contacted with metallic electrodes: electrodes dope the region underneath them. As a result, the potential μ_∞ of the region close to the graphene–electrode contact cannot be modulated much with the voltages.

10.4   Pseudodiffusion

Consider a strip of ballistic graphene attached between two electrodes as shown in Fig. 10.18. Let us calculate the transmission through such a ballistic graphene strip. The simplest model for the leads consists of assuming them to be made also from graphene but with a potential $\mu_\infty \ll \mu_g$.[23] This assumption allows describing many propagating modes inside the leads, coupled to propagating and evanescent modes of the wire. In particular, at the Dirac point $E = 0$ the density of states (eqn (10.14)) vanishes and all the modes inside the wire are evanescent. As shown in this section, the contribution of the evanescent waves on transport through a wide ballistic graphene flake results in a diffusive-like behaviour of the transmission eigenvalues. In particular, the conductance of the sample is inversely proportional to its length.

The specific assumption about the leads being also made from graphene can be justified by referring to the real measured systems. They typically contain a graphene sheet and metal electrodes fabricated on top of it, as in Fig. 10.19. These metal electrodes dope the graphene underneath by transferring charge between the two systems until the Fermi levels in the two systems match. This results into the potential distribution roughly in line with Fig. 10.18. To solve for the transmission amplitudes we can use the results of the previous section on Klein tunnelling, but replacing E by $E - \mu_\infty$ inside regions *a* and *c* and V_0 by μ_g inside the region *b* corresponding to the graphene wire. Taking $\mu_\infty \to -\infty$ yields

$\phi = 0$ and thereby

$$T(k_y) = \frac{2q_x^2}{2q_x^2 + k_y^2(1 - \cos(2q_x L))}, \qquad (10.43)$$

where $q_x = \sqrt{(E - \mu_g)^2/(\hbar v_F)^2 - k_y^2}$. The resultant conductance and the Fano factor describing shot noise (see Ch. 6) can be calculated from the usual formulae (see eqns (3.49) and (6.23))

$$G = \frac{2e^2}{h} \sum_n T_n \quad \text{and} \quad F = \frac{\sum_n T_n(1 - T_n)}{\sum_n T_n},$$

where the sums go over the transverse states inside the graphene wire acting as the lead. The details of these states depend on the boundary conditions at the edges of the wire (see Sec. 10.5 below). However, the results depend on the boundary conditions only when the width of the wire is of the order of or less than the length L. The conductance and the Fano factor are plotted as a function of gate voltage for structures with metallic armchair edges in Fig. 10.20.

If the wire is sufficiently wide (in practice, $W \gtrsim 3L$ is enough), we can calculate the total linear conductance by replacing the summation over the transverse modes by an integral,

$$G = \frac{4e^2}{h} \frac{W}{\pi} \int_0^\infty dk_y T(k_y). \qquad (10.44)$$

The prefactor W/π is due to the level spacing inside the wire and the 'conductance quantum' $2 \times 2e^2/h$ contains the spin and valley degeneracies. At the Dirac point $\mu_g = 0$ we obtain

$$G = \frac{4e^2}{\pi h} \frac{W}{L}. \qquad (10.45)$$

The conductance thus scales with the size as in an ohmic wire with conductivity $\sigma = 4e^2/(\pi h)$ (at the Dirac point). Analogously we can obtain the Fano factor for shot noise,

$$F = \frac{\int_0^\infty dk_y T(k_y)[1 - T(k_y)]}{\int_0^\infty dk_y T(k_y)}. \qquad (10.46)$$

At the Dirac point we obtain $F = 1/3$—again similar to what one finds for a diffusive wire (see Sec. 6.2.1), even though we are describing a wire without disorder! This is why the phenomenon has been dubbed 'pseudodiffusion'.

Graphene experiments have reached rather close to the ballistic prediction described above (see Fig. 10.21). The main differences between the ideal theory and experiments are typically found very close to the Dirac point, because even mild disorder changes the measured conductivity.[24] On the other hand, far away from the Dirac point graphene can often be described with the simple Boltzmann approach described in Ch. 2.

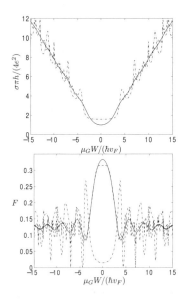

Fig. 10.20 (Top) Conductivity $\sigma = GL/W$ and (bottom) Fano factor of a graphene strip of different width-length ratios as a function of the gate-induced potential μ_g: $W/L = 30$ (solid), $W/L = 3$ (dashed) and $W/L = 1$ (dash-dotted). The smallest systems show resonant-like features at a non-zero gate voltage.

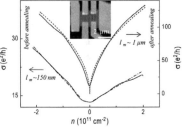

Fig. 10.21 Measured conductivity of suspended graphene (shown in the inset) as a function of the charge density n, which is proportional to the electrochemical potential. The lower solid line shows the conductivity of a disordered sample, and the upper solid line that of a pure sample. The dashed line is the prediction from the ballistic model, eqn (10.44), and it agrees well with the experiments elsewhere than at the Dirac point. (K.I. Bolotin *et al.*, *Phys. Rev. Lett.* **101**, 096802 (2008), Fig. 1.) © 2008 by the American Physical Society.

[24]For a more detailed description, see (Das Sarma *et al.*, 2011).

Fig. 10.22 Transmission electron microscope image of graphene nanoribbons with different types of edges. (X. Jia *et al.*, *Science* **323**, 1701 (2009), Fig. 1.) Reprinted with permission from AAAS.

Fig. 10.23 Graphene ribbons: Left: 'armchair' edges. Center: 'zigzag' edges. Right: 'reczag' edges. The ribbons are assumed to be infinite in the horizontal x-direction. The lower part of the figure shows the detailed structure of the edges.

[25]Controlled formation of different types of edges has been demonstrated in (Jia *et al.*, 2009), see Fig. 10.22.

10.5   Graphene nanoribbons

In Ch. 3 the scattering states in the leads are described in terms of the quantized transverse states. For ordinary metals these transverse states extend across the transverse direction of the leads, and are quantized as $\epsilon_n = \hbar^2/(2m)(n^2/w^2)$, where w is the width of the lead and n is an integer. These quantized states can be studied in graphene by making narrow and long ribbons. It turns out that the nature of the transverse states depends on the properties of the ribbon edges. In particular, for one (zigzag) type of the edges, one can find a low-energy edge state, which exists only at the edge of the ribbon. This edge state is interesting because of its specific (flat band) dispersion and because it breaks the valley symmetry. Moreover, manipulation of the ribbons allows to control an energy gap for transport through them (Han *et al.*, 2007).

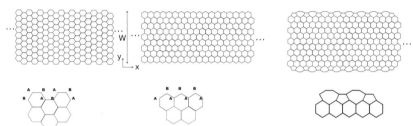

The most usual types of graphene ribbons are depicted in Fig. 10.23. The two first types have 'armchair' and 'zigzag' edges, whereas the rightmost one, with the 'reczag' edge contains a local reconstruction of the graphene lattice, where the hexagonal pattern has been locally deformed into pairs of pentagons and heptagons. It has been shown (Koskinen *et al.*, 2008) by density functional theory that the zigzag edge is unstable against the reconstruction into the reczag form.[25] These edges induce different types of boundary conditions for the Dirac equation (Brey and Fertig, 2006; van Ostaay *et al.*, 2011). Below, I describe these boundary conditions for the armchair and zigzag edges; the boundary condition for the reczag edge is somewhat more complicated (van Ostaay *et al.*, 2011).

To be able to derive the boundary conditions for the nanoribbons, we write the graphene electron wave function in the basis of the sublattices A and B. These wave functions can be written as

$$\Psi_{A/B}(\vec{r}) = e^{i\vec{K}\cdot\vec{r}}\psi_{A/B}(\vec{r}) + e^{i\vec{K}'\cdot\vec{r}}\psi'_{A/B}(\vec{r}). \tag{10.47}$$

Here $\psi_{A/B}(\vec{r})$ are the two components of the solution to eqn (10.11) with the upper sign in front of the x-derivative and $\psi'_{A/B}(\vec{r})$ are the solutions with the lower sign.

10.5.1   Zigzag ribbons

The fact that the zigzag ribbon contains only one type of pseudospin indices on each edge (A-type on one and B-type on the other), results in the low-energy boundary condition for the zigzag edges located at $x = 0$ and $x = W$:

$$\psi_A(x, y = 0) = \psi_B(x, y = W) = 0, \qquad (10.48)$$

similar for both valley indices. We can make an ansatz of the form

$$\psi(x, y) = \begin{pmatrix} \chi_A(y) \\ \chi_B(y) \end{pmatrix} e^{ikx} \qquad (10.49)$$

to the Dirac equation (10.11). Assuming further that

$$\chi_{A/B}(y) = c_{A/B} e^{iqy} + d_{A/B} e^{-iqy} \qquad (10.50)$$

we get $\epsilon(k, q) = \pm \hbar v_F \sqrt{k^2 + q^2}$ and

$$c_A = \frac{\pm k - iq}{\sqrt{k^2 + q^2}} c_B, \quad d_A = \frac{\pm k + iq}{\sqrt{k^2 + q^2}} d_B. \qquad (10.51)$$

Here and in the rest of the section the upper sign corresponds to valley $\vec{K}$ and the lower sign to valley $\vec{K}'$. From the boundary condition (10.48) we obtain a condition for the transverse wave vector q:

$$\frac{\pm k + iq}{\pm k - iq} = e^{-2iqW}. \qquad (10.52)$$

Besides the trivial solution $q = 0$ which would yield $\psi \equiv 0$, this has two types of solution for q, depending on whether $|k| < 1/W$ or $|k| > 1/W$. The first type of solution can be found by noting that for a real q we can take a logarithm from both sides of eqn (10.52), identifying the left-hand side with $i \arctan(\pm q/k)$. Hence, the real solutions for eqn (10.52) are obtained from

$$k = \mp \frac{q}{\tan(qW)} \qquad (10.53)$$

where now both sides are purely real. For $k \ll 1/W$ the right-hand side of eqn (10.53) tends to $\pm(2n + 1)\pi[q - (2n + 1)\pi/(2W)]/2$, where n is an integer. The solutions in this limit are of the form

$$q_n = \frac{(2n + 1)\pi}{2W} \pm \frac{2k}{(2n + 1)\pi}. \qquad (10.54)$$

These solutions depend on the valley index, but not very strongly.

The low-energy spectrum of graphene ribbons, solved from eqn (10.52), is plotted in Fig. 10.24. With $k > 1/W$ for valley K' (lower sign) and with $k < -1/W$ for valley K (upper sign) eqn (10.52) has one purely imaginary non-trivial solution with $q = iz$, $z \lesssim k$. Such a state corresponds to an edge state, which has a predominantly B contribution near $x = 0$ and A contribution near $x = W$. This state has a very low energy $\epsilon = \hbar v_F \sqrt{k^2 - z^2}$: for $|k| \gg 1/W$ we can expand eqn (10.52) in $\delta z = -z \mp k$ to linear order and get $\delta z = 2|k| e^{-2|k|W}$ and hence $\epsilon \approx 2\hbar v_F |k|^2 e^{-2|k|W} \approx 0$. This state thus has a vanishing dispersion at large enough ribbons.[26] A state with a vanishing dispersion has a very high density of states. Because of this, such a state is very susceptible

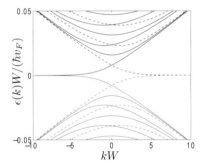

Fig. 10.24 Energy spectrum of a zigzag nanoribbon for longitudinal momentum k around the Dirac points. The solid line is for valley $\vec{K}$ and the dashed line for valley $\vec{K}'$. The zero-energy states, present for $k < -1/W$ for valley $\vec{K}$ and for $k > 1/W$ for valley $\vec{K}'$, are localized around the edges of the nanoribbon.

[26] The presence of this zero-energy edge state can be traced back to the special momentum space topology of bulk graphene, as argued in (Ryu and Hatsugai, 2002).

for different types of ordering. For example, it has been predicted that graphene nanoribbons with zigzag edges could be magnetic (Son *et al.*, 2006), but so far no convincing experimental evidence of this exists. Moreover, as the zigzag edge state exists only for negative (positive) longitudinal momentum k for valley $\vec{K}$ ($\vec{K}'$), zigzag nanoribbons have also been suggested for valley filtering (Rycerz *et al.*, 2007) similarly as the spin is filtered in ferromagnetic heterostructures (discussed in Sec. 2.7).

One possible reason for the lack of evidence of magnetism in zigzag nanoribbons may be that the zigzag edge is not stable, but reconstructs into other types of edges; for example, the reczag edge depicted in Fig. 10.23. This edge can also be described with an effective boundary condition for the Dirac equation (van Ostaay *et al.*, 2011). In this case, there is also an edge state, but it has a finite dispersion of the form $\epsilon(k) = \mp \hbar v_F k \sin(\vartheta)$, where the angle $\vartheta \approx 0.15$ describes the effective boundary condition.

10.5.2   Armchair ribbons

Armchair ribbons contain atoms from both sublattices at the ribbon edge. The boundary conditions at the ribbon edges are

$$\Psi_A(y=0) = \Psi_B(y=0) = \Psi_A(y=W) = \Psi_B(y=W) = 0, \quad (10.55)$$

where $\Psi_{A/B}$ is defined in eqn (10.47). Assuming translational invariance in the x-direction, we can again use ansatz (10.49) for the solutions of eqn (10.11). Combining this with eqn (10.55) leads to

$$e^{ikx}\chi_{A/B}(0) + e^{ikx}\chi_{A/B}(0)' = 0 \qquad (10.56a)$$
$$e^{iKW}e^{ikx}\chi_{A/B}(L) + e^{-iKW}e^{ikx}\chi'_{A/B}(L). \qquad (10.56b)$$

Here we have chosen the directions such that the Dirac points are at $\vec{K} = (0, K)$ and $\vec{K}' = (0, -K)$ where $K = 4\pi/(3a)$.[27] This equation is satisfied with

$$\chi_{A/B}(0) + \chi'_{A/B}(0) = 0 \qquad (10.57a)$$
$$e^{iKW}\chi_{A/B}(W) + e^{-iKW}\chi_{A/B}(W) = 0. \qquad (10.57b)$$

It is enough to consider the boundary conditions for one pseudospin index only, say A, as the other pseudospin is obtained from the solution of eqn (10.11), and the boundary conditions are symmetric in the pseudospin index. Let us use the ansatz

$$\chi_A(y) = Ae^{iqy} + Be^{-iqy}, \quad \chi'_A(y) = Ce^{iqy} + De^{-iqy}.$$

Then the boundary conditions yield

$$A + B + C + D = 0$$
$$Ae^{i(q+K)W} + De^{-i(q+K)W} + Be^{-i(q-K)W} + Ce^{i(q-K)L}.$$

[27]To get an armchair edge in the y-direction, we have to rotate the graphene by 90 degrees compared to the discussion in Sec. 10.1. As a result, the valleys in positions $(\pm K, 0)$ in Fig. 10.4 go to positions $(0, \pm K)$.

We can choose $C = B = 0$ and $A = -D$ to obtain the condition on the wave number q: $\sin[(q + K)W] = 0$. This is solved with

$$q_n = \frac{n\pi}{W} - \frac{4\pi}{3a}, \tag{10.58}$$

giving the eigenenergies $\epsilon(k, q) = \hbar v_F \sqrt{k^2 + q_n^2}$. In Exercise 10.8 you will analyze this to determine a condition on whether a given armchair nanoribbon is metallic or semiconducting.

Armchair ribbons do not contain any edge states, as all states have a real wave number.

Further reading

In this chapter, the reader should be acutely aware of the fact that research efforts on understanding the properties of graphene are enormous, and the field is advancing rapidly. The idea here is just to introduce some of the main graphene properties and provide a flavour of the expected—and in some cases already demonstrated—exotic effects. For up-to-date accounts of electronic transport properties of graphene, I would therefore like to refer to the extensive reviews on the topic, such as those by Castro Neto *et al.* (2009) and Das Sarma *et al.* (2011). The former focuses on the basic electronic properties, but also contains information on, for example, phonons, whereas the latter concentrates more on electron transport properties of graphene. Many more popular accounts on the topic also exist; for example the 'Advanced information' by the Nobel committee (NobelGraphene, 2010).

Exercises

(10.1) Analyze the dispersion of monolayer graphene by following these steps:

    (a) Justify the eigenenergies $\epsilon_{\vec{k}}$ in eqn (10.8).

    (b) Find the non-equivalent $\vec{k}$-vectors satisfying $\epsilon_{\vec{k}} = 0$, i.e., $|\gamma_{\vec{k}}| = 0$. These are typically denoted as $\vec{K}$ and $\vec{K}'$.

(10.2) Find the low-energy Hamiltonian with the following steps:

    (a) Linearize eqn (10.6) around $\vec{K}$ and $\vec{K}'$.

    (b) Express v_F in terms of t and a.

(10.3) Include the higher orders in momentum in the low energy monolayer graphene Hamiltonian to calculate the corrections to the linear dispersion. Do this by expanding $|\gamma_{\vec{k}}|^2$ to the third order in the wave vectors. This result reflects the trigonal form of the low-energy contours seen in Fig. 10.4.

(10.4) Calculate the density of states of monolayer graphene close to the Dirac point.

(10.5) Find the eigensolutions of bilayer graphene Hamiltonian, eqn (10.19).

(10.6) Derive eqns (10.33) for multilayered rhombohedral graphene.

(10.7) Analyze the electron guiding in a graphene np junction: a) Derive the angular dependence of the transmission, eqn (10.37). b) By using the fact that $|t|^2$ is a probability for transmission, find the region of phases ϕ_L for which electrons are totally reflected at the np junction. Alternatively, you can derive this condition from eqn (10.36).

(10.8) Using the spectrum of armchair nanoribbons, find out a condition to determine which nanoribbons are metallic, i.e., the lowest-energy transverse state has a vanishing energy for a vanishing longitudinal wave number k. Picking up armchair nanoribbons of different sizes, how big percentage of them are metallic?

(10.9) **Question on a scientific paper.** Take the paper K. Novoselov *et al.*, *Science* **306**, 666 (2004) and try to understand the following questions concerning its Fig. 2. Note that the measurement concerns the conductivity of a few layer graphene as a function of temperature and gate voltage.

a) In the presence of a gate voltage V_g, the experimentally measured resistivity ρ and conductivity σ are given in panels A and B. Why is there a peak in the resistivity? Why is the resistivity peak higher the lower the temperature?

b) Derive the relation between σ and gate voltage in panel B (obtained by inverting the 70K curve) (Hint: what is the relation between carrier density and conductivity given by the Drude formula?).

You can assume that the graphene chemical potential $\mu_g = cV_g$ is linearly proportional to V_g. What determines the position of the minimum in panel B? What majority carriers should be expected for $V_g = 0$? Compare your answer to the results from panel C where the Hall resistance is plotted. (Hint: the only information you need to know about R_H is that its sign depends on the type of carriers responsible for the current flow.)

c) Panel D gives the temperature dependence of the number of carriers for graphene with increasing thickness from top to bottom. Compare this result with the theoretical calculations for mono- and bilayer graphene. Evaluate the temperature dependence with respect to the value at an arbitrary reference temperature T^*, i.e., $n_0(T)/n_0(T^*)$. Hint: Use the identity

$$\int_0^\infty \frac{u^{n-1} du}{\exp{[u]} + 1} = \left(1 - 2^{-1/n}\right) \Gamma(1+1/n)\zeta(1+1/n),$$

where $\Gamma(1+x)$ and $\zeta(1+x)$ are continuous functions on the positive real axis.

Nanoelectromechanical systems

One important and emerging application of nanoelectronic systems is in probing mechanical vibrations in nanostructures. Suspending a wire or a sheet of material between two contacts or only from one of its ends (a cantilever) allows this system to vibrate. If the vibrating object conducts electricity it can be electrically coupled, for example, to nanoelectronic systems, say a single-electron transistor, or alternatively the vibrating wire may form the island of the SET. This coupling may then be used to *transduce* the vibration energy, i.e., convert the mechanical motion into other forms of energy, usually electronic signals. On the other hand, the electronic system may be used to apply a force in order to excite the vibrations. This process is called *actuation*.

Quite generally, the mechanical eigenfrequencies of the vibrations are inversely proportional to the size of the vibrating object (see Sec. 11.1 below). On the scale of nanomechanical resonators,[1] these eigenfrequencies typically range from a few tens to a few hundreds of MHz. Examples of nanoelectromechanical systems are shown in Fig. 11.1. Refrigerating the smallest resonators, with eigenfrequencies ω_m exceeding GHz, to low temperatures, the mode energy $\hbar\omega$ may become of the order of the thermal energy $k_B T$. In this case, only the ground state of the vibrations is occupied, and effects related to, for example, the superpositions of different vibration states can be observed. One of the present aims in the study of nanoelectromechanical systems is to cool the resonators close to their ground state and try to observe the remaining zero-point motion.[2] Such cooling is often very difficult or impossible directly by cooling the substrate, but it can also be achieved via electronic means (Naik *et al.*, 2006; Rocheleau *et al.*, 2009; Teufel *et al.*, 2011).

Besides the possibility of detecting quantum effects, there is another strong motivation for studying ever smaller mechanical resonators: Their resonance frequency depends on the total mass of the resonator. By measuring changes in this resonance, changes in the mass of the resonator can be measured, for example, when the resonators are loaded with extra particles. As the sensitivity of such measurements depends on the relative change in the total mass, making smaller (and thereby lighter) resonators allows the detection of smaller masses. At present the mass detection sensitivity of the smallest resonators is already in the yoctogram (10^{-24} g) range (Jensen *et al.*, 2008; Hüttel *et al.*, 2009; Chaste *et al.*, 2012) corresponding to the mass of a proton. The same applies for

---

[1] In this case the length of the resonators typically exceeds or is of the order of 1 μm. However, the width and thickness of the nanomechanical resonators is typically much less than 1 μm, down to even atomic thicknesses as in the cases of carbon nanotubes and graphene.

[2] The first observations of ground-state cooling of nanomechanical resonators has been very recently reported; see (O'Connell *et al.*, 2010). This experiment is discussed more in Sec. 11.4.1. In that case the resonator was made from piezoelectric material. The first ground-state cooling of a pure mechanical resonance was demonstrated by Teufel *et al.* (2011).

measuring forces applied on the resonators: the force measurement sensitivity improves with decreasing size of the resonators. Ultimately, this allows direct studies of quantum fluctuations in the forces coupling to the resonators. Finally, as discussed in Sec. 11.2.2, coupling the nanoresonator electrically to a gate capacitor allows changing of the resonance frequency by tuning the gate voltage. This then allows for a sensitive charge detection, with resolution in some cases even exceeding that of the radio-frequency SET discussed in Ch. 7.

Fig. 11.1 Examples of nanoelectromechanical systems. Top left: Scanning electron micrograph of an aluminum membrane resonator. (J.D. Teufel *et al.*, *Nature* **475**, 359 (2011), Fig. 1.) Reprinted by permission from Macmillan Publishers Ltd., © 2011. Top right: Suspended aluminum beam resonator cut with a focused ion beam from the Al pad on the left (J. Sulkko *et al.*, *Nano Lett* **10**, 4884 (2010), Fig. 1). In both figures on the top the motion of the resonator modulates the capacitance of a nearby superconducting microwave resonator (not shown), which is used to transduce the motion. Bottom left: Suspended graphene sheet between metal electrodes. (J.S. Bunch *et al.*, *Science* **315**, 490 (2007), Fig. 1) Reprinted with permission from AAAS. Bottom right: Suspended carbon nanotube (marked with the black arrow) between metal electrodes. (G.A. Steele *et al.*, *Science* **325**, 1103 (2009), Fig. 1.). Reprinted with permission from AAAS.

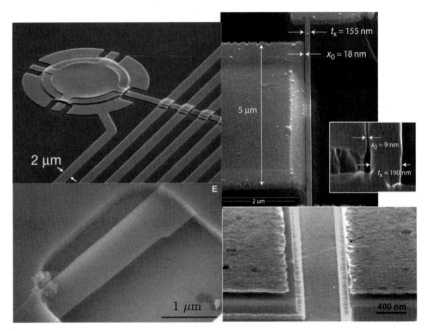

[3] A thorough account of this theory, including a derivation of the Euler–Bernoulli equation employed in Sec. 11.1, is given in (Landau and Lifshitz, 1985) and (Cleland, 2003).

This chapter starts by discussing briefly the basic elastic theory describing mechanical vibrations. The aim of this discussion is to provide the necessary details to describe the dependence of the resonant frequency on the size of the objects, and of the coupling to the electronic system. We do not dwell on the underlying elastic theory in detail.[3] We proceed by introducing two coupling schemes between the electrical and mechanical systems and discuss the resultant effective coupling strengths. Schemes related to non-linear electronic elements (SQUIDs and SETs) and resonant electronic circuits are then exemplified. The rest of the chapter is devoted to a description of quantum effects that have been recently measured in nanoelectromechanical systems. This text is by no means exhaustive, but its aim is to illustrate the possibilities of nanoelectromechanical systems via some selected examples. The reader must also note that this text has been written while the field is advancing rapidly. Therefore it may miss some concepts that turn out to be relevant while the research progresses.

11.1   Nanomechanical systems

This section introduces briefly the basic concepts of elasticity theory and then proceeds to study the properties of flexural vibrations of beams. I discuss only the specific yet relevant case of a doubly clamped beam. The flexural vibration modes in long macroscopic beams are described by the Euler–Bernoulli differential equation, and this is our starting point. In particular, the eigensolutions of this equation provide us with estimates of the vibration eigenfrequencies and waveforms. The vibrations of a single mode satisfy a harmonic oscillator equation, and hence its quantization can be readily achieved by introducing the harmonic oscillator creation and annihilation operators. In the subsequent sections, this allows to describe the quantum effects predicted for these oscillators.

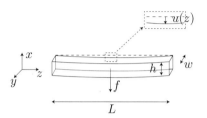

Fig. 11.2 The deformation of a solid beam as a response to a force f pointing in the negative x-direction can be described with the function $u(z)$. The beam on this picture is clamped at both its ends, and therefore $u = 0$ at the ends.

11.1.1   Basic elastic theory

Consider a solid beam suspended from one or both of its ends and subject to a force f pressing it downwards (see Fig. 11.2). The force stretches the beam, i.e., induces *strain*. The stiffness of the solid leads to a *stress* force that acts against the strain. In the bulk linear elastic theory, the latter is related to the former via *Young's modulus E*. In the presence of a time-dependent force, the resultant flexural vibrations of the beam are described by the *Euler–Bernoulli* equation (Landau and Lifshitz, 1985)

$$\rho A \frac{\partial^2 u}{\partial t^2} + \frac{\partial^2}{\partial z^2}\left[EI_y \frac{\partial^2 u}{\partial z^2}\right] = f(z,t), \qquad (11.1)$$

where $u(z,t)$ describes the deformation as depicted in Fig. 11.2, ρ is the mass density and A the cross-section of the beam, and I_y is the beam geometry-dependent bending modulus describing deformations in the x-direction. Equation (11.1) is valid for linear elastic beams which are long compared to their width and thickness, and in the absence of an additional tension.[4] The most relevant cases are that of a rectangular beam with width w in the direction opposite to the bending and to the long dimension of the beam and height h in the direction of the bending, and cylindrical wires with radius r. The bending moduli for these cases are (Cleland, 2003)

$$I_y^{\text{rectangular}} = \frac{wh^3}{12} \quad \text{and} \quad I_r^{\text{cylindrical}} = \frac{\pi r^4}{4}. \qquad (11.2)$$

In some cases the beam may be under a constant tension due to its contacts, or a constant tension may be applied through a time independent force. In this case another term should be added to the Euler–Bernoulli equation, resulting in

$$\rho A \frac{\partial^2 u}{\partial t^2} - T\frac{\partial^2 u}{\partial z^2} + EI_y \frac{\partial^4 u}{\partial z^4} = f(z,t), \qquad (11.3)$$

where T describes the tension. The bending-induced tension can be calculated by estimating the strain for a bent beam (see Fig. 11.3). A

[4]One interesting aspect of nanomechanics is to find out how small systems can be described with the bulk elastic theory. In all examples shown in this chapter this theory holds rather well.

small unbent wire segment of length dz stretches in the presence of bending to

$$dl = dz\sqrt{1 + u'(z)^2} \approx dz\left(1 + \frac{u'(z)^2}{2}\right), \tag{11.4}$$

where $u'(z) = \partial u/\partial z \ll 1$. The strain of the wire is hence $dl/dz - 1 \approx u'(z)^2/2$. If the beam is thin compared to its length, the strain is approximatively the same throughout its cross-section. The tensile stress equals Young's modulus times the strain, and therefore the average stress or tension within the wire is

$$T = \frac{EA}{2L}\int_0^L dz u'(z)^2. \tag{11.5}$$

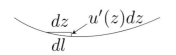

Fig. 11.3 The strain in a bent beam can be calculated by integrating the local strain $dl - dz$ over the length of the beam.

In the presence of a force inducing the tension, this needs to be calculated self-consistently by combining eqns (11.3) and (11.5).

The Euler–Bernoulli equation needs to be supplemented with boundary conditions describing how the beam is clamped at its ends. As it is a fourth-order equation in the position coordinate, we need four boundary conditions. For a doubly clamped beam, both the deformations and their slopes must vanish at the ends of the beam, i.e.,

$$u(0) = u(L) = u'(0) = u'(L) = 0, \tag{11.6}$$

where $u'(z)$ is the z-derivative of the deformation. In a cantilever, one of the ends is hanging freely. In this case the boundary conditions correspond to a vanishing of the transverse force and torque at the freely hanging end, say at $x = L$. The boundary conditions at $x = L$ thus become

$$u''(L) = u'''(L) = 0. \tag{11.7}$$

In the following, we concentrate on the doubly clamped beam, but the results for a cantilever follow from a similar procedure after the change of the boundary conditions.

11.1.2   Flexural eigenmodes of a doubly clamped beam without tension

Consider a doubly clamped isotropic (constant E, I_y) beam without tension and in the absence of an external force. Assume a harmonic time dependence $u(z,t) = u(z)e^{-i\omega t}$ for the deformations. In this case the deformations satisfy

$$\rho A \omega^2 u(z) = E I_y \frac{\partial^4 u(z)}{\partial z^4}. \tag{11.8}$$

This is an eigenvalue equation that can be solved with $u(z) = u_0 e^{\kappa z}$. There are solutions with $\kappa = \pm\beta, \pm i\beta$, where

$$\beta = \left(\frac{\rho A \omega^2}{E I_y}\right)^{1/4}. \tag{11.9}$$

Introducing the boundary conditions (11.6), let us try the ansatz

$$u(z) = \alpha_n \chi_n(z) = \alpha_n \{a_n[\sin(\beta_n z) - \sinh(\beta_n z)] - [\cos(\beta_n z) - \cosh(\beta_n z)]\},$$
(11.10)

which already takes into account the boundary conditions at $z = 0$. Requiring $\chi_n(L) = 0$ yields

$$a_n = \frac{\cos(\beta_n L) - \cosh(\beta_n L)}{\sin(\beta_n L) - \sinh(\beta_n L)}.$$
(11.11)

Moreover, requiring $\chi'_n(L) = 0$ leads to

$$\frac{2\beta_n(\cos(\beta_n L)\cosh(\beta_n L) - 1)}{\sinh(\beta_n L) - \sin(\beta_n L)} = 0.$$
(11.12)

Besides the trivial solution $\beta_0 = 0$, the wave numbers are hence obtained from the real solutions of

$$\cos(\beta_n L)\cosh(\beta_n L) = 1.$$
(11.13)

This can be solved numerically to yield $\beta_n = c_n/L$ with $c_1 = 4.73004$, $c_2 = 7.8532$, $c_3 = 10.9956$, $c_4 = 14.1372$, $c_5 = 17.2788$ and so on. Plugging these numbers into eqn (11.11) yields $a_1 = 0.982502$, $a_2 = 1.00078$, $a_3 = 0.999966$, $a_4 = 1.000001$ and $a_n \approx 1$ for $n > 4$.[5] The mode shapes of the first five non-trivial eigenfunctions of flexural vibrations are plotted in Fig. 11.4.

Note that functions $\chi_n(z)$ introduced in eqn (11.10) are solutions of a linear differential equation and therefore form an orthogonal basis. Furthermore, these modes are normalized. This means that

$$\frac{1}{L}\int_0^L dz\, \chi_n(z)\chi_m(z) = \delta_{nm}.$$
(11.14)

Hence, any function $u(z)$ satisfying the boundary conditions (11.6) can be written as a linear combination of $\chi_n(z)$, where the coefficients are projections of $u(z)$ into these eigenmodes.

The eigenfrequency of these modes is obtained by inverting eqn (11.9)

$$\omega_n = \sqrt{\frac{EI_y}{\rho A}}\frac{c_n^2}{L^2}.$$
(11.15)

Contrary to the usual solid vibrations (say, bulk three-dimensional phonons), where the eigenfrequency is linear in the wave number, for flexural modes it is quadratic. Let us estimate the dependence of the eigenfrequencies on the size of the resonator. For a beam with a square cross-section ($h = w$) we have $I_y = w^4/12$. In this case the eigenfrequencies are

$$\omega_n = \sqrt{\frac{E}{12\rho}}\frac{c_n^2 w}{L^2}.$$
(11.16)

They hence scale roughly inversely with the size of the system.

[5] Within six decimals

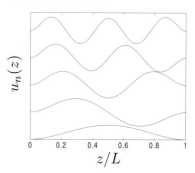

$u_n(z)$

0   0.2   0.4   0.6   0.8   1

z/L

Fig. 11.4 Mode shapes of the lowest five non-trivial eigenfunctions of flexural vibrations. For clarity, the vertical axis has been shifted for the different modes. Due to the boundary conditions, all of the modes vanish at the edges.

[6]Note that the usage of a Young's modulus assumes the validity of continuum theory even for the single- or a few-atom-thick systems. This is a widely used assumption, and the quoted value for the Young's modulus is approximatively valid in the case of a monolayer graphene provided the thickness 0.34 nm corresponding to the interlayer separation in graphite is used. See the discussion in (Qian *et al.*, 2002) for the Young's modulus of carbon nanotubes.

In Exercise 11.1 you will estimate the typical eigenfrequencies of a few example systems. The Young's modulus in our three example systems depicted in Fig. 11.1 is 70 GPa for Al beams (Sulkko *et al.*, 2010), of the order of 1.0 TPa for graphene (Bunch *et al.*, 2007) and carbon nanotubes.[6] Moreover, the mass density is $\rho = 2330$ kg/m³ for Si and 1400 kg/m³ for carbon nanostructures.

11.1.3 Effect of tension on the vibration modes

In the presence of tension the eigenfrequency equation becomes somewhat more complicated, but yet still tractable. Inserting the ansatz $u(z,t) = u_0 e^{\kappa z} e^{i\omega t}$ in eqn (11.3) in the absence of a force yields

$$(-\rho A \omega^2 - T\kappa^2 + EI_y \kappa^4)u = 0 \qquad (11.17)$$

This may be solved with four possible values of κ,

$$\kappa = \pm i\sqrt{\sqrt{\xi^2 + \lambda\omega^2} - \xi} \equiv \pm i\eta \qquad (11.18\text{a})$$

$$\kappa = \pm\sqrt{\sqrt{\xi^2 + \lambda\omega^2} + \xi} \equiv \pm\beta, \qquad (11.18\text{b})$$

where $\xi = T/(2EI_y)$ and $\lambda = \rho A/(EI_y)$. The eigensolutions are then superpositions of $e^{\kappa_n z}$ satisfying the boundary conditions. We start from the ansatz

$$u(z) = \alpha_n \chi_n(z) = \alpha_n \{a_n[\sin(\eta_n z) - \frac{\eta_n}{\beta_n}\sinh(\beta_n z)] + \cos(\eta_n z) - \cosh(\beta_n z)\}, \qquad (11.19)$$

where we take into account the boundary conditions $\chi_n(0) = \chi_n'(0) = 0$. The other two boundary conditions $\chi_n(L) = \chi_n'(L) = 0$ for a doubly clamped beam yield

$$\cos(\eta_n L)\cosh(\beta_n L) - 1 + \frac{(\eta_n^2 - \beta_n^2)}{2\eta_n\beta_n}\sin(\eta_n L)\sinh(\beta_n L)$$
$$\equiv \cos(\eta_n L)\cosh(\beta_n L) - 1 - \frac{\xi}{\sqrt{\lambda}\omega_n}\sin(\eta_n L)\sinh(\beta_n L) = 0. \qquad (11.20)$$

In general, this equation has to be solved numerically to find the eigenfrequencies corresponding to some given value of ξ and λ.

In the limit of strong tension ($\xi \gg \sqrt{\lambda}\omega_n$), the last term in eqn (11.20) starts to dominate and the condition becomes $\sin(\eta_n L) = 0$. Using $\eta_n \approx \sqrt{A\rho/T}\omega_n$ gives the typical eigenfrequencies of a wave equation,

$$\omega_n \approx \sqrt{\frac{T}{\rho A}}\frac{n\pi}{L}. \qquad (11.21)$$

This is the limit of a 'guitar string', in which the resonance frequency can be tuned with the tension.

11.1.4   Driving and dissipation

Assume we drive the beam with a force of the form $f(z,t) = f(z)e^{-i\omega_d t}$. In the stationary situation the resultant deformation follows the time dependence of the force, i.e., $u(z,t) = u(z)e^{-i\omega_d t}$. Let us write the deformation $u(z)$ in terms of the eigenfunctions $\chi_n(z)$,

$$u(z) = \sum_n \alpha_n \chi_n(z). \tag{11.22}$$

The Euler–Bernoulli equation (without induced tension) can hence be written as

$$\sum_n \alpha_n(-\omega_d^2 + \omega_n^2)\chi_n(z) = \frac{1}{m}f(z), \tag{11.23}$$

where $m = \rho A L$ is the mass of the resonator. Multiply this equation by u_m and average over the position. Using the orthonormality condition (11.14) we get

$$\alpha_m(-\omega_d^2 + \omega_m^2) = \frac{1}{mL}\int dz \chi_m(z)f(z). \tag{11.24}$$

For the mode amplitude α_m this is the same equation as for a harmonic oscillator. It can thus readily be quantized in terms of harmonic oscillator creation and annihilation operators. This is discussed more in Sec. 11.4.

Typically it is enough to concentrate on a single mode with ω_m closest to ω_d, as that is the mode that couples to the driving the strongest. Exactly at $\omega_d = \omega_n$ the amplitude α_m diverges based on eqn (11.24). This is due to the fact that we ignored the mechanical dissipation of the beams. The latter is typically described by adding a small imaginary part, $\omega_m \to \omega_m - i\gamma_m$ in the time-dependent response (see Sec. 11.4.2). With this substitution, eqn (11.24) in the case $\gamma_m \ll \omega_m$ transforms to[7]

$$\alpha_m(-\omega_d^2 - i\omega_m\gamma_m + \omega_m^2) = \frac{1}{mL}\int dz \chi_m(z)f(z) \equiv \frac{f_0 \eta_m}{m}. \tag{11.25}$$

Here f_0 is the amplitude of the force and

$$\eta_m \equiv \frac{1}{L}\int_0^L dz \chi_n(z)\frac{f(z)}{f_0} \tag{11.26}$$

is the projection of the force on the mode with eigenfunction $\chi_n(z)$. In the case of resonators, the 'strength' of resonance is often described by the quality factor Q_m satisfying

$$Q_m^{-1} \equiv \frac{\text{Im}(\omega_m)}{|\omega_m|} = \frac{\gamma^2}{\sqrt{\omega_m^2 + \gamma_m^2}} \overset{\gamma \ll \omega_m}{\approx} \frac{\gamma}{\omega_m}. \tag{11.27}$$

The amplitude response of the resonator eigenmode as a function of the driving frequency ω_d is described by the imaginary part of α_m (see

[7]This is valid for $\omega_d \approx \omega_m$. See Example 11.1 for the more general approach.

for example Sec. 11.2.1). It has a characteristic Lorentzian shape for $\omega_m \approx \omega_d$,

$$\text{Im}(\alpha_m) = \frac{\eta_m f}{m} \frac{Q_m \omega_m^2}{\omega_m^4 + (\omega_d^2 - \omega_m^2)^2 Q_m^2} \overset{\omega_d \approx \omega_m}{\approx} \frac{\eta_m f}{m} \frac{Q_m}{\omega_m^2 + 4Q_m^2(\omega_d - \omega_m)^2}. \tag{11.28}$$

For a uniform (z-independent) force, the projection of the force on the lowest flexural eigenmodes is $\eta_1 = 0.830862$, $\eta_2 = 0$, $\eta_3 = 0.363769$, $\eta_4 = 0$ and $\eta_5 = 0.231498$. Modes with an even order hence do not couple to position independent forces.[8]

The dissipation in resonators comes from various sources, such as the presence of mobile defects that can absorb energy from the vibrations, dynamic impurities at the surfaces, clamping losses, and in the case of nanoelectromechanical systems, also from the dissipation in the electronic system. Until quite recently the Q-factors of the mechanical resonators were found to scale with the size of the resonator such that the smallest resonators have the smallest Q-factors.[9] This has been thought to be due to the increasing impact of the surface effects in small resonators. However, recent experiments on ultraclean carbon nanotubes[10] have shown that high Q-values can be obtained also in nanomechanical resonators.[11]

Example 11.1 Langevin and Fokker–Planck equations for a harmonic oscillator

The effect of thermal or driven fluctuations on the motion of a harmonic oscillator (such as one of the flexural modes of a mechanical resonator) may be described by assuming a fluctuating force $\delta f(t)$ coupled to the resonator. The effect of this force on the resonator can be described by the *Langevin equation*, which in this case has the form (see eqn (11.25), in the following we absorb the matrix element η_m to the forces)

$$\ddot{\alpha}_m + \gamma_m \dot{\alpha}_m + \omega_m \alpha_m = \frac{f_m + \delta f(t)}{m}, \tag{11.29}$$

where f_m describes the effect of a steady force and $\delta f(t)$ its fluctuations. As we describe the average force with f_m, we can assume $\langle \delta f(t) \rangle = 0$, where $\langle \cdot \rangle$ denotes a time average. We aim to describe thermal noise in the limit $\omega_m \ll k_B T/\hbar$. In this case $\delta f(t)$ is described by the correlator (see Appendix C)

$$\langle \delta f(t) \delta f(t') \rangle = 2k_B T \gamma_m m \delta(t - t'), \tag{11.30}$$

satisfying the fluctuation–dissipation theorem.

Sometimes it may be advantageous to consider the probability distribution $P(\alpha_m, v_m; t)$ of the vibration amplitude α_m and its time derivative $v_m = \dot{\alpha}_m$, induced by the fluctuating force. If the force fluctuations are described by a Gaussian probability distribution, the time evolution of $P(\alpha_m, v_m; t)$ is described by the *Fokker–Planck equation*,[12]

$$\frac{\partial P}{\partial t} = -v_m \frac{\partial P}{\partial \alpha_m} + \frac{\partial}{\partial v_m}\left[\left(\gamma_m v_m + \omega_m^2 \alpha_m - \frac{f_m}{m}\right)P\right] + \frac{k_B T \gamma_m}{m} \frac{\partial^2 P}{\partial v_m^2}. \tag{11.31}$$

In the stationary limit where both sides of this equation evaluate to zero, this is solved by the Boltzmann distribution

$$P(\alpha_m, v_m) = P_0 \exp\left(-\frac{E_m}{k_B T}\right), \tag{11.32}$$

[8]Or in general, modes of even order do not couple to forces which are symmetric with respect to the centre of the beam.

[9]See especially Fig. 3 in (Ekinci and Roukes, 2005).

[10]See for example (Steele *et al.*, 2009).

[11]These devices show also characteristic *non-linear damping* (Eichler *et al.*, 2011), where the damping strength depends on the amplitude of the vibrations.

[12]See for example (Reichl, 2004).

where P_0 is a normalization constant and $E_m = m(v_m^2 + \omega_m \alpha_m^2)$ is the total energy of the oscillator.

As discussed in Sec. 11.2.4 and Example 11.1 these equations can be generalized to describe the effects of a coupling of the mechanical resonator to an external electronic component driving it.

11.2 Coupling to nanoelectronics

The above discussion concentrates on the properties of nanomechanical systems under an applied force. Such forces are typically applied either electronically or optically. In this chapter we concentrate on the previous form of *actuating* the mechanical motion. There are a variety of different types of schemes, but we discuss here two of them: a) magnetomotive actuation, and b) capacitive actuation.[13] In addition, the actuation methods have to be appended by methods to transduce the mechanical motion into electrical energy, such that the former can be detected. Traditionally this is done by similar methods as actuation.[14] In the magnetomotive actuation/detection scheme an alternating current in a magnetic field produces a force on the mechanical oscillator, and the motion of the mechanical oscillator gives rise to an alternating voltage which is used for the detection. In the capacitive actuation an alternating voltage in a capacitor gives rise to a the force on the beam, and the vibrations are detected by measuring the resultant change in the capacitance. However, there is also another way to transduce the motion, by converting it directly into a change in a current or voltage, by using a non-linear system, like a SQUID for magnetic read-out or single-electron transistor or quantum dot for detecting the vibrations induced by a gate voltage. Generally, such non-linear systems act as amplifiers of the signal and allow for more sensitive detection than the linear direct detectors. However, their back-action on the vibrations is typically also stronger.

A further alternative electronic actuation/transduction scheme is to couple the mechanical resonator to an electronic (LC) resonance circuit typically operating in the microwave regime. This type of a scheme is discussion in Sec. 11.3.

Besides actuation and transduction of the mechanical motion, strongly driven detectors can in some cases be used for *cooling* the nanomechanical resonators.[15]

11.2.1 Magnetomotive actuation and detection

Consider the setup shown in Fig. 11.5. Applying a magnetic field B perpendicular to the resonator beam, and a current I through it produces a Lorentz force (per unit length) $f = BI$. For a uniform magnetic field, the total force acting on the beam is thus $F = BIL$. If the current oscillates in time as $I(t) = I_0 e^{-i\omega_d t}$, so does the force.[16] Neglecting the screening of the magnetic field due to the current, to a good approximation the force can be considered as position independent. This way the

[13]In addition, piezoelectric coupling has been used; see for example (O'Connell *et al.*, 2010).

[14]These linear schemes of actuation and transduction do not depend on the size of the resonator, and hence they can be used to access vibrations of larger objects as well.

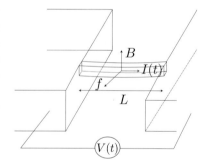

Fig. 11.5 Magnetomotive actuation and detection of mechanical motion.

[15]See for example (Naik *et al.*, 2006), (Rocheleau *et al.*, 2009) and (Teufel *et al.*, 2011).

[16]For an example of magnetomotive actuation and detection of nanoelectromechanical systems, see (Cleland and Roukes, 1996).

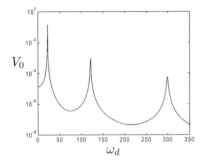

Fig. 11.6 Amplitude of the part of voltage in phase with the driving current as a function of driving frequency. The frequencies are in units of $\omega_0 = \sqrt{EI_y/(\rho A)}L^{-2}$ and the voltage in units of $I_0 B^2 L/(M\omega_0^2)$. The peaks correspond to the first, third and fifth flexural eigenmodes of the beam, and the voltage has been calculated by assuming $Q_m = 100$ for each mode.

[17] This includes the plain charging energy and the work done by the gate voltage, and corresponds to the limit $C_L = C_R = 0$ of a single-electron transistor. See also Exercise 11.2.

force couples to the odd-order eigenmodes of the flexural vibrations, as discussed above.

The simplest way of detecting the vibrations in this scheme is to measure the ac voltage through the device. Namely, the vibrations of the beam in the magnetic field generate an electromotive force, which according to Faraday's law equals the rate of change of the flux Φ through the region spanned by the beam. This voltage is

$$V_{\text{emf}} = \frac{d\Phi}{dt} = B \int_0^L \frac{\partial u(z,t)}{\partial t} dz = -i\omega_d B L e^{-i\omega_d t} \sum_n \alpha_n \eta_n. \quad (11.33)$$

Close to the resonance of mode m, the real part of the voltage oscillating in phase with the current is a Lorentzian function of the driving frequency ω_d:

$$\text{Re}(V_{\text{emf}} e^{i\omega_d t}) \approx \frac{\eta_m^2 B^2 I_0 L}{m} \frac{Q_m}{\omega_m^2 + 4Q_m^2(\omega_d - \omega_m)^2}. \quad (11.34)$$

This function is plotted in Fig. 11.6 for a high-quality ($Q_m = 100$) resonator.

Besides a direct readout of the oscillating voltage, more advanced readout schemes involve the use of radio frequency reflection measurements, similar to those described in Sec. 7.6.2.

11.2.2 Capacitive actuation and detection

Another way to actuate the mechanical motion is by applying an alternating gate voltage as shown schematically in Fig. 11.7. For a capacitance C, the polarization charge on the capacitor is $q = CV_g$ and the stored charging energy is $E_{\text{ch}} = q^2/(2C) - qV_g = -CV_g^2/2$.[17] The capacitance depends on the distance $d(z)$ between the beam and the gate. The gate voltage hence induces a force

$$f = -\frac{\partial E_{\text{ch}}}{\partial x} = \frac{V_g^2}{2} \frac{\partial C}{\partial d(z)} \quad (11.35)$$

on the beam. The sign of the force can be estimated by requiring that the effect of the force is to reduce the charging energy. Since the capacitance increases at a decreasing distance d, the force pulls the capacitor

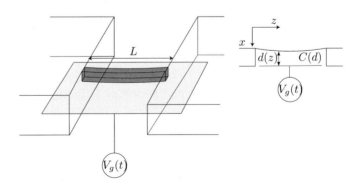

Fig. 11.7 Capacitive actuation of mechanical oscillations.

towards the gate. The distance $d(z)$ can be written in terms of the vibration amplitude $u(z)$ via $d(z) = d_0 - u(z)$, where $d_0 > u(z)$ is the distance between the beam and the gate in the absence of the gate voltage. Applying a voltage of the form $V_g(t) = V_{g0} + \delta V_g e^{-i\omega_d t}$ with $\delta V_g \ll V_{g0}$ induces an average stress force on the beam, proportional to $(V_{g0})^2$, and an alternating stress force, proportional to $2V_{g0}\delta V_g e^{-i\omega_d t}$. In the limit $\delta V_g \ll V_{g0}$ we may disregard the part oscillating with the double frequency. As shown in the following, the average dc voltage can be used to tune the vibration eigenfrequencies.[18]

As an example, let us consider a parallel-plate capacitor model,[19]

$$C(d) = \epsilon S/d, \qquad (11.36)$$

where ϵ is the permittivity of the space between the gate and the vibrating object and $S = wL$ is the surface area of the wire. The force per unit length at position z acting on the wire is

$$f(z) = \frac{\epsilon S V_{g0}^2}{d(z)^2} + \frac{2\epsilon S V_{g0}\delta V_g}{d(z)^2} e^{-i\omega_d t}, \qquad (11.37)$$

where the sign of the force is such that positive force is towards the gate capacitor.

For small gate voltages the bending of the tube is small, $u(z) \ll d_0$, and we can approximate $d(z) \approx d_0$. In this case we can proceed as in the magnetomotive scheme above. For example, the resultant force is independent of position and therefore only odd modes are actuated.

For higher gate voltages we have to take into account the dependence of the force on the induced oscillation amplitude $u(z,t)$. In this case we can write the time dependence of the oscillation amplitude as a Fourier series,

$$u(z,t) = \sum_n u_n(z)e^{-i\omega_d t}. \qquad (11.38)$$

This form is imposed by the fact that the induced oscillation amplitude has to have the same period as the applied force. Assuming that the ac voltage amplitude is much smaller than the dc voltage, $\delta V_g \ll V_{g0}$ we also get that $u_n \ll u_0$. This allows us to write the Euler–Bernoulli equation for the dc deflection as

$$EI_y \frac{\partial^4 u_0}{\partial z^4} - T\frac{\partial^2 u_0}{\partial z^2} = \frac{\epsilon S V_{g0}^2}{(d_0 - u_0(z))^2}. \qquad (11.39)$$

Solving this numerically for $u_0(z)$[20] yields for the ac part

$$\omega u_1(z) - EI_y \frac{\partial^4 u_1(z)}{\partial z^4} - T_{dc}\frac{\partial^2 u_1}{\partial z^2} - T_1\frac{\partial^2 u_0}{\partial z^2} = \frac{2\epsilon S V_{g0}^2}{(d_0 - u_0)^3}u_1(z) + \frac{2\epsilon S V_{g0}\delta V_g}{(d_0 - u_0)^2}. \qquad (11.40)$$

Here T_{dc} is the dc part of the tension and T_1 its first harmonic, resulting from the induced tension, eqn (11.5).

Solving eqn (11.40) allows us to find the voltage-induced changes in the eigenfrequencies. These are plotted numerically as a function of the

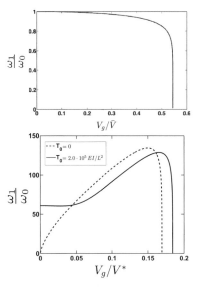

Fig. 11.8 Top: Frequency of the fundamental flexular mode of a thick membrane ($h \gtrsim d_0$) as a a function of the gate voltage. The voltage scale is here given by $\bar{V} = \sqrt{m\omega_0^2 d_0^3 \epsilon_0 w L}$, where $\omega_0 = \omega_1(V_g = 0, T_0 = 0)$, obtained by equating the charging energy with the bending energy of a beam bent to amplitude d. Bottom: The same for a thin resonator, $d \gg h$ (here $d = 283h$) with (solid line) and without (dashed line) initial tension T_0. Now the voltage scale is $V^* = Ed_0^5 \nu^2/(4L^3\epsilon_0)$, where ν is the average of $\chi_1''(z)$ along the wire. This is obtained by equating the charging energy with the stress energy of a beam bent to amplitude d_0. Courtesy of Raphaël Khan.

[18]See an example in (Sapmaz et al., 2003) for the theory and (Sazonova et al., 2004) for an experimental demonstration of this effect.

[19]This assumes that the gate has roughly the same size as the vibrating wire and $d \ll w, L$. An alternative model for a one-dimensional wire next to an infinite plate is used for example in (Sapmaz et al., 2003).

[20]This can be done, for example, by using the Galerkin method: write $u(z)$ in the form (11.22), including some N basis functions, and invert the resultant matrix equation for the prefactors.

dc voltage in Fig. 11.8. There are two regimes of parameters: when the thickness of the wire is of the order of the distance d_0 to the gate, the bending-induced tension is negligible and the only effect of the voltage is to reduce the frequency. At a certain critical voltage the beam collapses into the gate, 'pulls in'. This is signalled by the eigenfrequency tending to zero. On the other hand, for a thin beam (thickness $h \ll d_0$), the eigenfrequencies of the flexular modes first increase due to an induced tension on the beam, before collapsing quickly to zero at the pull-in point (Sillanpää *et al.*, 2011).

The tuning of the resonant frequency with the gate voltage allows one to use this device as an alternative charge detection scheme to the single electron transistor, as the gate voltage can be replaced by the charge one aims to measure. The best charge sensitivities reached with this a method can rival those of the single electron devices (Bunch *et al.*, 2007).

Besides the choice made above for the time dependence of the gate, we could also have chosen a time dependence of the form $V_g(t) = V_{g1}e^{-i\omega_P t} + V_{g2}e^{-i\omega_d t}$. In this case both the effective spring constant and the driving force oscillate in time and one can realize the phenomenon of *parametric resonance*. We do not discuss this effect here, but direct the interested reader to the References.[21]

The resultant motion of the beam can be detected in many ways. Typical schemes are based on measuring the changes in the capacitance due to the motion of the beam. This can be done extremely accurately by coupling the resonator with an electomagnetic resonator, in which case the vibration-induced changes in the capacitance tune the electrical resonance frequency (see Sec. 11.3), or exploiting the non-linear single-electron effects and converting the capacitance modulation into a modulation of the current, either through a single electron transistor or through a quantum dot. Such non-linear methods are discussed next.[22]

11.2.3 SQUID detection

One suggested non-linear generalization of the magnetomotive detection technique is to replace one arm of a superconducting quantum interference device (SQUID) by the nanomechanical resonator (see Fig. 11.9 for an example of an actual device). Here the idea is based on the fact that upon a constant magnetic field B, the flux threading the loop changes as a function of the deformation $u(z)$ of the resonator. This can then be used for the measurement of the vibrations in the resonator. In practice, the SQUID is operated in the dissipative mode, by driving a bias current through it and monitoring the voltage that depends on the flux via the SQUID critical current (see Sec. 9.3.1). The motion of the resonator gives rise to a time-dependent output voltage oscillating at the eigenfrequency of the resonator. This allows for the detection of the mechanical oscillations.

For a detailed description of SQUIDs as detectors of mechanical motion, see for example (Buks and Blencowe, 2006).

---

[21] See Sec. 9.3 in (Cleland, 2003).

[22] One further interesting effect combining charge dynamics with a mechanical resonance is *charge shuttling* (Gorelik *et al.*, 1998), where the mechanical resonance of a vibrating SET island is self-excited by charges transferred through it. As a result, the average current through the SET is directly proportional to the mechanical vibration frequency of the island.

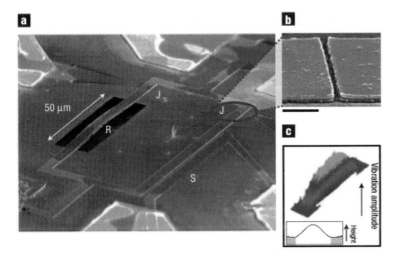

Fig. 11.9 SQUID coupled to a mechanical resonator. a) Scanning electron micrograph of the device. The resonator is marked with the letter R. (b) Closeup of the Josephson junctions. (c) Three-dimensional image of the vibration amplitude of the driven beam resonance. (S. Etaki *et al.*, *Nature Phys.* **4**, 785 (2008), Fig. 1.) Reprinted by permission from Macmillan Publishers Ltd., © 2008.

11.2.4 Detection through single-electron effects

Another accurate read-out scheme of nanoelectromechanical devices relies on coupling the mechanical vibrations with single-electron effects. The idea is based on the fact that in structures discussed in Sec. 11.2.2 the mechanical vibrations tune the gate capacitance C_g, inducing a time-dependent oscillation $C_g(t) = C_{g0} + \delta C_g(t)$, oscillating at the driving frequency. Therefore, the gate charge on the single-electron devices oscillates as $Q_G(t) = Q_{G0} + \delta Q_G(t)$, where

$$\delta Q_G(t) = \delta(C_g(t)V_g(t)) \approx C_{g0}\delta V_g(t) + V_{g0}\delta C_g(t). \qquad (11.41)$$

The first term is a direct consequence of the fact that the gate voltage actuating the mechanical motion oscillates, whereas the second term results from the mechanical motion.

Typical eigenfrequencies of the mechanical oscillations are below the tunnelling rates of the single-electron devices. In this case the current through these devices follows the change in the gate adiabatically,

$$\delta I(t) = \frac{\partial G}{\partial V_g}\left(\delta V_g(t) + V_{g0}\frac{\delta C_g(t)}{C_{g0}}\right)V_b, \qquad (11.42)$$

where V_b is the bias voltage over the single-electron device and $G(V_g)$ is its linear conductance. The conversion of the mechanical motion to electronic signal is most efficient at the Coulomb blockade threshold. Alternatively, using the fact that the conductance is a non-linear function of V_g, the signal can also be read at the dc current. Assuming a gate charge modulation of the form $\delta Q_G = \delta Q_{G0}\cos(\omega_d t)$, the induced change in the dc current is

$$\delta I_{dc} = \frac{\partial^2 G}{\partial Q_G^2}\frac{\delta Q_{G0}^2}{2}. \qquad (11.43)$$

Figure 11.10 shows an example of the detected dc current vs. the driving frequency in the case where the nanomechanical oscillator was a single-wall carbon nanotube in the quantum dot regime. In this case the carbon

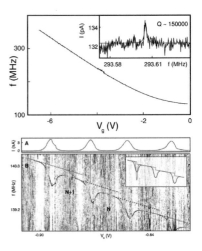

Fig. 11.10 (Top): Measurement of a mechanical resonance in a single-wall carbon nanotube. The inset shows the modulation of the dc current as the frequency of the driving was varied. The main figure shows how in this case the resonant frequency can be greatly tuned with the gate voltage. (Bottom): Addition of a single charge into the carbon nanotube results in strong changes in its vibration eigenfrequency. At the points of charge degeneracy the current through the nanotube is increased, which also results in dips in the mechanical resonance frequencies. (G.A. Steele *et al.*, *Science* **325**, 1103 (2009), Figs. 1 and 2.) Reprinted with permission from AAAS.

nanotube was used both as the mechanical resonator and as the read-out device. The theoretical description of this device combines the master equation for the charge state of the SET (see Sec. 7.3) with the Fokker–Planck equation for the mechanics (Example 11.1). This also allows us to describe, besides the direct effect of the mechanical motion on the ac current, the changes in the damping and effective temperature of the mechanical resonator, and the resultant effects on the current noise through the SET (Isacsson and Nord, 2004; Armour, 2004; Usmani *et al.*, 2007).

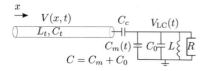

Fig. 11.11 Equivalent circuit model of a mechanical resonator (the part dependent on the vibration amplitude $u(z)$ denoted with the variable capacitance C_m) as part of an electrical LC resonator (microwave cavity) coupled capacitively to a transmission line. Resistance R models the internal losses of the electrical resonator, so that the limit $R \to \infty$ represents the case of a lossless resonator.

[23] Vibrations corresponding to single resonator quanta have been measured with such setups; see (Teufel *et al.*, 2011).

[24] In the latter case of optomechanics the mechanical resonator forms one of the ends of an optical cavity, and thereby the mechanical motion modulates the optical properties of the cavity. For a general discussion, see (Marquardt and Girvin, 2009).

11.3 Coupling to microwave resonant circuits

An efficient way to actuate and transduce mechanical motion is to couple the mechanical resonator to modulate the capacitance of a resonant LC circuit. An equivalent model of such a circuit is shown in Fig. 11.11 and an example realization in Fig. 11.12. As explained in Complement 11.1, in this case the mechanical motion can be measured via measuring side-bands of probe microwaves sent into the cavity. Besides allowing for an extremely accurate measurement of the mechanical motion,[23] this scheme allows for cooling of the mechanical vibrations. Moreover, this setup was recently realized to make a mechanical microwave amplifier (Massel *et al.*, 2011), where the mechanical motion was exploited to amplify electronic signals with frequency close to that of the LC resonance.

In principle this scheme can be realized by choosing the electrical resonance to any desired frequency. In practice it has been mostly used in the microwave regime, using superconducting resonators, and in the optical regime, using optical cavities.[24]

Consider the setup of a transmission line coupled capacitively to a microwave cavity (Fig. 11.11), the latter described within the lumped element model valid in the case when the wavelength of incoming radiation is at most of the order of the length of the cavity. The voltage and the current in the transmission line satisfy a wave equation (see Appendix E),

$$\partial_t^2 V(x,t) = c\partial_x^2 V(x,t), \quad I(x,t) = \frac{1}{L_t}\int^t dt' \partial_x V(x,t), \qquad (11.44)$$

where $c = 1/(L_t C_t)$ is the speed of light in the transmission line with inductance/unit length L_t and capacitance/unit length C_t. At the capacitive contact to the LC resonator (say, at $x = 0$), we can write for the (stray) current through the capacitor

$$I_c = C_c(\dot{V}_{\mathrm{LC}}(t) - \dot{V}(0,t)), \qquad (11.45)$$

where $\dot{V} = \partial_t V$. Moreover, the current through the LC resonator is

$$I_{\mathrm{LC}} = C\dot{V}_{\mathrm{LC}}(t) + \frac{1}{L}\int^t dt' V_{\mathrm{LC}}(t') + V_{\mathrm{LC}}/R. \qquad (11.46)$$

Let us consider a wave of the form

$$V(x,t) = V_0 e^{i\omega t}(e^{i\omega/cx} + re^{-i\omega/cx}), \tag{11.47}$$

consisting of the incoming wave transmitted towards the LC resonator and a reflected wave (with reflection coefficient r) travelling away from it. Solving the current conservation condition $I(0,t) = I_c = -I_{\rm LC}$ yields

$$\sqrt{\frac{C_t}{L_t}}(1-r) = i\omega C_c\left(\frac{V_{\rm LC}}{V_0} - 1 - r\right) = \frac{\omega^2 - \tilde\omega_c^2 - i\omega\tilde\gamma_c}{i\omega/C}\frac{V_{\rm LC}}{V_0} \tag{11.48}$$

for the voltage $V_{\rm LC}(t) = V_{\rm LC}e^{i\omega t}$. Here $\tilde\omega_c^2 = 1/(LC)$ and $\tilde\gamma_c = 1/(RC)$ of the bare LC resonator, with capacitance $C = C_0 + C_m$. Solving these equations in the typical limit $\omega C_c \ll \sqrt{C_t/L_t}$ gives

$$V_{\rm LC} \approx V_0\frac{2\omega^2 C_c}{\omega^2(C_c + C) - \tilde\omega_c^2 + i\tilde\omega/(RC)} = V_0\frac{2\omega^2 C_c/(C_c + C)}{\omega^2 - \omega_c^2 + i\omega\gamma_c}, \tag{11.49}$$

where $\omega_c^2 = C\tilde\omega_c^2/(C+C_c)$ and $\gamma_c = C\tilde\gamma_c/(C+C_c)$ contain the renormalization of the LC oscillator capacitance due to the coupling capacitance.

The total grand canonical electromagnetic energy $E_{\rm LC}$ inside the LC resonator is a sum of four parts, the two charging energies across the capacitors C and C_c, the magnetic energy in the inductor L and the energy $V_0 Q_c$ needed to charge the capacitor C_c with charge Q_c. To describe the effect on the moving capacitance in the limit $\omega C_c \ll \sqrt{C_t/L_t}$, it is enough to consider the voltage $V(0,t) \approx V_0 e^{i\omega t}$. Then, we get

$$E_{\rm LC} = \frac{C}{4}|V_{\rm LC}(\omega)|^2 + \frac{C_c}{4}(|V_{\rm LC}|^2 + |V(0)|^2 - 2\mathrm{Re}(V_{\rm LC}(\omega))V(0))$$
$$+ \frac{L}{4}\frac{|V_{\rm LC}(\omega)|^2}{L^2\omega^2} + C_c\frac{V(0)(\mathrm{Re}[V_{\rm LC}(\omega)] - V(0))}{2}. \tag{11.50}$$

Here the overall prefactor $1/2$ comes from using the amplitude of voltage oscillations instead of their rms magnitude. Moreover, the real part of $V_{\rm LC}(\omega)$ in two of the terms reflects the effect of a phase shift in $V_{\rm LC}(\omega)$ compared to the driving voltage $V(0)$. Combining all these terms yields

$$E_{\rm LC} = \frac{C + C_c}{4}\left(1 + \frac{\omega_c^2}{\omega^2}\right)|V_{\rm LC}(\omega)|^2. \tag{11.51}$$

The two remaining terms reflect the total charging and magnetic energies of the resonator, and at resonance $\omega = \omega_c$ the two are equal.

If part of the capacitance is due to a moving mechanical resonator, i.e., capacitance depends on the time-dependent displacement $u(z,t) = \sum_m \alpha_m \chi_m$, this shows up as an energy that depends on the amplitude of deformation. The corresponding prefactor of this dependence is a mechanical force on the resonator. This force is called the *radiation pressure* force, and it results from the momentum change of the electromagnetic field upon reflection from the capacitance plate. Its magnitude should be calculated in the presence of a constant voltage $V_{\rm LC}$. We get[25]

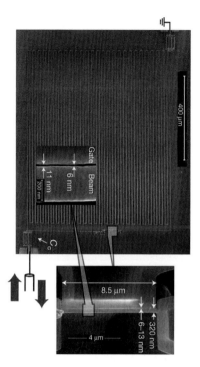

Fig. 11.12 Superconducting microwave resonator (top) coupled at its end to the vibrating Al beam (bottom and zoom-up on top of the superconducting resonator). (F. Massel *et al.*, *Nature* **480**, 351 (2011), Fig. 1.) Reprinted by permission from Macmillan Publishers Ltd., © 2011.

[25]Note that we also have to take the derivative of the ω_c-term, which depends on the capacitance.

$$f = \frac{|V_{\mathrm{LC}}|^2}{4} \frac{dC}{du(z)} = V_0^2 \left(\frac{C_c}{C_c + C_0}\right)^2 \frac{\omega^4}{(\omega^2 - \omega_c^2)^2 + \omega^2 \gamma_c^2} \frac{dC}{du(z)}.$$
(11.52)

The force is thus greatest when the driving is close to the resonance of the cavity: $\omega - \omega_c \lesssim \gamma_c$. For example, in a parallel plate model $C_{\mathrm{mech}} = \epsilon S/(d - u(z))$ with area S of the capacitor plates held at average distance d from each other we get $dC/du(z) \approx \epsilon S u(z)/d^2$, valid for small deformations $u(z) \ll d$. In what follows, we project this on mode m with amplitude α_m, obtaining an overall prefactor η_m as defined in eqn (11.26).

The dependence of the force on the moving capacitance C_m through $|V_{\mathrm{LC}}|$ gives a shift of the mechanical eigenfrequency of oscillations. This is straightforward to show by using the characteristic equation of a harmonic oscillator,

$$\ddot{\alpha}_m + \gamma_m \dot{\alpha}_m + \omega_m^2 \alpha_m = \frac{f(\alpha_m)}{m} \approx \eta_m \frac{f_0(t)}{m} + \frac{\eta_m}{m} \frac{\partial f}{\partial \alpha_m}\Big|_{\alpha_m = 0} \alpha_m.$$

$$\ddot{\alpha}_m + \gamma_m \dot{\alpha}_m + \underbrace{\left(\omega_m^2 - \frac{\eta_m}{m} \frac{\partial f}{\partial \alpha_m}\right)}_{\delta(\omega^2)} \alpha_m = \frac{\eta_m f_0}{m}.$$
(11.53)

This frequency shift is the *electromagnetic spring effect* on the mechanical oscillations due to the electromagnetic force acting on the vibrations. Similar *optical spring effects* have been observed in mechanical resonators coupled to optical cavities.

For the LC resonator, the derivative of the force is[26]

$$\frac{\partial f}{\partial \alpha_m} = \left(\frac{dC}{d\alpha_m}\right)^2 \partial_{C_m} \left(\frac{|V_{\mathrm{LC}}|^2}{4}\right) + \frac{|V_{\mathrm{LC}}|^2}{4} \frac{d^2 C}{d^2 \alpha_m},$$
(11.54)

Using the parallel-plate capacitor model gives

$$\frac{\partial f}{\partial \alpha} = V_0^2 \underbrace{\frac{\epsilon S}{d^3}}_{C_{\mathrm{mech}}/d^2} \frac{8 C_c^2 C_0}{(C_c + C_0)^3} \frac{\omega^6 (\omega_c^2 - \omega^2)}{(\omega^2 - \omega_c^2)^2 + \gamma_c^2 \omega^2}$$
$$+ V_0^2 \underbrace{\frac{\epsilon^2 S^2}{d^4 C_0}}_{C_{\mathrm{mech}}^2/(d^2 C_0)} \frac{C_c^2}{(C_c + C_0)^2} \frac{\omega^4}{(\omega^2 - \omega_c^2)^2 + \gamma_c^2 \omega^2}.$$
(11.55)

Note that although the prefactor of the second-order term is $C_{\mathrm{mech}}/C_0 \ll 1$ times smaller than that of the first term, the first term vanishes at $\omega = \omega_c$. Therefore, close to resonance both terms are relevant.

Equation (11.53) assumes that the force responds instantaneously to the motion. This would correspond to the case where the cavity linewidth is much larger than the mechanical frequency of oscillation ($\gamma_c \gg \omega_m$). Including the delayed response on the force due to the cavity linewidth γ_c, this equation should be replaced by (Metzger and Karrai, 2004)[27]

[26] Note that the second term is disregarded in (quantum-mechanical) approaches starting from the Hamiltonian describing a radiation pressure coupling of the form $\hat{n}_c \hat{x}$ between the LC resonator with $\hat{n}_c$ photons coupled to the oscillations with amplitude $\hat{x}$. The second term would require a coupling of the form $\hat{n}_c \hat{x}^2$.

[27] Note that this force delay model is only a qualitatively correct description of the effect, whereas more accurate theory requires solving the equations of motion for the cavity and the mechanics simultaneously, as in (Marquardt et al., 2007) and the Supplementary information of (Massel et al., 2011). The two approaches agree quantitatively only in the 'non-resolved sideband limit' where $\gamma_c \gg \omega_m$ provided that γ_c is replaced by $\gamma_c/2$.

$$\ddot{\alpha}_m + \gamma_m \dot{\alpha}_m + \omega_m \alpha_m = \frac{\eta_m f_0(t)}{m} + \frac{\eta_m}{m} \frac{df}{d\alpha_m} \int_{-\infty}^{t} \alpha_m(t') \gamma_c e^{-\gamma_c(t-t')} dt'.$$

$$(11.56)$$

The second part of the force at time t thus represents the information about the position at an earlier time, with a delay time $1/\gamma_c$. If $\alpha_m(t)$ does not depend on time on this scale (i.e., if $\gamma_c \gg \omega_m$), it is straightforward to see that the integral over time gives simply $\alpha_m(t)$, consistent with eqn (11.53).

Now, inserting a force $f(t) = f_0 e^{i\omega_f t}$ gives us resonator oscillations $\alpha_m(t) = \alpha_{m0} e^{i\omega_f t}$, where

$$\alpha_{m0} = \frac{\eta_m f_0}{m(\omega_{\text{eff}}^2 - \omega_f^2 + i\gamma_{\text{eff}}\omega)}, \qquad (11.57)$$

where the effective frequency now contains the optical spring effect,

$$\omega_{\text{eff}}^2 = \omega_m^2 - \frac{\eta_m}{m} \frac{df}{d\alpha_m} \frac{\gamma_c^2}{\omega_f^2 + \gamma_c^2} = \omega_m^2 \left(1 - \frac{df}{d\alpha_m} \frac{\eta_m \gamma_c^2}{m\omega_m^2(\omega_f^2 + \gamma_c^2)} \right)$$

$$(11.58)$$

and the effective damping is given by

$$\gamma_{\text{eff}} = \gamma_m + \frac{\eta_m}{m} \frac{df}{d\alpha_m} \frac{\gamma_c}{\omega_f^2 + \gamma_c^2} = \gamma_m \left(1 + Q_m \eta_m \frac{df}{d\alpha_m} \frac{\gamma_c}{m\omega_m(\omega_f^2 + \gamma_c^2)} \right).$$

$$(11.59)$$

Here $Q_m = \omega_m/\gamma_m$ is the mechanical quality factor. Note that in the non-resolved side-band limit $\gamma_c \gg \omega$ of instantaneous response we get the optical spring effect described in eqn (11.53), and no change in the damping. However, in the more typical case $\gamma_c \approx \omega_m$ the change in the damping is larger than the shift in the bare frequency by the mechanical quality factor Q_m.

As seen in eqn (11.55), $df/d\alpha_m$ may have either sign, depending on the value of ω with respect to ω_c. The case $df/d\alpha_m > 0$ obtained with $\omega \lesssim \omega_c$ (red detuning) hence corresponds to an increased damping, whereas the case $df/d\alpha_m < 0$ obtained for $\omega \gtrsim \omega_c$ (blue detuning) describes decreased damping. When $\hbar\omega_c \gg k_B T$, the optical driving does not add any fluctuations while the damping is changed. Because of this, these regimes also correspond to cooling and heating of the mechanical vibrations. Experiments probing these regimes are discussed in Exercises 11.7 and 11.8.

Complement 11.1 Effect of mechanical vibrations on a driven LC resonance

To determine the effect of mechanical vibrations on the oscillations of the electronic degrees of freedom in the LC resonator, let us write the equations of motion for the relevant coupled oscillator degrees of freedom in the setup. For the mechanical resonator, this is obviously the amplitude α_m of motion. For the electrical resonator we could choose the voltage $V_{\text{LC}}(t)$, but typically a more convenient choice is the phase

$$\phi(t) = \frac{e}{\hbar} \int_0^t V(t') dt' + \phi(0). \qquad (11.60)$$

The prefactor makes the phase dimensionless and facilitates the discussion of the quantum regime of electronic systems, as discussed in Sec. 9.4. Note, however, that the Dirac constant $\hbar$ is chosen above simply to set the scale of the phase, as all the effects described in this complement are classical, and the measurable voltages or currents are independent of the value chosen for $\hbar$.

The phase is driven with the time-dependent voltage $V_0(t)$ inside the transmission line. Its equation of motion can be readily written down from the Fourier transform of eqn (11.48),

$$\ddot{\phi}(t) + \gamma_c \dot{\phi}(t) + \omega_c^2 \phi = \frac{e}{\hbar} \frac{C_c}{C + C_c} \dot{V}(0,t), \tag{11.61}$$

where $\omega_c = 1/\sqrt{L(C_0 + C_c + C_m)}$ and $\gamma_c = 1/\sqrt{R(C_0 + C_c + C_m)}$. On the other hand, the force driving the mechanical mode m is given by $f = (\hbar^2 \dot{\phi}^2)/(2e^2)\partial_{\alpha_m} C_m$ (see eqn (11.35)).

Taking into account the mechanical vibrations, the coupled equations of motion for the phase and the vibration amplitude are thus[28]

[28]This is a classical treatment of the problem. The quantum calculation is very similar, but the typical approach is to use the *input/output* formalism of quantum optics; see (Walls and Milburn, 2010).

$$\ddot{\phi} + \omega_c^2[1 - g_m \alpha_m(t)]\phi + \gamma_c \dot{\phi} = \frac{e}{\hbar} \frac{C_c}{C_0 + C_c} \dot{V}(0,t) \tag{11.62a}$$

$$\ddot{\alpha}_m + \omega_m^2 \alpha_m + \gamma_m \dot{\alpha}_m = \frac{\hbar^2 g_m \dot{\phi}(t)^2 (C_0 + C_c)}{2e^2 m} + \frac{\delta f_m(t)}{m}. \tag{11.62b}$$

Here we only take into account the lowest-order term in the force, not including higher order corrections from the varying capacitance. To describe the non-zero temperature T_m of the resonator, we can include the (Langevin) force $\delta f_m(t)$ related with the damping of the resonator as discussed in Example 11.1. The coupling strength between the phase and mechanical motion in a beam of length L_0 is described by[29]

[29]Do not confuse the length L_0 and inductance L!

$$g_m = \frac{1}{(C_0 + C_c)L_0} \int_0^{L_0} \frac{\delta C}{\delta u(z)} \chi_m(z) dz \tag{11.63}$$

and it has a dimension 1/length. Typically this coupling is rather weak, so that the frequency shift of the LC resonator resulting from this term is much smaller than ω_c.

Note that often $\omega_c \gg \omega_m$ and the two resonators are far from a common resonance. However, a situation close to resonant can be realized by driving the LC resonator with a voltage oscillating with frequency $\omega_p \approx \omega_c \pm \omega_m$ ('blue/red detuned driving'), in which case the two systems can be made to exchange energy almost resonantly.

It is convenient to describe the resonances via the (dimensionless) response functions of the type

$$r_{c/m}(\omega) = \frac{\omega_{c/m}^2}{\omega_{c/m}^2 - \omega^2 - i\gamma_{c/m}\omega}. \tag{11.64a}$$

$$R_{c/m}(\omega) = r_{c/m}(\omega)r_{c/m}(-\omega) = |r_{c/m}(\omega)|^2 = \frac{\omega_{c/m}^4}{(\omega^2 - \omega_{c/m}^2)^2 + \omega^2\gamma_{c/m}^2}. \tag{11.64b}$$

[30]The corresponding time-domain response is $\phi_0(t) = eV_0 C_c \omega_p / [(C_c + C_0)\hbar\omega_c^2]\{\text{Re}[r_c(\omega_p)]\cos(\omega_p t) + \text{Im}[r_c(\omega_p)]\sin(\omega_p t)\}$.

Consider driving the system with a time-dependent voltage $V(t) = V_0 \sin(\omega_p t)$. In the absence of the mechanics, this gives rise to a coherent response $\phi_0(t)$ oscillating at the drive frequency. The latter can be evaluated in the frequency domain to yield[30]

$$\phi_0(\omega) = \frac{\pi e C_c \omega_p V_0 r_c(\omega)}{\hbar (C_0 + C_c) \omega_c^2} [\delta(\omega - \omega_p) + \delta(\omega + \omega_p)]. \qquad (11.65)$$

In the presence of the mechanical motion, $\phi(t)$ obtains another, incoherent part, i.e., $\phi(t) = \phi_0(t) + \delta\phi(t)$. For low coupling strength g_m, we may evaluate $\delta\phi(t)$ and the amplitude of the mechanical motion from the coupled linear equations in Fourier space:

$$\frac{\omega_c^2 \delta\phi(\omega)}{r_c(\omega)} - \frac{g_m e \omega_p C_c V_0}{2\hbar (C_0 + C_c)} [r_c(\omega_p)\alpha_m(\omega - \omega_p) + r_c(-\omega_p)\alpha_m(\omega + \omega_p)] = \delta f_\phi(\omega)$$

$$(11.66a)$$

$$\frac{\omega_m^2 \alpha_m(\omega)}{r_m(\omega)} - \frac{g_m \hbar \omega_p C_c V_0}{2 e m \omega_0^2} [r_c(\omega_p)\delta\phi(\omega - \omega_p) + r_c(-\omega_p)\delta\phi(\omega + \omega_p)] = \frac{\delta f(\omega)}{m}.$$

$$(11.66b)$$

Here the second term on the second equation describes the first-order force term, $\phi(t)^2 \approx \phi_0(t)^2 + 2\phi(t)\delta q(t)$: the first of these terms contains a non-oscillating part and a part oscillating at $2\omega_p$. Here we assume that both of these frequencies are far from the mechanical resonance frequency, and therefore we may disregard the driving of the resonator due to the $\phi_0(t)$ term. Moreover, I have included a possible further (small) probing signal $\delta f_\phi(\omega) = e C_c / (C_0 + C_c) \delta V(\omega) / (\hbar\omega)$. It is set to zero in the following, but its effects are analyzed in Exercise 11.5.

In order to solve eqn (11.66) it is convenient to make another approximation: let us assume that $\omega_c \approx \omega_p \gg \omega_m$ and consider frequencies $\omega \approx \pm\omega_p$. Then, depending on the sign of ω we may drop either of the two terms in square brackets in eqn (11.66). This allows for a full solution of the problem via an inversion of the coefficient matrix.

Choosing $\omega_p \approx \omega_c \pm \omega_m$ (red/blue sidebands) then allows us to describe the side-band cooling/heating of the resonator (Marquardt *et al.*, 2007; Massel *et al.*, 2011) and the resultant (measurable) phase fluctuations. In the following I only consider the latter and calculate the terms linear in g_m. Such linear terms may be obtained by solving $\alpha_m(\omega)$ for $g_m = 0$ and inserting this to the equation for $\delta\phi(\omega)$. We get

$$\delta\phi_1(\omega) = \frac{e V_0 g_m \omega_p r_c(\omega)}{2 m \hbar \omega_c^2 \omega_m^2} \frac{C_c}{C_c + C_0} \sum_{\sigma = \pm 1} [\delta f(\omega - \sigma\omega_p) r_c(\sigma\omega_p) r_m(\omega - \sigma\omega_p)].$$

$$(11.67)$$

This function describes the Fourier amplitude of the phase oscillations in the LC resonator due to the mechanical motion driven by the force δf. This is not yet the observable measured in a typical experiment. Rather, what is measured is the amplitude (squared) of the resultant wave propagating out of the system inside the transmission line, of the form $V_{out}(x, t) = V_{out}(t) e^{-i(\omega/v)x}$. The amplitude $V_{out}(t)$ of this wave can be calculated by noting that a time-dependent voltage (or phase) on the LC resonator gives a boundary condition for the transmission line: current $I_{out}(0, t)$, related to V_{out} by eqn (11.44), equals $C_c(\dot{V}_{LC} - \dot{V}_{out}(t))$. From this we may obtain a relation for the Fourier amplitudes $V_{out}(\omega)$ and $\delta\phi(\omega)$,

$$V_{out}(\omega) = \frac{\hbar\omega^2 C_c}{e\sqrt{C_t/L_t}} \delta\phi(\omega). \qquad (11.68)$$

Here, for simplicity, we neglect a small factor $i\omega C_c / \sqrt{C_t/L_t} \ll 1$.

Note that for thermal forces the average $\langle \delta f \rangle = 0$, and therefore also the related average amplitude $\langle V_{out} \rangle$ vanishes. However, utilizing eqn (11.30) we can

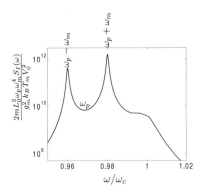

Fig. 11.13 In the presence of the mechanical oscillations, the output voltage noise correlator of the transmission line shows side-bands at $\omega = \omega_p \pm \omega_m$. Here $\omega_m = \gamma_c = \omega_c/100$, $\omega_p = 0.97\omega_c$, $\gamma_m = 0.001$ and the scale of the voltage correlator is given by the prefactor of the $R(\omega)$-terms in eqn (11.69). Note the logarithmic scale!

[31] For example, in (Teufel *et al.*, 2011) the total mass of the resonator was 48 pg and $\omega_m \approx 10$ MHz, which resulted in $x_{\mathrm{ZPM}} \approx 4.1$ fm.

[32] In reality this is slightly more complicated due to the combination of facts that also the coupling to the detecting electrical system may be made stronger in larger systems, and larger resonators have typically a higher quality factor.

[33] The correspondence principle states that for linear systems the quantum-mechanical equation of motion has to be the same as the classical one.

[34] Nanoscale systems are difficult to cool much below 10 mK. In frequency units, 10 mK corresponds to $h \cdot 210$ MHz, where h is Planck's constant.

[35] Demonstrated in (O'Connell *et al.*, 2010) and (Teufel *et al.*, 2011). However, this as such is not yet a quantum effect, unless the zero-point vibrations are accessed.

[36] The latter has been recently demonstrated in resonators coupled to an optical cavity (Safavi-Naeini *et al.*, 2012) by measuring the asymmetry between the emission and absorption spectra of the vibrations, as discussed in Sec. 6.1.

calculate the average of the square of V_{out}, i.e., the voltage–voltage correlator. This is

$$\langle V_{\mathrm{out}}(\omega)V_{\mathrm{out}}(-\omega)\rangle =$$
$$\frac{L_t C_c^2 V_0^2}{C_t} \frac{C_c^2}{(C_c + C_0)^2} \frac{g_m^2 \omega^4 \omega_p^2}{2 m \omega_c^4 \omega_m^4} k_B T \gamma_m R_c(\omega) R_c(\omega_p)[R_m(\omega - \omega_p) + R_m(\omega + \omega_p)].$$
$$(11.69)$$

Contrary to the coherent response of the LC cavity that oscillates at the frequency ω_p of the pump, this 'thermal response' contains many frequencies. The response functions $R_{c/m}(\omega)$ are peaked at $\omega \approx \pm\omega_{c,m}$. This shows up as side-band peaks in the voltage–voltage correlator, residing at frequencies $\omega = \omega_p \pm \omega_m$ (see Fig. 11.13). Measuring the positions of these side bands hence allows determination of the eigenfrequency of the mechanical oscillations with great accuracy. Moreover, the amplitude of the peaks is proportional to the temperature of the resonator, and can therefore be used to measure it.

11.4 Quantum effects

One of the recent trends in the studies of nanoelectromechanical systems is the search for true quantum-mechanical behaviour related to the macroscopic mechanical vibrations. Curiously, such macroscopic quantum effects have been demonstrated in optical and electronic setups, but quantum-mechanical mechanical (with intentional repetition) motion remained unaccessible until only very recently.

One of the outstanding questions is how large systems can exhibit quantum effects. Making the systems larger makes the observation of such effects generally more difficult. This is partially because the eigenfrequencies of mechanical motion typically are inversely proportional to the size of the resonator (see eqns (11.16) and (11.21)), and it is hence difficult to cool the systems to their mechanical ground state. Secondly, at the ground state the resonators exhibit quantum zero-point vibrations with amplitude

$$x_{\mathrm{ZPM}} = \sqrt{\frac{\hbar}{2 m \omega_m}}, \qquad (11.70)$$

which also quite generally becomes smaller at larger resonators.[31] Therefore, measurement of zero-point vibrations becomes increasingly more difficult as the system size becomes larger.[32] Recently the experiments on mechanical resonators with one dimension in excess of micrometers have demonstrated the necessary accuracy to measure the zero-point vibrations (Teufel *et al.*, 2011) with electrical setups.

The main problems in finding quantum behaviour in nanomechanical systems are related with the linearity of the oscillators[33] and their low eigenfrequencies compared to the achievable temperatures.[34] Suggested quantum effects include cooling the resonators close to their ground state,[35] creating superpositions between different Fock states of the vibration modes (see Sec. 11.4.1 below), direct detection of the zero-point vibrations[36] and macroscopic quantum tunnelling of nanomechanical beams (Sillanpää *et al.*, 2011). Moreover, it was recently suggested (Massel *et al.*, 2011) that coupling mechanical resonators to microwave resonators allows the use of this system as a quantum-limited amplifier of microwaves—provided the mechanical system can first be cooled to its ground state.

The quantization of the equations of motion can be carried out as in any quantum-mechanics textbook: we introduce the creation and annihilation operators $\hat{a}$ and $\hat{a}^\dagger$, satisfying the bosonic commutation relations (A.5), and relate them to the amplitude α_m of oscillations and the corresponding momentum $p_m = m\dot{\alpha}_m$ via

$$\alpha_m = x_{\mathrm{ZPM}}(\hat{a}^\dagger + \hat{a}), \quad p_m = ip_{\mathrm{ZPM}}(\hat{a}^\dagger - \hat{a}), \tag{11.71}$$

where $p_{\mathrm{ZPM}} = \sqrt{\hbar m \omega_m / 2}$. As a result, the Hamiltonian describing the energy of the harmonic oscillator can be written as[37]

$$H_{\mathrm{ho}} = \hbar\omega_m(\hat{a}^\dagger \hat{a} + \frac{1}{2}). \tag{11.72}$$

This Hamiltonian thus describes the amplitude of motion of a single flexural mode:[38] to describe many modes, we should introduce the corresponding operators for each mode, and sum over the mode index.

11.4.1   Creating a quantum superposition of vibration states in an oscillator–qubit system

One way to demonstrate true quantum effects in a mechanical resonator is to couple it to a structure working as a qubit.[39] In this case, a superposition state created in the qubit can be transferred to a superposition state of the resonator.[40] This was achieved recently by coupling a superconducting phase qubit to a mechanical resonator made of a piezoelectric material, see (O'Connell *et al.*, 2010). Such systems and the scheme of transferring superpositions are outlined in this section.

As discussed in Sec. 9.4.3, the Hamiltonian of a qubit can in general be written in terms of the Pauli spin matrices[41]

$$H_q = -\frac{B_z}{2}\hat{\sigma}_z - \frac{B_x}{2}\hat{\sigma}_x, \tag{11.73}$$

where the components of the effective 'magnetic field' $B_{z,x}$ are related to the macroscopic control fields, such as a gate voltage or an external magnectic field.

Coupling the qubit to a mechanical resonator allows the state of the resonator to affect the fields $B_{z,x}$: for example, in a superconducting charge qubit (eqn (9.44) in Sec. 9.4.2) B_z depends on the value of the gate charge, which on the other hand depends on the size of the gate capacitance and therefore on the vibrations. We can characterize the effect of the nanomechanical motion on the field B_z as in Sec. 11.2.4 above. By tuning the average gate charge suitably, the total Hamiltonian can be written without the average B_z-term,

$$H = \underbrace{\hbar\omega_m \hat{a}^\dagger \hat{a}}_{\text{oscillator}} \underbrace{- \frac{B_x}{2}\hat{\sigma}_z}_{\text{qubit}} + \underbrace{g(\hat{a} + \hat{a}^\dagger)\hat{\sigma}_x}_{\text{coupling}}. \tag{11.74}$$

[37] I disregard the 1/2 term in the following, as it there has no observable consequences.

[38] See Exercise A.2.

[39] See, for example, the description of spin qubits in Sec. 8.5 and superconducting qubits in Sec. 9.4.3.

[40] Note that most of these types of scheme involve superpositions of the Fock states of the resonators, labelled by the occupations of the flexular modes. The resonators cannot therefore be said to be in a superposition of 'here' and 'there', but rather in a more abstract superposition of 'this' and 'that' amplitude of oscillations.

[41] See Appendix A.6.

For convenience I also rotate the qubit Hamiltonian by 90 degrees, converting σ_x to $-\sigma_z$ and σ_z to σ_x. Here we focus only on one eigenmode of the oscillator, characterized by the bosonic annihilation (creation) operator $\hat{a}^{(\dagger)}$. This is usually the $n = 1$ mode with the lowest eigenfrequency. Moreover, g is the coupling energy related with the addition of one phonon into the resonator.

The idea of the quantum manipulation of the resonator is now the following. For most of the manipulations the qubit and the resonator are far from resonance, and therefore weakly coupled. First, while the systems are out of resonance, the qubit and the resonator are initialized into their ground states. The qubit is excited into its excited state by applying a suitably pulsed control field with frequency equal to the level separation.[42] After this the systems are brought into resonance by varying the qubit field B_x quickly to the value $B_x = \omega_m$. As discussed in Example 11.2 below, the excited state of the qubit can now be transferred to a phonon in the resonator, after which the systems are again brought out of resonance. Repeating this cycle allows the addition of phonons one by one to the resonator.

[42] How this is done is explained in books on quantum computing, such as (Nielsen and Chuang, 2000) and (Nakahara and Ohmi, 2008).

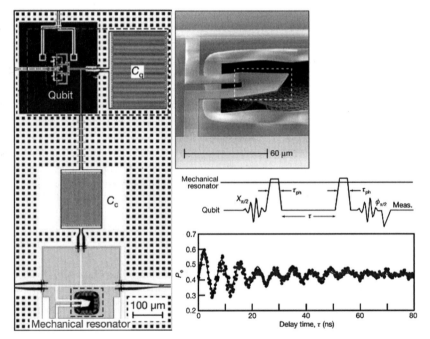

Fig. 11.14 The first experimental setup studying quantum coherent effects in qubit-mechanical resonator systems. Left: Optical micrograph of the mechanical resonator coupled to a superconducting phase qubit. Right top: scanning electron micrograph of the suspended film bulk acoustic resonator, made of a piezoelectric material. Right centre: Control field pulse sequence used to create and measure the superposition of the resonator Fock states and their decoherence. Right bottom: Coherent oscillations of the resonator superposition state, eqn (11.75), accompanied by their decoherence. What is plotted is the qubit excited state probability once the phonon excitation is transferred back to the qubit after delay time τ. (A.D. O'Connell *et al.*, *Nature* **464**, 697 (2010), Figs. 1, 2 and 6.) Reprinted by permission from Macmillan Publishers Ltd., © 2010.

A superposition of different Fock states of the resonator, i.e., states with different occupation numbers, can be realized by replacing the second step above, exciting the qubit, to the transformation of the qubit state to the superposition

$$\psi_{\text{qubit}} = (|\downarrow\rangle + |\uparrow\rangle)/\sqrt{2}.$$

Bringing the qubit again to resonance with the resonator for a suitable time allows transformation of this superposition state to the mechanical resonator, bringing the system to the state

$$\psi = (|\downarrow 0\rangle + i|\downarrow 1\rangle)/\sqrt{2} = |\downarrow\rangle(|0\rangle + i|1\rangle)/\sqrt{2}. \qquad (11.75)$$

This is therefore a state involving a superposition of different vibration states of a macroscopic mechanical resonator. Its creation and measurement was demonstrated recently by O'Connell *et al.* (2010) (see Fig. 11.14).

Example 11.2 Description of the qubit–oscillator system at resonance

The effective coupling between the qubit and the resonator in eqn (11.74) depends strongly on the relative magnitudes of the oscillation frequencies B_x and ω_0. If one is much larger than the other, the dynamics of the subsystem with a higher eigenfrequency can be calculated treating the other system parameters as constant in time. However, the dynamics is strongly coupled if the systems are at resonance. In this case only the 'resonant' terms in the coupling are typically retained, transforming the total Hamiltonian to[43]

$$H = \hbar\omega_m \hat{a}^\dagger \hat{a} + \frac{B_x}{2}\hat{\sigma}_z + g(\hat{a}\hat{\sigma}^\dagger + \hat{a}^\dagger\hat{\sigma}_\downarrow). \qquad (11.76)$$

This is frequently called the Jaynes–Cummings Hamiltonian, traditionally describing the coupling between light and matter (atoms). The state space of this system consists of an outer product of spin states $|\uparrow\rangle, |\downarrow\rangle$ and states $|n\rangle$ with $n = 0, 1, 2, \ldots$ phonons occupying the fundamental flexular mode described by the operators a and $a^\dagger$, i.e., $a^\dagger a|n\rangle = n|n\rangle$. For brevity we denote these outer product states below by $|\sigma n\rangle$ where $\sigma \in \{\uparrow, \downarrow\}$. In the absence of a coupling, these outer product states are eigenstates of the system with energy

$$\epsilon_{\uparrow n} = n\hbar\omega_m + \frac{B_x}{2}, \quad \epsilon_{\downarrow n} = n\hbar\omega_m - \frac{B_x}{2}. \qquad (11.77)$$

For $B_x = \hbar\omega_m$, the states $|\uparrow n\rangle$ are degenerate with states $|\downarrow n+1\rangle$. Close to this resonance, when the system is at a temperature $k_B T \ll \min(\hbar\omega_m, B_x)$, it is enough to discuss the effect of the coupling g in the subspace of the total Hilbert space spanned by the vectors $|\uparrow 0\rangle$ and $|\downarrow 1\rangle$. Projecting eqn (11.76) into this subspace can be described by the effective Hamiltonian

$$H_{\text{coupled}} = \frac{B_x - \hbar\omega_m}{2}\sigma_z + g\sigma_x. \qquad (11.78)$$

The time evolution of a given state with this Hamiltonian (in the Schrödinger picture) is given by

$$\psi(t) = e^{-\frac{i}{\hbar}H_{\text{coupled}}t}\psi(0). \qquad (11.79)$$

Let us consider what happens to an initial state $\psi(0) = |\uparrow 0\rangle$, i.e., the qubit in the excited state and the resonator in its ground state, as a function of time, when the two systems are at resonance, i.e., $B_x = \hbar\omega_m$. In this case

$$\begin{aligned}
\psi(t) &= e^{-\frac{i}{\hbar}g\sigma_x t}\psi(0) = [\cos(gt/\hbar)\sigma_0 + i\sin(gt/\hbar)\sigma_x]\psi(0) \\
&= \cos(gt/\hbar)|\uparrow 0\rangle + i\sin(gt/\hbar)|\downarrow 1\rangle.
\end{aligned} \qquad (11.80)$$

At $t = \hbar/g$ the system contains a qubit in its ground state and a resonator containing a single phonon. The excitation energy has thus been moved from the qubit to the resonator in a coherent way. This can be used for the coherent manipulation of the resonator states once the qubit can be separately (i.e., off-resonance) manipulated.

[43] The resonant terms couple degenerate states $|\downarrow n\rangle$ and $|\uparrow n+1\rangle$, whereas the off-resonant terms $\hat{a}\hat{\sigma}_\downarrow$ and $\hat{a}\hat{\sigma}_\uparrow$ couple states with a large energy difference $\omega_0 + B_x \approx 2\omega_0$, resulting in a weaker effect.

11.4.2   Describing dissipation

As discussed in Sec. 6.6, isolated quantum-mechanical systems show time-reversible dynamics, and, for example, there is no dissipation of energy. For a description of such effects, the isolated system has to be coupled to an environment. Most often this environment is modelled as an infinite set of harmonic oscillators describing the excitations in the environment. For example, the Hamiltonian describing a single harmonic oscillator (say, a vibration state in a nanomechanical beam) coupled to a dissipative environment is of the form

$$H = \hbar\omega_0 \hat{a}^\dagger \hat{a} + \hbar \sum_\alpha \left[\gamma_\alpha \hat{x}\hat{x}_\alpha + \frac{\omega_\alpha}{4}\left(\hat{p}_\alpha^2 + \hat{x}_\alpha^2\right) + \frac{1}{\omega_\alpha}\gamma_\alpha^2 \hat{x}^2 \right], \quad (11.81)$$

where $\hat{a}^{(\dagger)}, \hat{a}_\alpha^{(\dagger)}$ are bosonic annihilation (creation) operators, $\hat{x}_{(\alpha)} = \hat{a}_{(\alpha)}^\dagger + \hat{a}_{(\alpha)}$ is the normalized position operator and $\hat{p}_{(\alpha)} = i(\hat{a}_{(\alpha)}^\dagger - \hat{a}_{(\alpha)})$ the normalized momentum operator. The first term describes the oscillator of interest, with resonance frequency ω_0, the first term in the square brackets is a coupling between the main oscillator and the environment oscillators, the second term in the brackets is the resonance energy of the bath oscillator, and the last term is needed to balance an average effect from the bath on the system.

Applying the Heisenberg equation of motion for $\hat{x}$ and $\hat{x}_\alpha$ once gives

$$-i\hbar\dot{\hat{x}} = [H, \hat{x}] = -i\hbar\omega_0\hat{p} \qquad (11.82a)$$

$$-i\hbar\dot{\hat{x}}_\alpha = [H, \hat{x}_\alpha] = -i\hbar\omega_\alpha\hat{p}_\alpha. \qquad (11.82b)$$

The second derivative is

$$-i\hbar\ddot{\hat{x}} = [H, \omega_0\hat{p}] = -i\hbar\omega_0^2\hat{x} + 2i\hbar\omega_0\left(\sum_\alpha \gamma_\alpha \hat{x}_\alpha + \frac{2\gamma_\alpha^2}{\omega_\alpha}\hat{x}\right) \qquad (11.83a)$$

$$-i\hbar\ddot{\hat{x}}_\alpha = [H, \omega_\alpha\hat{p}_\alpha] = -i\hbar\omega_\alpha^2\hat{x}_\alpha + 2i\hbar\omega_\alpha\gamma_\alpha\hat{x}. \qquad (11.83b)$$

Let us now average this equation over the state of the system, converting operators $\hat{x}, \hat{x}_\alpha$ to their averages $x = \langle\hat{x}\rangle, x_\alpha = \langle\hat{x}_\alpha\rangle$. Fourier transforming $(\partial_t^2 \mapsto -\omega^2)$, we can solve for $x_\alpha(\omega)$,

$$x_\alpha(\omega) = \frac{2\omega_\alpha\gamma_\alpha x}{\omega^2 - \omega_\alpha^2} \qquad (11.84)$$

and insert this in the equation for x, yielding

$$(\omega^2 - \omega_0^2)x = 4\omega_0 x \sum_\alpha \gamma_\alpha^2 \left[\frac{\omega_\alpha}{\omega^2 - \omega_\alpha^2} + \frac{1}{\omega_\alpha}\right] \qquad (11.85)$$

$$\equiv -ix\omega\eta,$$

where

$$\eta = \sum_\alpha \gamma_\alpha^2 \frac{4i\omega_0\omega}{\omega_\alpha(\omega^2 - \omega_0^2)}. \qquad (11.86)$$

In the following I aim to show that under some assumptions η can be made real and independent of ω.

Let us assume that there are many oscillators and that their level spacing $\delta\omega$ is small compared to ω_0. This allows us to convert the sum into an integral. Moreover, let us assume that the couplings are of the form $\gamma_\alpha^2 = a\omega_\alpha$. Therefore[44]

$$\eta = \frac{4i}{\delta\omega} \int_0^\infty \frac{a\omega}{\omega^2 - \omega_\alpha^2} d\omega_\alpha = \frac{2\pi a}{\delta\omega}, \qquad (11.87)$$

which is real and independent of ω as desired.

As a result, the equation of motion for x,

$$(\omega^2 + i\omega\eta - \omega_0^2)x = 0, \qquad (11.88)$$

describes damped harmonic oscillation with a damping time $1/\eta$. This trick thus allows us to describe damping (dissipation) of the harmonic oscillator even when starting from a full quantum-mechanical model.

[44]Strictly speaking, this integral needs a convergence factor, a tiny imaginary part in ω, which can be taken to zero at the end of the calculation.

Outlook

Microelectromechanical systems are today found in many everyday applications, ranging from accelerometers and pressure sensors to microscopes, to name only a few. As shown in this chapter, miniaturization of the resonators towards the nanomechanical regime allows the increase of the characteristic frequencies and simultaneously improve the force, mass and charge sensitivities of these devices. However, the study of the combination of non-linear mesoscopic transport effects and controlled (nano)mechanical motion is still in its infancy, and may well reveal new and surprising phenomena. Simultaneously, this combination also opens the possibilities for studying macroscopic quantum effects related to mechanical motion. This chapter provides only a glimpse of the possible effects. An example of a new combination of systems is that of nanomechanics and spintronics; see (Kovalev *et al.*, 2011). A lot more discoveries can be expected in the coming years.

Further reading

- Foundations of nanomechanics, along with some of the coupling schemes presented here are discussed in (Cleland, 2003). This book also includes a detailed description of the linear elastic theory.
- A review of nanoelectromechanical systems has been given by (Ekinci and Roukes, 2005).
- (Marquardt and Girvin, 2009) contains a short review on optomechanics, and also includes systems combining electronic *LC* resonators to nanomechanics.

Exercises

(11.1) Estimate the fundamental flexural mode eigenfrequencies of (a) 1 μm long Al beam with height $h = 0.5$ μm and width $w = 0.3$ μm, (b) 1 μm long carbon nanotube with diameter 1.6 nm, and (c) 1 μm long and 500 nm wide graphene sheet with thickness (height) 0.3 nm. Estimate the mass detection sensitivity of the carbon nanotube NEMS, assuming that its Q-value is 10000. Use the fact that the resonance frequency can be measured with an accuracy ω_m/Q.[45]

(11.2) Calculate the force due to the gate voltage in the case when the island of a single-electron box is suspended (see Fig. 11.15 and Sec. 7.1.1).

(11.3) By summing the currents in eqn (7.24) and the current $I_{\cot} = e\Gamma_{\cot}$ with $\Gamma_{\cot}$ given by eqn (7.27), estimate the sensitivity $\partial I/\partial\alpha_n$ of the ac single-electron tunnelling current to mechanical vibrations at the resonance $\omega_d = \omega_1$, assuming some Q-value for the mechanical resonance. You can assume a two-plane capacitor model $C(d) = \epsilon_0 A/d$. Assume the SET is biased with a low voltage $V \ll e/C$ and gated close to the resonance point $Q_G = 1/2 - x$ with some small deviation x.

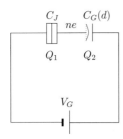

Fig. 11.15 Circuit for a suspended single-electron box with a deformation-dependent gate capacitance.

(11.4) Using eqns (11.59) and the first term of (11.55) with $\omega = \omega_c + \omega_m$ and $\omega_f = \omega_m$, estimate the voltage amplitude V_0 required to reach the threshold of a parametric instability of oscillations where $\gamma_{\mathrm{eff}} = 0$.

(11.5) Consider the probed response of the mechanical vibrations coupled to an LC resonator, described by eqns (11.66) but neglecting the fluctuating force $\delta f(\omega)$ and including the force δf_ϕ. Following a similar calculation as for the thermal fluctuations, calculate the reflection probability of the signal $\delta V(t)$ oscillating at frequency ω_δ.

(11.6) The description of dissipation in Sec. 11.4.2 relies on time-reversible Hamiltonian dynamics, but leads to a time-irreversible equation, eqn (11.88). Can you determine which assumption in the calculation leads to this result?

(11.7) **Question on a scientific paper.** (Probed resonance): Consult the paper Massel *et al.*, *Nature* **480**, 351 (2011) and try to understand what was measured there. (a) Figure 3a shows the measured reflection amplitude (see Appendix E) of the superconducting cavity (LC resonator)–mechanical resonator system. What can you say about the properties of the superconducting cavity resonance based on Fig. 3a? (b) Figure 3b shows the same quantity but zoomed inside the cavity absorption dip. The mechanical eigenfrequency was 32.5 MHz. Why can the measured peak, residing at around 6.98 GHz, contain information about the mechanical resonance? (c) Fig. 3b also shows that with $n_c \gtrsim 3.6 \times 10^5$, the probe signal is amplified (reflection probability exceeds unity). How is this possible: where does the excess energy come from?

(11.8) **Question on a scientific paper.** (Thermal response): Consult the paper Teufel *et al.*, *Nature* **475**, 359 (2011), presenting results on cooling mechanical resonators to the ground state. Try to understand how the mechanical occupation number was measured. The schematic of the measurement is explained in Fig. 1b: coupling the driving with frequency ω_d with the thermal mechanical oscillations at frequency ω_m creates side-bands at $\omega_d \pm \omega_m$. (a) How can you deduce the number of thermal quanta from the results presented in Fig. 3a? (b) Why are n_m not quantized to integer values? (c) Can you understand why zero-point fluctuations do not show up in this measurement?

---

[45]In fact, this is not a fundamental limitation, and typically this accuracy depends on many things—primarily the noise in the measurement and the averaging time. However, it can be regarded as a useful rule of thumb.

Important technical tools

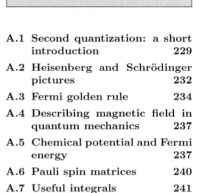

Most of the calculations carried out in this book require only the knowledge of the most basic technical tools taught typically in the first quantum mechanics courses. As a reminder for the students, I detail here some of them: a short introduction to second quantized operators for bosons and fermions, to Schrödinger, Heisenberg and Interaction pictures, and to the Fermi golden rule. I also explain briefly how the orbital effect of a magnetic field is included in the Schrödinger equation via minimal substitution, elaborate the relation between the chemical potential and the Fermi energy, and introduce the Pauli spin matrices useful for describing not only spin but any quantum two-level system. Finally, Sec. A.7 lists some of the most common integrals used in the description of electronic properties of open quantum systems: integrals over different combinations of Fermi functions together with some lowest powers of energy.

A.1 Second quantization: a short introduction

Various parts of this book employ the formalism of second quantization. I introduce it here separately for bosons and fermions.

A.1.1 Bosons

Consider a system of N bosons occupying quantum states with normalized wave functions $\phi_i(x)$, $i = 1, \ldots$. The state of such a system is described by a set of numbers N_i of particles in state i. The normalized total wave function for the occupation numbers $N_1, N_2, \ldots$ is given by

$$\Phi_{N_1, N_2, \ldots} = \left(\frac{N_1! N_2! \ldots}{N!} \right)^{1/2} \sum_P \phi_{j_1}(x_1) \phi_{j_2}(x_2) \ldots \phi_{j_N}(x_N). \quad (A.1)$$

Here j_m are the labels of the states, for example $j_1, j_2, \ldots, j_{N_1} = 1$, $j_{N_1+1}, j_{N_1+2}, \ldots, j_{N_1+N_2} = 2$, etc. The sum goes over all possible permutations of the labels and the prefactor takes care of the normalization. For Bose particles this wave function is symmetric on interchanging the labels.

For what follows, we use the bra–ket notation for this state,

$$|N_1, N_2, \ldots\rangle = \Phi_{N_1, N_2, \ldots},$$

i.e., labelling the different many-particle states by the occupation numbers of the different states. For example, for a three-particle system the state with $N_1 = 2$ and $N_2 = 1$ is

$$|2, 1\rangle =$$
$$\frac{1}{\sqrt{3}}[\phi_1(x_1)\phi_1(x_2)\phi_2(x_3) + \phi_1(x_1)\phi_2(x_2)\phi_1(x_3) + \phi_2(x_1)\phi_1(x_2)\phi_1(x_3)].$$

Normalization corresponds to requiring that the integral over the coordinates x_i gives unity.

To introduce any linear operator acting on a quantum state, it is sufficient to specify its matrix elements between the states that span the Hilbert space of interest. So let us introduce an operator a_i that decreases the occupation number in state i by 1. We define it such that

$$\langle N_1, N_2, \ldots, N_i - 1, \ldots |a_i| N_1, N_2, \ldots, N_i, \ldots \rangle = \sqrt{N_i}. \tag{A.2}$$

The matrix element is naturally finite only for this type of combination of states, and for other combinations it vanishes. The conjugate operator is defined such that

$$\langle N_1, N_2, \ldots, N_i, \ldots |a_i^\dagger| N_1, N_2, \ldots, N_i - 1, \ldots \rangle = \sqrt{N_i}. \tag{A.3}$$

Combining the two operators then gives the properties

$$a_i^\dagger a_i = N_i, \quad a_i a_i^\dagger = N_i + 1, \tag{A.4}$$

i.e., the combined operator is proportional to the unit operator. For different states $i \neq j$, any diagonal matrix element of $a_i a_j^\dagger$ (i.e., bracketed between the same states) has to vanish, and for the non-diagonal matrix elements their order does not matter: a_i and $a_j^\dagger$ with $i \neq j$ commute. The same holds for the matrix elements of $a_i a_j$ and $a_i^\dagger a_j^\dagger$ with all combinations of i and j. This result can be stated in a simple form:

$$[a_i, a_j^\dagger] \equiv a_i a_j^\dagger - a_j^\dagger a_i = \delta_{ij} \tag{A.5a}$$
$$a_i a_j - a_j a_i = a_i^\dagger a_j^\dagger - a_j^\dagger a_i^\dagger = 0. \tag{A.5b}$$

These are the bosonic commutation relations.

To derive the average occupation number of state i with energy ϵ_i in a grand canonical ensemble relevant for open systems, we use the definition of the grand partition function,

$$Z_G = \prod_i \sum_{n_i=0}^{\infty} \exp[-n_i \beta(\epsilon_i - \mu)] = \prod_i \{1 - \exp[-\beta(\epsilon_i - \mu)]\}^{-1}, \tag{A.6}$$

where $\beta = 1/(k_B T)$ is the inverse temperature fixing the average energy and μ is the chemical potential fixing the average number of particles. The grand potential is obtained as $\Omega_G = -k_B T \ln Z_G$, from which we get the average number of particles,

$$N = -\left(\frac{\partial \Omega_G}{\partial \mu}\right) = \sum_i n_i.$$

The contribution from each state i gives the average number of bosons occupying state i,

$$n_i = \frac{1}{\exp[\beta(\epsilon_i - \mu)] - 1}. \tag{A.7}$$

This is the Bose–Einstein distribution.

A.1.2 Fermions

For fermions, the total wave function has to be antisymmetric with respect to interchanging any two particles,

$$|N_1, N_2, \dots\rangle = \frac{1}{\sqrt{N!}} \sum_P \delta_P \phi_{j_1}(x_1)\phi_{j_2}(x_2)\cdots\phi_{j_N}(x_N). \tag{A.8}$$

Here δ_P is $+1$ or -1 depending on whether the state $j_1, j_2, \dots$ is obtained after an even or odd number of transpositions from some initial configuration. For example, we have the three-particle state

$$|1, 0, 1, 1\rangle =$$
$$\frac{1}{\sqrt{6}}[\phi_1(x_1)\phi_3(x_2)\phi_4(x_3) - \phi_1(x_1)\phi_4(x_2)\phi_3(x_3) - \phi_3(x_1)\phi_1(x_2)\phi_4(x_3)$$
$$- \phi_4(x_1)\phi_3(x_2)\phi_1(x_3) + \phi_3(x_1)\phi_4(x_2)\phi_1(x_3) + \phi_4(x_1)\phi_1(x_2)\phi_3(x_3)]. \tag{A.9}$$

The antisymmetry of the wave function implies the *Pauli exclusion principle* that there cannot be more than one particle occupying each state. Namely, consider the possibility of having the state

$$|2, 0, 1\rangle =$$
$$\frac{1}{\sqrt{6}}[\phi_1(x_1)\phi_1(x_2)\phi_3(x_3) - \phi_1(x_1)\phi_3(x_2)\phi_1(x_3) - \phi_1(x_1)\phi_1(x_2)\phi_3(x_3)$$
$$- \phi_3(x_1)\phi_1(x_2)\phi_1(x_3) + \phi_1(x_1)\phi_3(x_2)\phi_1(x_3) + \phi_3(x_1)\phi_1(x_2)\phi_1(x_3)] = 0.$$

It is straightforward to see that the same holds for an arbitrary wave function with more than a single particle occupying the same state.

The creation and annihilation operators are defined through the matrix elements

$$\langle N_1, N_2, \dots, N_i + 1, \dots | a_i^\dagger | N_1, N_2, \dots, N_i, \dots\rangle = (-1)^{\sum_{k=1}^{i-1} N_k} \delta_{N_i, 0} \tag{A.10}$$

and

$$\langle N_1, N_2, \dots, N_i - 1, \dots | a_i | N_1, N_2, \dots, N_i, \dots\rangle = (-1)^{\sum_{k=1}^{i-1} N_k} \delta_{N_i, 1}. \tag{A.11}$$

These satisfy (again for diagonal matrix elements) (see Exercise A.1)

$$a_i^\dagger a_i = N_i, \quad a_i a_i^\dagger = 1 - N_i \tag{A.12}$$

and

$$a_i^\dagger a_j + a_j a_i^\dagger = 0 \tag{A.13}$$

for $i \neq j$. Thus, we find the fermionic anticommutation relations

$$\{a_i, a_j^\dagger\} \equiv a_i a_j^\dagger + a_j^\dagger a_i = \delta_{ij} \tag{A.14a}$$

$$a_i a_j + a_j a_i = a_i^\dagger a_j^\dagger + a_j^\dagger a_i^\dagger = 0. \tag{A.14b}$$

The grand partition function is obtained by exploiting the Pauli exclusion principle,

$$Z_G = \prod_i \sum_{n_i=0}^{1} \exp[-n_i \beta(\epsilon_i - \mu)] = \prod_i 1 + \exp[-\beta(\epsilon_i - \mu)]. \tag{A.15}$$

The average particle number, obtained similarly as for the bosons, is $N = \sum_i n_i$ with

$$n_i = \frac{1}{\exp[\beta(\epsilon_i - \mu)] + 1}. \tag{A.16}$$

This is the Fermi–Dirac distribution.

A.2   Heisenberg and Schrödinger pictures

In this book, many of the phenomena are discussed using the Heisenberg picture of quantum mechanics, where the operators rather than the wave functions are time dependent. The two pictures are connected as follows. Let the Hamiltonian $\hat{H}$ be time independent. The wave function $\psi(t)$ in the Schrödinger picture obeys the Schrödinger equation

$$i\hbar \frac{\partial \psi}{\partial t} = \hat{H}\psi. \tag{A.17}$$

Its solution can be written formally as

$$\psi(t) = e^{-i\hat{H}t/\hbar} \psi_H, \tag{A.18}$$

where ψ_H is time independent. The time dependence of an observable O is calculated as a function of the matrix elements of the corresponding operator $\hat{O}$,

$$O_{mn} = \langle \psi_m(t) | \hat{O} | \psi_n(t) \rangle = \langle \psi_{Hm} | e^{i\hat{H}t/\hbar} \hat{O} e^{-i\hat{H}t/\hbar} | \psi_{Hn} \rangle. \tag{A.19}$$

This is thus the same as the matrix element of a time-dependent (Heisenberg) operator

$$\hat{O}_H(t) \equiv e^{i\hat{H}t/\hbar} \hat{O} e^{-i\hat{H}t/\hbar}, \tag{A.20}$$

calculated over the time-independent basis ψ_{Hn}. This operator satisfies the Heisenberg equation

$$\frac{\partial \hat{O}_H}{\partial t} = \frac{i}{\hbar} [\hat{H}, \hat{O}_H]. \tag{A.21}$$

Typically, if not stated otherwise, any time-dependent operator refers to the Heisenberg operator.

A.2.1   Interaction picture

Many problems in quantum mechanics deal with an application of a force on a given system and calculating the response to this force. Let us denote the total Hamiltonian in this case by $\hat{H} = \hat{H}_0 + \hat{V}$. Typically the system Hamiltonian $\hat{H}_0$ is somehow manageable, i.e., its eigenstates and eigenvalues can be determined, whereas the effect of a (possibly time-dependent) perturbation $\hat{V}(t)$ is harder to calculate. In this case the interaction picture is of great help. In the interaction picture, both states and operators are time dependent, defined with respect to the Schrödinger picture by

$$|\psi_I(t)\rangle = e^{i\hat{H}_0 t/\hbar}|\psi_S(t)\rangle \tag{A.22a}$$

$$\hat{O}_I(t) = e^{i\hat{H}_0 t/\hbar}\hat{O}_S e^{-i\hat{H}_0 t/\hbar}. \tag{A.22b}$$

The operator $\hat{H}_0$ does not depend on time. The idea of the interaction picture can be seen by taking the time derivative of the wave function:

$$
\begin{aligned}
i\hbar\partial_t|\psi_I(t)\rangle &= e^{i\hat{H}_0 t/\hbar}(-\hat{H}_0 + i\hbar\partial_t)|\psi_S(t)\rangle \\
&= \exp(i\hat{H}_0 t/\hbar)(-\hat{H}_0 + \hat{H})\exp(-i\hat{H}_0 t/\hbar)|\psi_I(t)\rangle \quad \text{(A.23)} \\
&= \hat{V}_I(t)|\psi_I(t)\rangle.
\end{aligned}
$$

That means that the time evolution of a state in the interaction picture is governed only by the perturbation $\hat{V}_I(t) = \exp(i\hat{H}_0 t/\hbar)\hat{V}\exp(-i\hat{H}_0 t/\hbar)$.

Let us write the general solution to eqn (A.23) by introducing a time-evolution operator $\hat{U}(t, t_0)$:

$$|\psi_I(t)\rangle = \hat{U}(t, t_0)|\psi_I(t_0)\rangle. \tag{A.24}$$

Clearly $\hat{U}(t_0, t_0) = \hat{1}$, the identity operator. From eqn (A.23) we obtain a differential equation for $\hat{U}(t, t_0)$:

$$i\hbar\partial_t\hat{U}(t, t_0) = \hat{V}_I(t)\hat{U}(t, t_0). \tag{A.25}$$

This can be integrated to yield an integral equation:

$$\hat{U}(t, t_0) = \frac{1}{i\hbar}\int_{t_0}^{t}\hat{V}_I(t')\hat{U}(t', t_0)dt'. \tag{A.26}$$

Using the fact that $\hat{U}(t_0, t_0) = \hat{1}$, we can write a solution to this as a series expansion by iteration: the first-order term is obtained by substituting the zeroth-order term $\hat{U}_0(t, t_0) = \hat{1}$ to the right-hand side of eqn (A.26), the second-order term by substituting the first-order term, and so on. As a result, we obtain the series expansion of the form

$$\hat{U}(t, t_0) = 1 + \frac{1}{i\hbar}\int_{t_0}^{t}\hat{V}_I(t_1)dt_1 + \left(\frac{1}{i\hbar}\right)^2\int_{t_0}^{t}\hat{V}_I(t_1)dt_1\int_{t_0}^{t_1}\hat{V}_I(t_2)dt_2 + \dots \tag{A.27}$$

Note that each subsequent integral goes from t_0 to the argument of the previous integral. As the operators $\hat{V}_I(t)$ do not necessarily commute with each other at different times, their order has to be strictly respected. But since the t_i are dummy arguments, we can write any nth-order term as a symmetric combination of $n!$ terms where the dummy indices have been reordered. For example, the second-order term can be written as

$$\left(\frac{1}{i\hbar}\right)^2 \frac{1}{2!} \left[\int_{t_0}^t \hat{V}_I(t_1)dt_1 \int_{t_0}^{t_1} \hat{V}_I(t_2)dt_2 + \int_{t_0}^t \hat{V}_I(t_2)dt_2 \int_{t_0}^{t_2} \hat{V}_I(t_1)dt_1 \right] \tag{A.28}$$

In the first term $\hat{V}_I(t_1)$ precedes $\hat{V}_I(t_2)$, while in the latter it is the opposite. But in both cases the operator evaluated at the later time goes to the left. Let us thus define a time-ordering operator $\mathcal{T}$ such that

$$\mathcal{T}[\hat{V}_I(t_1)\hat{V}_I(t_2)] = \hat{V}_I(t_1)\hat{V}_I(t_2)\theta(t_1-t_2) + \hat{V}_I(t_2)\hat{V}_I(t_1)\theta(t_2-t_1), \tag{A.29}$$

where $\theta(x)$ is a Heaviside step function: $\theta(x) = 1$ for $x \geq 0$ and $\theta(x) = 0$ for $x < 0$. Using this operator we can extend the integrals in eqn (A.28) to the full time range $t_0 \ldots t$ and the second-order term becomes

$$\left(\frac{1}{i\hbar}\right)^2 \frac{1}{2!} \int_{t_0}^t \int_{t_0}^t \mathcal{T}[\hat{V}_I(t_1)\hat{V}_I(t_2)]dt_1 dt_2. \tag{A.30}$$

The definition of $\mathcal{T}$ can be extended to multiple operators: the operators under it are always organized so that the operator with the latest time is ordered to the left. Therefore, we can write the full time development of $\hat{U}(t, t_0)$ as

$$\hat{U}(t, t_0) = \sum_{n=0}^{\infty} \frac{1}{n!} \left(\frac{1}{i\hbar}\right)^n \mathcal{T}\left[\int_{t_0}^t \hat{V}_I(t')dt' \right]^n \equiv \mathcal{T}\exp\left(-\frac{i}{\hbar}\int_{t_0}^t \hat{V}_I(t')dt' \right). \tag{A.31}$$

The inverse of this operator is given by using the anti-time-ordering operator $\tilde{\mathcal{T}}$,

$$\hat{U}(t, t_0)^{-1} = \hat{U}(t, t_0)^\dagger = \tilde{\mathcal{T}}\exp\left(\frac{i}{\hbar}\int_{t_0}^t \hat{V}_I(t')dt' \right). \tag{A.32}$$

These expressions are used as a starting point for the most general perturbation theories used in quantum mechanics. An example of such a calculation is given in the next section and in the exercises.

A.3   Fermi golden rule

A typical task in quantum mechanics is to calculate a rate for a transition induced by an external force between two energy eigenstates of a known system (Hamiltonian $\hat{H}_0$). Let us denote such a force with an interaction term $\hat{V}_S(t) = \lambda \hat{f}_s e^{\eta t}$. The last factor, $e^{\eta t}$, indicates that this force is adiabatically 'turned on'. At a later stage of the calculation we take the limit $\eta \to 0$. In the following, our aim is to calculate the rate of change of the probability for the system to be in the final state $|f_S(t)\rangle$, provided it was initially in the state $|i_S(t_0)\rangle$. For this calculation we assume that the force is weak and take only the lowest-order term in λ. We can express the time evolution of the initial state by using eqns (A.22) and (A.24),

$$|i_S(t)\rangle = e^{-i\hat{H}_0 t/\hbar} \hat{U}(t, t_0) e^{i\hat{H}_0 t_0/\hbar} |i_S(t_0)\rangle. \qquad (A.33)$$

Assuming a weak force, let us in the following retain only the first-order term of expansion (A.27). Now the overlap between state $|i_S(t)\rangle$ and some final state $|f\rangle$ (assumed orthogonal with $|i_S(t_0)\rangle$) is

$$
\begin{aligned}
\langle f|i_S(t)\rangle &\approx \frac{i}{\hbar}\lambda\langle f|e^{-i\hat{H}_0 t/\hbar} \int_{t_0}^{t} dt' e^{i\hat{H}_0 t'/\hbar} \hat{f} e^{\eta t'} e^{-i\hat{H}_0 t'/\hbar} e^{i\hat{H}_0 t_0/\hbar} |i_S(t_0)\rangle \\
&= \frac{i}{\hbar}\lambda\langle f|\hat{f}|i_S(t_0)\rangle e^{-i\epsilon_f t/\hbar} \int_{t_0}^{t} dt' e^{i\epsilon_f t'/\hbar} e^{\eta t'} e^{-i\epsilon_i(t'-t_0)/\hbar} \rangle \\
&= \lambda\frac{\langle f|\hat{f}|i_S(t_0)\rangle}{\epsilon_f - \epsilon_i - i\eta\hbar} e^{i\epsilon_i t_0/\hbar - i\epsilon_f t/\hbar}\Big|_{t_0}^{t} e^{i(\epsilon_f - \epsilon_i - i\eta\hbar)t'/\hbar}. \qquad (A.34)
\end{aligned}
$$

The second line of this equation makes the assumption that the initial and final states are eigenstates of $\hat{H}_0$ with eigenvalues ϵ_f and ϵ_i:

$$\hat{H}_0|i\rangle = \epsilon_i|i\rangle, \quad \hat{H}_0|f\rangle = \epsilon_f|f\rangle.$$

Now let us take the initial time t_0 to $-\infty$. In this case we get

$$\langle f|i_S(t)\rangle = \lambda\langle f|\hat{f}|i_S(t_0)\rangle \frac{e^{-i\epsilon_i(t-t_0)/\hbar} e^{\eta t}}{\epsilon_f - \epsilon_i - i\eta\hbar}. \qquad (A.35)$$

The conditional probability of occupying the final state $|f\rangle$, given that the initial state was $|i(t_0)\rangle$, is

$$P_{fi} = |\langle f|i_S(t)\rangle|^2 = |\lambda|^2|\langle f|\hat{f}|i_S(t_0)\rangle|^2 \frac{e^{2\eta t}}{(\epsilon_f - \epsilon_i)^2 + \eta^2\hbar^2}. \qquad (A.36)$$

The transition rate between these states is the rate of change of P_{fi}. In the limit $\eta \to 0$, this equals (see Exercise A.3)

$$\Gamma_{fi} = \partial_t P_{fi} = \frac{2\pi|\lambda|^2}{\hbar}|\langle f|\hat{f}|i_S(t_0)\rangle|^2\delta(\epsilon_f - \epsilon_i). \qquad (A.37)$$

This is the *Fermi golden rule*. In the case when there are many degenerate final states, these are weighted by their density of states in the final expression.

As the above calculation makes no assumption about the particle number, the Fermi golden rule can be used also in many-particle systems. An example of this is given in Ch. 7, where the tunnelling rate of a single-electron transistor is calculated with the Fermi golden rule.

A.3.1   Higher order: generalized Fermi golden rule

Conceptually, the generalization of the Fermi golden rule to include higher orders in λ is not much more complicated than the lowest-order calculation. Including all terms of eqn (A.27) in the calculation of the overlap between the final and the initial states, eqn (A.34) yields

$$\langle f|i_S(t)\rangle = \langle f|e^{-i\hat{H}_0 t/\hbar} \left[\frac{1}{i\hbar} \int_{t_0}^{t} \hat{V}_I(t_1) dt_1 \right.$$
$$\left. + \left(\frac{1}{i\hbar}\right)^2 \int_{t_0}^{t} \hat{V}_I(t_1) dt_1 \int_{t_0}^{t_1} \hat{V}_I(t_2) dt_2 + \ldots \right] e^{i\hat{H}_0 t_0/\hbar} |i_S(t_0)\rangle.$$

Now we have to be rather more careful how we 'turn on' the interaction. Let us choose the interaction term to be of the form $\hat{V}_I(t) = \lambda \hat{f}_I(t) g(t)$, where $g(t)$ vanishes at $t \ll t_0$ and tends to unity at some time scale $1/\eta \ll (t - t_0)$ around t_0. We hence *first* take the time difference $t - t_0$ to infinity and *then* take η to zero. Taking t_0 to $-\infty$, the nth-order term in the expansion has integrals of the form

$$\int_{-\infty}^{t} dt_1 \hat{f}_I(t_1) g(t_1) \int_{-\infty}^{t_1} dt_2 \hat{f}_I(t_2) g(t_2) \ldots \int_{-\infty}^{t_{n-1}} dt_n \hat{f}_I(t_n) g(t_n) |i_S(t_0)\rangle.$$
$$\text{(A.38)}$$

We thus have $t_1 > t_2 > \ldots t_{n-1} > t_n > t_0 \to -\infty$. With a small error that becomes negligible in the limit $\eta \to 0$ we can hence assume that all other $g(t_{k<n}) \approx 1$ and only $g(t_n) = e^{\eta t}$. Let us include the time dependence of $\hat{f}_I(t) = e^{i\hat{H}_0 t/\hbar} \hat{f}_S e^{-i\hat{H}_0 t/\hbar}$ and calculate the rightmost integral of eqn (A.38). It yields

$$\int_{-\infty}^{t_{n-1}} dt_n e^{i(\hat{H}_0 - \epsilon_0 - i\hbar\eta) t_n/\hbar} \hat{f}_S |i_S(t_0)\rangle = \frac{\hbar e^{i(\hat{H}_0 - \epsilon_0 - i\hbar\eta) t_{n-1}/\hbar}}{i(\hat{H}_0 - \epsilon_0 - i\hbar\eta)} \hat{f}_S |i_S(t_0)\rangle.$$

The above equation uses the fact that $e^{i\hat{H}_0 t_n/\hbar} |i_S(t)\rangle = e^{i\epsilon_i t_n/\hbar} |i_S(t)\rangle$, which allows us to commute the (scalar) exponential factor $e^{i\epsilon_i t_n/\hbar}$ with $\hat{f}_S$. The next integral looks exactly the same, so we may perform all the integrals to get

$$|\langle f|i_S(t)\rangle| = \left| \frac{e^{\eta t}}{\epsilon_f - \epsilon_i - i\eta} \langle f|\hat{T}|i\rangle \right|, \quad \text{(A.39)}$$

where the *T-matrix* satisfies the *Dyson equation*

$$\hat{T} = \lambda \hat{f}_S + \lambda^2 \hat{f}_S \frac{1}{(\hat{H}_0 - \epsilon_i - i\hbar\eta)} \hat{f}_S$$
$$+ \lambda^3 \hat{f}_S \frac{1}{(\hat{H}_0 - \epsilon_i - i\hbar\eta)} \hat{f}_S \frac{1}{(\hat{H}_0 - \epsilon_i - i\hbar\eta)} \hat{f}_S + \ldots \quad \text{(A.40)}$$
$$= \lambda \hat{f}_S + \lambda \hat{f}_S \frac{1}{(\hat{H}_0 - \epsilon_i - i\hbar\eta)} \hat{T}.$$

The latter form can be checked by recursion to equal the infinite series on the first two lines.

Following the same steps as in the lowest-order theory, the transition rate can be written as

$$\Gamma_{fi} = \frac{2\pi}{\hbar} |\langle f|\hat{T}|i_S(t_0)\rangle|^2 \delta(\epsilon_f - \epsilon_i). \quad \text{(A.41)}$$

This equation can be used, for example, in calculating the cotunnelling rate through a single electron transistor, Sec. 7.4.

A.4 Describing magnetic field in quantum mechanics

A charged particle in an electromagnetic field in quantum mechanics is described analogously to that in analytical mechanics. The electric and magnetic fields $\vec{E}$ and $\vec{B}$ are described by the scalar and vector potentials φ and $\vec{A}$ so that

$$\vec{E} = -\nabla\varphi - \frac{\partial \vec{A}}{\partial t}, \quad \vec{B} = \nabla \times \vec{A}. \tag{A.42}$$

The Lagrangian including these fields and describing the dynamics of a particle with charge q is

$$L = \frac{1}{2}mv^2 - q\varphi + q\vec{v} \cdot \vec{A}. \tag{A.43}$$

It has been chosen so that the corresponding Euler–Lagrange equation yields the Lorentz force,

$$\frac{d}{dt}(m\vec{v}) = -q(\nabla\varphi + \partial_t\vec{A}) + q\vec{v} \times (\nabla \times \vec{A}) = q\vec{E} + q\vec{v} \times \vec{B}, \tag{A.44}$$

The canonical momentum is

$$\vec{p} = \frac{\partial L}{\partial \vec{v}} = m\vec{v} + q\vec{A}. \tag{A.45}$$

Thus, the Hamiltonian is

$$H = \vec{p} \cdot \vec{v} - L = \frac{1}{2}mv^2 + \phi = \frac{1}{2m}(\vec{p} - q\vec{A})^2 + \phi. \tag{A.46}$$

The effect of the magnetic field can hence be described by the substituting the canonical momentum with the kinetic momentum,

$$\vec{p} \mapsto \vec{p} - q\vec{A}. \tag{A.47}$$

In addition to this *orbital* effect of the magnetic field, there is a *Zeeman* effect due to coupling to the spin $\vec{S}$, described by an additional term $H_Z = -\mu_B g \vec{S} \cdot \vec{B}/\hbar$ in the Hamiltonian. Here $\mu_B = e\hbar/(2m)$ is the Bohr magneton and $g \approx 2$ is the (electron) g-factor. The resultant Zeeman splitting of electron eigenstates with different spins can be measured for example by measuring the conductance of quantum dots (Ch. 8) as a function of magnetic field. To explain such measurements, however, the effective g-factor has to be typically renormalized to include, for example, effects from spin-orbit scattering. The measured effective g-factor may then range from values of smaller than one to several tens.

A.5 Chemical potential and Fermi energy

Essentially all of the physics described in this book concentrates on low energies close to the Fermi level of the metal electrodes typically connected to the measured small conductor in which one is interested. Discussions of the transport physics then necessarily deal with the concepts of the chemical and electrical potentials and the Fermi energy. Here I define these quantities in detail.

Because of the Pauli exclusion principle, many electrons cannot occupy the same momentum states. Rather, at a vanishing temperature, they are filled at the order of increasing energies up to the *Fermi energy* E_F. In a finite system with volume V, the latter can be connected to the particle density n by

$$n(T=0) = \frac{1}{V} \sum_{\text{spin}} \sum_{\vec{k} \in \Omega_F} 1 \approx 2 \times \frac{1}{(2\pi)^3} \int_{\Omega_F} d^3k, \qquad (A.48)$$

where the sum and the integral are over the Fermi sphere Ω_F. In a system where the energy for a given momentum state is independent of the momentum direction, we can write the integral in spherical coordinates

$$n(T=0) = \frac{1}{4\pi^3} \int_0^{k_F} k^2 dk \int_0^{2\pi} d\phi \int_0^{\pi} \sin(\theta)d\theta = \frac{1}{3\pi^2} k_F^3, \qquad (A.49)$$

where k_F is the Fermi momentum; the momentum of the electrons residing at the Fermi energy. On the other hand, the total energy density of the system is

$$E_{\text{tot}} = \frac{1}{V} \sum_{\text{spin}} \sum_{\vec{k}} \epsilon_k = \frac{1}{\pi^2} \int_0^{k_F} \epsilon_k k^2 dk, \qquad (A.50)$$

where ϵ_k is the energy for momentum state labelled by k. Now the *chemical potential* is defined as the energy required to add or remove an electron into/from the electron system, i.e.,

$$\mu = \frac{dE_{\text{tot}}}{dn} = \frac{dE_{\text{tot}}}{dk_F}\frac{dk_F}{dn} = \epsilon_{k_F} \equiv E_F. \qquad (A.51)$$

This equality is valid only at $T = 0$. At a non-zero temperature eqn (A.49) is replaced by

$$\begin{aligned}
n(T) &= \frac{1}{\pi^2} \int_0^{\infty} \frac{k^2 dk}{e^{(\epsilon_k - \mu(T))/(k_B T)} + 1} \\
&= \int_0^{\infty} \frac{N(\epsilon)d\epsilon}{e^{(\epsilon - \mu(T))/(k_B T)} + 1} = \int_0^{E_F} N(\epsilon)d\epsilon,
\end{aligned} \qquad (A.52)$$

where $N(\epsilon)$ is the density of states. The last equality reflects the fact that the total particle number does not change when temperature is increased. Therefore, we have to include a temperature dependence in μ. This is typically a weak dependence. Let us denote $\mu(T) = E_F + \delta\mu(T)$, and $\epsilon = E_F + \delta\epsilon$. For $k_B T \ll E_F$, the differences between the two integrands reside mostly around E_F. We may hence expand $N(\epsilon) \approx N_F + N'(E_F)\delta\epsilon$. After some manipulations, (A.52) goes into the form

$$\int_0^{E_F} d\delta\epsilon \frac{N_F - N'(E_F)\delta\epsilon}{e^{(\delta\epsilon + \delta\mu(T))/(k_B T)} + 1} = \int_0^{\infty} d\delta\epsilon \frac{N_F + N'(E_F)\delta\epsilon}{e^{(\delta\epsilon - \delta\mu(T))/(k_B T)} + 1}. \qquad (A.53)$$

Now the integrand is non-zero mostly for $\delta\epsilon \lesssim k_B T$, so that we may extend the upper limit of the first integral also to infinity. These integrals can be evaluated in terms of special functions, but most purposes it is enough to expand them to the first order in $\delta\mu$. We then get integrals of the type

$$\int_0^\infty \frac{1}{e^{\epsilon/(k_B T)} + 1} = k_B T \ln 2 \tag{A.54a}$$

$$\int_0^\infty \frac{\epsilon}{e^{\epsilon/(k_B T)} + 1} = (k_B T)^2 \frac{\pi^2}{12} \tag{A.54b}$$

$$\partial_\mu \int_0^\infty \frac{1}{e^{(\epsilon+\mu)/(k_B T)} + 1} = -\frac{1}{2} \tag{A.54c}$$

$$\partial_\mu \int_0^\infty \frac{\epsilon}{e^{(\epsilon+\mu)/(k_B T)} + 1} = -k_B T \ln 2. \tag{A.54d}$$

Using these, we can write for the temperature dependence of the chemical potential

$$N_F(k_B T \ln 2 - \frac{\delta\mu}{2}) - N'(E_F)((k_B T)^2 \frac{\pi^2}{12} - \delta\mu k_B T \ln 2) = \\ N_F(k_B T \ln 2 + \frac{\delta\mu}{2}) + N'(E_F)((k_B T)^2 \frac{\pi^2}{12} + \delta\mu k_B T \ln 2). \tag{A.55}$$

This is readily solved to yield

$$\delta\mu(T) = -\frac{\pi^2}{6} \frac{N'(E_F)}{N_F} (k_B T)^2 \sim -\frac{(k_B T)^2}{E_F}, \tag{A.56}$$

where the latter equation may be shown, for example, in the free-electron model (see Sec. 1.3). We hence get that for most practical purposes in this book, the deviations between the chemical potential and the Fermi energy are small, and these terms can hence be used interchangably. To take into account the fact that the two can have a small difference, the term *Fermi level* is also often used for the chemical potential.

Because of strong screening in metals, the *electrostatic potential $e\phi$* in them typically follows the chemical potential. This can be seen by considering the case when there is a non-zero charge density $\delta n(\vec{r}) = n(\vec{r}) - n_0$ at some point in the metal. Here n_0 is the uniform density of the positive background created by the ion lattice, described by the chemical potential as in eqn (A.52). The electron density can be obtained from the same equation by replacing μ by $\mu - e\phi(\vec{r})$. Linearizing, we then get[1]

$$\delta n(\vec{r}) \approx -\partial_\mu n\phi(\vec{r}) e\phi(\vec{r}) \approx \frac{N_F}{2} e\phi(\vec{r}), \tag{A.57}$$

where the last approximate equality comes from the calculation of the μ derivative of eqn (A.52) with a constant density of states. Inserting this result into the Poisson equation yields

$$\nabla^2 \phi = \frac{2\pi N_F e^2}{\epsilon_0} \phi. \tag{A.58}$$

[1]This is called the (linearized) Thomas–Fermi theory of screening.

Its solution describes an exponentially damped potential with a screening length given by

$$\ell_{\text{TF}} = \sqrt{\frac{\epsilon_0}{2\pi N_F e^2}}. \tag{A.59}$$

For metals ℓ_{TF} is of the order of inter-atomic distance, so that for most purposes metals can be viewed as charge neutral. However, in semiconductors and graphene screening length is often larger, and in some cases the incomplete screening has to be taken into account especially when analyzing non-linear current–voltage characteristics.

A.6   Pauli spin matrices

The Hamiltonian of (quantum) two-level systems, such as qubits or (pseudo)spins (which are in fact the same thing in an abstract sense), can typically be specified in terms of 2×2 matrices. In this case, Pauli spin matrices allow for a simplification of such a description. These matrices are

$$\sigma_x = \begin{pmatrix} 0 & 1 \\ 1 & 0 \end{pmatrix}, \quad \sigma_y = \begin{pmatrix} 0 & -i \\ i & 0 \end{pmatrix}, \quad \sigma_z = \begin{pmatrix} 1 & 0 \\ 0 & -1 \end{pmatrix}. \tag{A.60a}$$

Any 2×2 matrix can be specified as a linear combination of the unit matrix (often denoted as σ_0) and the above spin matrices. One of the reasons for writing matrices in terms of σ_i are their commutation properties. First of all, their mutual anticommutators vanish:

$$\{\sigma_i, \sigma_j\} \equiv \sigma_i \sigma_j + \sigma_j \sigma_i = \delta_{ij} \sigma_0, \tag{A.61}$$

where $i, j \in \{x, y, z\}$. Secondly, their commutators are cyclic,

$$[\sigma_x, \sigma_y] \equiv \sigma_x \sigma_y - \sigma_y \sigma_x = 2i\sigma_z \tag{A.62a}$$
$$[\sigma_y, \sigma_z] \equiv \sigma_y \sigma_z - \sigma_z \sigma_y = 2i\sigma_x \tag{A.62b}$$
$$[\sigma_z, \sigma_x] \equiv \sigma_z \sigma_x - \sigma_x \sigma_z = 2i\sigma_y, \tag{A.62c}$$

which helps manipulations of their products enormously. A further useful property of the Pauli matrices is their vanishing trace, $\text{Tr}\sigma_{x,y,z} = 0$

Besides the Pauli matrices, so-called ladder operators are often used. They are defined as

$$\sigma_\uparrow \equiv \frac{1}{2}(\sigma_x + i\sigma_y) = \begin{pmatrix} 0 & 1 \\ 0 & 0 \end{pmatrix} \tag{A.63a}$$

$$\sigma_\downarrow \equiv \frac{1}{2}(\sigma_x - i\sigma_y) = \begin{pmatrix} 0 & 0 \\ 1 & 0 \end{pmatrix}. \tag{A.63b}$$

The ladder operators are similar to the fermionic creation and annihilation operators discussed in Sec. A.1: $\sigma_\uparrow$ turns a down-spin component of a wave function to an up-spin component and nullifies the up-spin component, and vice versa for $\sigma_\downarrow$.

A.7 Useful integrals

In almost all transport problems concerning mesoscopic objects, the current is calculated for a system connecting two or more reservoirs. These reservoirs are assumed to be in internal equilibrium, and hence the energy distribution function of electrons in them is a Fermi function. Therefore, almost all transport problems involve the calculation of one of the following integrals

$$\int_{-\infty}^{\infty} (f^0(E;\mu_1,T_1) - f^0(E;\mu_2,T_2))dE = \mu_1 - \mu_2 \quad \text{(A.64a)}$$

$$\int_{-\infty}^{\infty} E(f^0(E;\mu_1,T_1) - f^0(E;\mu_2,T_2))dE =$$
$$\frac{\pi^2}{6}k_B^2(T_1^2 - T_2^2) + \frac{1}{2}(\mu_1^2 - \mu_2^2) \quad \text{(A.64b)}$$

$$\int_{-\infty}^{\infty} E^2(f^0(E;\mu_1,T_1) - f^0(E;\mu_2,T_2))dE =$$
$$\frac{1}{3}(\mu_1^3 - \mu_2^3) + \frac{\pi^2}{3}k_B^2(\mu_1 T_1^2 - \mu_2 T_2^2) \quad \text{(A.64c)}$$

$$\int_{-\infty}^{\infty} dE\partial_\mu f^0(E;\mu,T) = 1 \quad \text{(A.64d)}$$

$$\int_{-\infty}^{\infty} dE\partial_T f^0(E;\mu,T) = 0 \quad \text{(A.64e)}$$

$$\int_{-\infty}^{\infty} dEE\partial_\mu f^0(E;\mu,T) = \mu \quad \text{(A.64f)}$$

$$\int_{-\infty}^{\infty} dEE\partial_T f^0(E;\mu,T) = \frac{k_B^2 T\pi^2}{3} \quad \text{(A.64g)}$$

$$\int_{-\infty}^{\infty} dE f^0(E;\mu,T)(1 - f^0(E - x;\mu,T))$$
$$= n^0(x,T)\int_{-\infty}^{\infty} dE[f^0(E - x;\mu,T) - f^0(E;\mu,T)] = xn^0(x,T). \quad \text{(A.64h)}$$

Here $f^0(E;\mu,T)$ are Fermi functions and $n^0(x,T)$ are Bose functions.

Exercises

(A.1) Prove eqns (A.12) and (A.14) by using the definitions (A.10) and (A.11).

(A.2) The Hamiltonian for a harmonic oscillator is

$$\hat{H}_{ho} = \hbar\omega_0(\hat{a}^\dagger\hat{a} + \frac{1}{2}), \quad \text{(A.65)}$$

where $\omega_0 = \sqrt{k/m}$ is the resonance frequency for mass m and spring constant k characterizing the potential, and $\hat{a}$ is the bosonic annihilation operator. With these operators, the (Heisenberg) operators for position $\hat{x}$ and and momentum $\hat{p}$ are $\hat{x} = \sqrt{\hbar/(2m\omega)}(\hat{a}+\hat{a}^\dagger)$ and $\hat{p} = i\sqrt{\hbar m\omega/2}(\hat{a}^\dagger - \hat{a})$.

From the Heisenberg equation of motion,

$$i\hbar \frac{d\hat{O}}{dt} = [\hat{O}, H], \qquad (A.66)$$

valid for any operator $\hat{O}$ without external time dependence, derive the equations of motion for $\hat{x}$ and $\hat{p}$.

(A.3) Show that the Fermi golden rule, eqn (A.37), follows from eqn (A.36).

(A.4) Consider the harmonic oscillator Hamiltonian $\hat{H}_{ho}$ in Exercise A.2. Its energy eigenstates satisfy $\hat{H}_{ho}|n\rangle = (n + \frac{1}{2})|n\rangle$. Assume a time-dependent force $f(t) = f_0 \cos(\omega_0 t)$ that acts on the x-coordinate of the oscillator, i.e., the full Hamiltonian of the system is $\hat{H}_{ho} + f(t)\hat{x}$. By a slight modification of the Fermi golden rule, calculate the lowest-order transition rate between oscillator eigenstates, induced by the force. Hint: assume a force that is adiabatically turned on.

Current operator for the scattering theory

The derivation of the current operator in the scattering theory can be carried out by using the quantum-mechanical expression for the current operator in terms of the field operators $\hat{\Psi}_\alpha(\vec{r}, t)$, $\hat{\Psi}_\alpha^\dagger(\vec{r}, t)$, annihilating or creating an electron in position $\vec{r}$ at time t in lead α,[1]

$$\hat{I}_\alpha(z, t) = \frac{\hbar e}{2im} \int dx dy \left[\hat{\Psi}_\alpha^\dagger(\vec{r}, t) \frac{\partial}{\partial z} \hat{\Psi}_\alpha(\vec{r}, t) - \left(\frac{\partial}{\partial z} \hat{\Psi}_\alpha^\dagger(\vec{r}, t) \right) \hat{\Psi}_\alpha(\vec{r}, t) \right].$$

(B.1)

[1] This derivation follows mostly (Büttiker, 1992).

Using the Heisenberg equation for the field operators it can be shown that this satisfies the continuity equation $\dot{\rho}_\alpha + \nabla \cdot \hat{I}_\alpha = 0$ between the current and the time-dependent charge density[2]

[2] Show this as an exercise!

$$\rho_\alpha(z, t) = e \int dx dy \hat{\Psi}_\alpha^\dagger(\vec{r}, t) \hat{\Psi}_\alpha(\vec{r}, t).$$

(B.2)

The annihilation operator may be written in terms of the wave function of the lead and using the energy representation instead of time,

$$\hat{\Psi}_\alpha(\vec{r}, t) = \int dE e^{-iEt/\hbar}$$

$$\times \sum_{n=1}^{M_\alpha(E)} \frac{\chi_{\alpha n}(x, y)}{\sqrt{2\pi \hbar v_{\alpha n}(E)}} [\hat{a}_{\alpha n}(E) e^{ik_{\alpha n}(E)z} + \hat{b}_{\alpha n}(E) e^{-ik_{\alpha n}(E)z}]$$

(B.3)

and likewise for the creation operator $\hat{\Psi}_\alpha^\dagger$. Inserting this into eqn (B.1) we find (dropping some of the arguments E, E')

$$\hat{I}_\alpha(z, t) = \frac{\hbar e}{2mi} \int dx dy dE dE' \sum_{nm} e^{i(E-E')t/\hbar} \frac{\chi_{\alpha n}^*(x, y) \chi_{\alpha m}(x, y)}{2\pi \hbar \sqrt{v_{\alpha n}(E) v_{\alpha m}(E')}} \times$$

$$\left\{ ik_{\alpha m}(E') \left[\hat{a}_{\alpha n}^\dagger e^{-ik_{\alpha n}(E)z} + \hat{b}_{\alpha n}^\dagger e^{ik_{\alpha n}(E)z} \right] \times \right.$$

$$\left[\hat{a}_{\alpha m} e^{ik_{\alpha m}(E')z} - \hat{b}_{\alpha m} e^{-ik_{\alpha m}(E')z} \right]$$

$$+ ik_{\alpha n}(E) \left[\hat{a}_{\alpha n}^\dagger e^{-ik_{\alpha n}(E)z} - \hat{b}_{\alpha n}^\dagger e^{ik_{\alpha n}(E)z} \right] \times$$

$$\left. \left[\hat{a}_{\alpha m} e^{ik_{\alpha m}(E')z} + \hat{b}_{\alpha m} e^{-ik_{\alpha m}(E')z} \right] \right\}.$$

(B.4)

Using orthogonality of the transverse wave functions we get

$$
\begin{aligned}
\hat{I}_\alpha(z,t) =& \frac{\hbar e}{2m} \int dE dE' \sum_n e^{i(E-E')t/\hbar} \frac{1}{2\pi\hbar\sqrt{v_{\alpha n}(E)v_{\alpha n}(E')}} \\
& \left\{ (k_{\alpha n}(E)+k_{\alpha n}(E')) \left[\hat{a}^\dagger_{\alpha n}\hat{a}_{\alpha n} e^{i(k_{\alpha n}(E')-k_{\alpha n}(E))z} \right. \right. \\
& \left. \qquad -\hat{b}^\dagger_{\alpha n}\hat{b}_{\alpha n} e^{i(k_{\alpha n}(E)-k_{\alpha n}(E'))z} \right] \\
& +(k_{\alpha n}(E)-k_{\alpha n}(E')) \left[\hat{a}^\dagger_{\alpha n}\hat{b}_{\alpha n} e^{-i(k_{\alpha n}(E)+k_{\alpha n}(E'))z} \right. \\
& \left. \left. \qquad -\hat{b}^\dagger_{\alpha n}\hat{a}_{\alpha n} e^{i(k_{\alpha n}(E)+k_{\alpha n}(E'))z} \right] \right\}.
\end{aligned}
$$

(B.5)

Let us assume that in the relevant energy regime, $k_{\alpha n}(E)$ and $v_{\alpha n}(E)$ are essentially independent of energy. Noting that $v = \hbar k/m$, we have

$$
\hat{I}_\alpha(t) = \frac{e}{2\pi\hbar} \int dE dE' \sum_n e^{i(E-E')t/\hbar} \left[\hat{a}^\dagger_{\alpha n}(E)\hat{a}_{\alpha n}(E') - \hat{b}^\dagger_{\alpha n}(E)\hat{b}_{\alpha n}(E') \right],
$$

which is now independent of z. Using eqn (3.52) we can write this in the form

$$
\begin{aligned}
\hat{I}_\alpha(t) =& \frac{e}{h} \int dE dE' \sum_j e^{i(E-E')t/\hbar} \left[\hat{a}^\dagger_{\alpha j}(E)\hat{a}_{\alpha j}(E') \right. \\
& \left. - \sum_{\beta\gamma}\sum_{mn}(s^{\alpha\beta}_{jm})^*(E)s^{\alpha\gamma}_{jn}(E')\hat{a}^\dagger_{\beta m}(E)\hat{a}_{\gamma n}(E') \right], \\
=& \frac{e}{h} \int dE dE' \sum_{\beta\gamma}\sum_{mn} e^{i(E-E')t/\hbar} \left[\delta_{\beta\alpha}\delta_{\gamma\alpha}\delta_{mn} - \sum_j (s^{\alpha\beta}_{jm})^*(E)s^{\alpha\gamma}_{jn}(E') \right] \\
& \times \hat{a}^\dagger_{\beta m}(E)\hat{a}_{\gamma n}(E'), \\
=& \frac{e}{h} \int dE dE' \sum_{\beta\gamma}\sum_{mn} e^{i(E-E')t/\hbar} \hat{a}^\dagger_{\beta m}(E)A^{\beta\gamma}_{mn}(\alpha,E,E')\hat{a}_{\gamma n}(E'),
\end{aligned}
$$

(B.6)

which gives eqn (3.54) after we multiply by 2 due to spin.

Fluctuation–dissipation theorem

This appendix presents the derivation of the fluctuation–dissipation theorem, describing, for example, equilibrium noise in electronic circuits. However, the theorem itself is much more general, and applies to any dissipative process in a linear (or linearized) system at equilibrium. The derivation starts by defining the quantum linear response theory, the linear response coefficients (susceptibilities) and the associated noise. With these definitions the actual derivation of the theorem is rather straightforward.

C.1 Linear response theory and susceptibility

Linear response theory describes a system characterized by a time-independent Hamiltonian H_0. Observables of this system are denoted by A_i and the corresponding operators by $\hat{A}_i$, such that

$$A_i = \langle \hat{A}_i \rangle = \text{Tr}[\rho(t)\hat{A}_i] = \sum_n \rho_n \langle n_S(t)|\hat{A}|n_S(t)\rangle, \qquad (C.1)$$

where $|n_S(t)\rangle$ are the energy eigenstates of the system in the Schrödinger picture, and $\rho_n = \exp(-\epsilon_n/k_B T) = \langle n_S(t)|\rho|n_S(t)\rangle$ are the diagonal elements of the equilibrium density matrix.

Assume that at time $t = t_0$ we start to apply a force $f(t)$ that acts on the observable A_j. The resultant Hamiltonian becomes

$$H = H_0 - f(t)\hat{A}_j. \qquad (C.2)$$

The force $f(t)$ is assumed to act in this system as a scalar function,[1] such that it commutes with the operators $\hat{A}_i$. The resultant change in the observable $A_i(t)$ can be conveniently characterized in the interaction picture (see Appendix A.2.1), where $|n_I(t)\rangle = U(t, t_0)|n_I(t_0)\rangle = U(t, t_0)|n_S(t_0)\rangle$, because the states of the interaction and the Schrödinger pictures coincide at the initial time t_0. Here the time-evolution operator is

$$U(t, t_0) = \mathcal{T} \exp\left(\frac{i}{\hbar} \int_{t_0}^{t} f(t')\hat{A}_{jI}(t')dt'\right), \qquad (C.3)$$

and the operator $\hat{A}_{jI}(t)$ is the operator in the interaction picture.

[1] However, it can still be an operator of some other system, and hence one should be careful when trying to commute $f(t)$ with itself at different times.

Let us consider the linear response of observable A_i to force f. In this case the time-evolution operator is

$$U(t, t_0) = 1 + \frac{i}{\hbar} \int_{t_0}^{t} f(t') \hat{A}_{jI}(t') dt'. \tag{C.4}$$

As the time evolution of the states in the Schrödinger picture can be obtained from $|n_S(t)\rangle = \exp(-iH_0 t/\hbar)|n_I(t)\rangle$, we get for the time dependence of operator A_i

$$A_i(t) = \langle A_{iI}^0(t) \rangle + \frac{i}{\hbar} \int_{-\infty}^{t} dt' \sum_n \rho_n \langle n_S(t_0) | [\hat{A}_{iI}(t), \hat{A}_{jI}(t')] | n_S(t_0) \rangle f(t')$$

$$\equiv \langle A_{iI}(t) \rangle + \frac{i}{\hbar} \int_{t_0}^{t} dt' \langle [\hat{A}_{iI}(t), \hat{A}_{jI}(t')] \rangle f(t'). \tag{C.5}$$

Here $\langle A_{iI}^0(t) \rangle$ is the (Heisenberg) time dependence in the absence of the force.

A similar procedure for higher orders in $f(t)$ would yield commutators of the form $[\hat{A}_{jI}(t), [\hat{A}_{jI}(t), \hat{A}_{iI}^0(t)]]$, and so on. Now, a linear quantum system is characterized by the property that the commutators between the operators $[\hat{A}_{jI}(t), \hat{A}_{iI}(t)]$ are scalars,[2] and thus their further commutators vanish. For such systems we may terminate the series to first order, and eqn (C.5) is exact. Otherwise we have to require that $f(t)$ is small such that higher orders do not contribute much.

Defining a response coefficient (or susceptibility)

$$\chi_{ij}(t, t') \equiv \frac{i}{\hbar} \langle [\hat{A}_{iI}(t), \hat{A}_{jI}(t')] \rangle \theta(t - t') \tag{C.6}$$

allows us to represent the time dependence of observable A_i by

$$A_i(t) = A_i^0(t) + \int_{-\infty}^{\infty} dt' \chi_{ij}(t, t') f(t'). \tag{C.7}$$

In a stationary system $\chi(t, t') = \chi(t - t')$. In that case eqn (C.7) is a convolution. Hence, in the Fourier transformed space it is

$$A_i(\omega) = A_i^0(\omega) + \chi_{ij}(\omega) f(\omega), \tag{C.8}$$

where $A_i(\omega)$, $\chi_{ij}(\omega)$ and $f(\omega)$ are the Fourier transforms of $A_i(t)$, $\chi_{ij}(t - t')$ and $f(t)$. Thus, the change in observable $A_i(\omega)$ due to the applied force $f(\omega)$ is described through the susceptibility $\chi(\omega)$. To be precise, only the imaginary part of $\chi(\omega)$ describes a dissipative process, and the real part is the reactive part.

[2] A prime example is harmonic oscillator with $\hat{A}_1 = \hat{x} \propto \hat{a} + \hat{a}^\dagger$ and $\hat{A}_2 = \hat{p} \propto i(\hat{a}^\dagger - \hat{a})$.

C.2   Derivation of the fluctuation–dissipation theorem

Now let us consider the noise correlation function,[3]

$$S_{ij}(t) \equiv \langle \hat{A}_i(t)\hat{A}_j(0)\rangle. \tag{C.9}$$

[3]Here the operators are written in the interaction picture, but I drop the subscript I for convenience.

This is not yet the symmetrized version introduced above, but we may symmetrize it at the end of the calculation. For simplicity, let us prove the fluctuation–dissipation relation in the case $i = j$, as this is what we mostly need, and then the proof is slightly simpler than in the general case. We may then drop the indices from S_{ij} and χ_{ij} altogether.

First, note that in the stationary case $S(t)$ follows the symmetry $S(t)^* = \langle \hat{A}_i(0)\hat{A}_i(t)\rangle = \langle \hat{A}_i(-t)\hat{A}_i(0)\rangle = S(-t)$. Now separate the real and imaginary parts of the correlator,

$$S(t) = B(t) + iC(t) \tag{C.10a}$$

$$B(t) = B(-t) = \frac{1}{2}\left(S(t) + S(-t)\right) \tag{C.10b}$$

$$C(t) = -C(-t) = \frac{1}{2i}(S(t) - S(-t)) = \frac{1}{2i}\langle[A_i(t), A_i(0)]\rangle. \tag{C.10c}$$

From the last equality we get $\chi(t) = -\frac{2}{\hbar}\theta(t)C(t)$. The Fourier transform of χ satisfies

$$\chi(\omega) = -\frac{2}{\hbar}\int_0^\infty dt e^{i\omega t}C(t) = -\frac{2}{\hbar}\int_{-\infty}^0 dt e^{-i\omega t}C(-t) = \frac{2}{\hbar}\int_{-\infty}^0 dt e^{-i\omega t}C(t). \tag{C.11}$$

We also have

$$C(\omega) = \int_{-\infty}^\infty dt e^{i\omega t}C(t) = -\frac{\hbar}{2}\left[\chi(\omega) - \chi(-\omega)\right] = -i\hbar\text{Im}[\chi(\omega)]. \tag{C.12}$$

The last relation follows from the fact that $\chi(t)$ is real (which in turn comes from the reality of $C(t)$).

For an equilibrium state, the correlator $S(t)$ satisfies the KMS relations (Kubo, 1957; Martin and Schwinger, 1959)

$$\begin{aligned}
S(t) &= \text{Tr}[e^{-\beta H_0}e^{iH_0 t/\hbar}\hat{A}_i(0)e^{-iH_0 t/\hbar}\hat{A}_i(0)]\\
&= \text{Tr}[\hat{A}_i(0)e^{-iH_0 t/\hbar}\hat{A}_i(0)e^{-\beta H_0}e^{iH_0 t/\hbar}]\\
&= \text{Tr}[\hat{A}_i(0)e^{-\beta H_0}e^{iH_0(-t-i\hbar\beta)/\hbar}\hat{A}_i(0)e^{-iH_0(-t-i\hbar\beta)/\hbar}]\\
&= \langle A_i(-t-i\hbar\beta)A_i(0)\rangle = S(-t - i\hbar\beta).
\end{aligned} \tag{C.13}$$

This relation uses the cyclic property of the trace and a simple reordering of the exponents. Correspondingly, the Fourier transform yields the detailed balance relation

$$S(\omega) = \int_{-\infty}^\infty dt e^{i\omega t}S(-t - i\hbar\beta) = \int_{-\infty}^\infty ds e^{-i\omega s}e^{\beta\hbar\omega}S(s) = e^{\beta\hbar\omega}S(-\omega). \tag{C.14}$$

Finally, we get for $C(\omega)$

$$C(\omega) = \frac{1}{2i}(S(\omega) - S(-\omega)) = \frac{1 - e^{-\beta\hbar\omega}}{2i}S(\omega) = \frac{-iS(\omega)}{\coth(\beta\hbar\omega/2) + 1}.$$
(C.15)

Combining eqns (C.12) and (C.15) yields the fluctuation–dissipation theorem,

$$S(\omega) = \hbar\mathrm{Im}[\chi(\omega)][\coth(\beta\hbar\omega/2) + 1].$$
(C.16)

The proof is slightly longer if $i \neq j$, but a similar theorem holds also in that case (see Exercise 6.2).

In an electrical system the admittance $Y(\omega) = Z^{-1}(\omega)$ can be defined through the response of the current to a time-dependent vector potential $\vec{A}$:[4] it is the vector potential that enters the Hamiltonian rather than the electric field $\vec{E}(\omega) = i\omega\vec{A}$. Using the fact that the current density follows the vector potential via $\vec{j}(\omega) = \sigma(\omega)\vec{E}(\omega) = i\omega\sigma(\omega)\vec{A}(\omega) = \chi(\omega)\vec{A}(\omega)$, we hence find $\sigma(\omega) = \chi(\omega)/(i\omega)$, or in the case of full current $Y(\omega) = \chi(\omega)/(i\omega)$. Therefore, the fluctuation–dissipation relation for current noise reads (see Fig. 6.4)

$$S_I(\omega) = 2\hbar\omega\mathrm{Re}[Y(\omega)][\coth(\beta\hbar\omega/2) + 1].$$
(6.7)

Here the prefactor 2 comes from the definition of the noise correlator. This correlator consists of two parts, symmetric and antisymmetric in the frequency. As $\mathrm{Re}[Y(\omega)] = \mathrm{Re}[Y(-\omega)]$, these are the first and second terms in the square brackets, respectively. The symmetrized correlator of eqn (6.1) is the one usually accessed in experiments; it captures only the $\coth(\cdot)$-part of the above expression. However, the non-symmetrized correlator can also be viewed differently: as discussed in Sec. 6.6, the negative-frequency noise corresponds to the 'emitted' and the positive-frequency noise to the 'absorbed' fluctuations.

[4]In this case one typically chooses the gauge where the scalar potential vanishes.

Derivation of the Boltzmann–Langevin noise formula

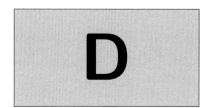

D

In this appendix, we derive eqn (6.39) for the noise by adding to the Boltzmann equation (2.6) a Langevin force $\xi(t)$ which describes the stochasticity of scattering. In the stationary case, concentrating on low frequencies, we can neglect the time dependence of both the average and the fluctuating part of the distribution function. In this case eqn (2.6) should be replaced by[1]

$$\vec{v} \cdot \nabla f(\vec{r}, \hat{p}, E, t) = I_{\text{coll}}[f] + \xi(\vec{r}, \vec{p}, t), \qquad (D.1)$$

where the Langevin forces ξ are

$$\xi(\vec{r}, \vec{p}, t) = \Omega \int \frac{d^3 \vec{p'}}{(2\pi\hbar)^3} \left[\delta J_{\vec{p'}, \vec{p}} - \delta J_{\vec{p}, \vec{p'}} \right]. \qquad (D.2)$$

Here Ω is the volume of the system, which cancels out at the end of the calculation. This term comes directly from the fluctuating part of $I_{\text{coll}}[f]$.

Let us consider what happens in the diffusive limit, where f can be expanded in spherical harmonics in the $\hat{p}$-dependence; see eqn (2.14). Now the two included harmonics contain both the average and the fluctuating parts. Proceeding as in Sec. 2.4, we first note that the angular average of the Langevin source terms vanishes,

$$\int d\hat{p}\, \xi(\vec{r}, \hat{p}, t) = 0. \qquad (D.3)$$

This reflects the fact that the number of electrons is conserved also in the presence of fluctuations. Therefore, eqn (2.19) is unaltered by the Langevin term. However, we get an extra term for the p-wave part compared to eqn (2.20). In the static case this is

$$\delta\vec{f} = -v\tau \nabla f_0 + 3\tau \int \hat{p}\xi\, d\hat{p}. \qquad (D.4)$$

Therefore, the full distribution function (average + fluctuations) satisfies an equation analogous to eqn (2.22),

$$D\nabla^2 f_0 = \ell_{\text{el}} \nabla \cdot \int \hat{p}\xi\, d\hat{p} + I_{\text{inel}}. \qquad (D.5)$$

[1] Here we neglect the fluctuation of the electric field, which would add a term on the left-hand side. This is valid if we can ignore the Coulomb-blockade -type effects, and concentrate on the limit of low frequencies.

Assuming that inelastic scattering is much weaker than elastic, we may ignore the collision integral for the fluctuating part, and get

$$D\nabla^2 \delta f_0 = \ell_{\text{el}} \nabla \cdot \int \hat{p}\xi d\hat{p}. \tag{D.6}$$

We can calculate the current density by integrating $\delta \vec{f}$ over the energy as in eqns (2.8), (2.24), and using the fluctuating distribution function instead of the average one. We get for the fluctuating part (see eqn (D.4))

$$\delta j = -eN_F v_F \int dE\delta(\delta\vec{f}) \int d\hat{p}\hat{p}^2 = \underbrace{eN_F \int dE(-D\nabla \delta f_0}_{\sigma\delta\mu/e} + \underbrace{\ell_{\text{el}} \int d\hat{p}\hat{p}\xi)}_{\delta j^s} \tag{D.7}$$

Here $\sigma = e^2 N_F D$ is the Drude conductivity. Thus, the total local current fluctuations consist of local potential fluctuations $\sigma\delta\mu/e$, and the fluctuations δj^s coming from elastic scattering.

In a quasi-one-dimensional geometry (where the distribution function f only changes in one direction, say, x), we can solve eqn (D.6) and get

$$\delta f_0 = -\frac{c_1}{D}x + \frac{\ell_{\text{el}}}{D}\int_0^x dx \int d\hat{p}\hat{p}\xi + c_2, \tag{D.8}$$

where c_1 and c_2 are integration constants. Assuming that the sample is purely voltage-biased, the fluctuations of the distribution function may be set to vanish at the contacts, say at $x = 0$ and $x = L$. This implies $c_2 = 0$ and

$$c_1 = \frac{\ell_{\text{el}}}{L}\int_0^L dx \int d\hat{p}\hat{p}\xi. \tag{D.9}$$

Comparing eqn (D.7) and (D.9), we find that the total current fluctuations are obtained as

$$\delta j = -eN_F \int dE c_1 = -\frac{1}{L}\int_0^L \delta j^s dx. \tag{D.10}$$

In the following, we derive an equation for the correlator of δj^s and thereby for δj.

Correlator of fluctuations

Now we should make an assumption about the correlator of the fluctuations. Typically one takes the currents $J_{\vec{p},\vec{p}'}$ as independent Poisson processes. This means that they are correlated only when the initial and final states and times are the same, and if they are evaluated at the same point. Moreover, in a Poisson process the second-order correlator ($\sim$ variance) is directly proportional to the average. Therefore, we may write

$$\langle \delta J_{\vec{p}_1,\vec{p}_1'}(\vec{r}_1,t_1)\delta J_{\vec{p}_2,\vec{p}_2'}(\vec{r}_2,t_2)\rangle$$
$$= \frac{(2\pi\hbar)^6}{\Omega}\delta(\vec{p}_1 - \vec{p}_2)\delta(\vec{p}_1' - \vec{p}_2')\delta(\vec{r}_1 - \vec{r}_2)\delta(t_1 - t_2)\bar{J}_{\vec{p},\vec{p}'}(\vec{r}_1,t_1). \tag{D.11}$$

Using this, we can write a relation for the correlator of the Langevin forces,

$$\langle \xi(\vec{r}, \vec{p}, t)\xi(\vec{r}', \vec{p}', t')\rangle =$$
$$\delta(\vec{r} - \vec{r}')\delta(t - t')\Omega \left\{ \delta(\vec{p} - \vec{p}')\int d\vec{p}'' \left[\bar{J}_{\vec{p}'',\vec{p}} + \bar{J}_{\vec{p},\vec{p}''} \right] - \bar{J}_{\vec{p},\vec{p}'} - \bar{J}_{\vec{p}',\vec{p}} \right\},$$
$$(D.12)$$

where in the latter term the position and time arguments have been omitted.

For elastic scattering it is enough to consider the case when the states lie on the same energy and only the angle $\hat{p}$ changes in the scattering. We thus assume

$$\bar{J}_{\vec{p},\vec{p}'} = \frac{1}{N_F}\delta(E_p - E_{p'})J_{\hat{p},\hat{p}'}, \tag{D.13}$$

where the density of states N_F at the Fermi level has been used for proper normalization. In this case we get

$$\langle \xi(\vec{r}, \hat{p}, E, t)\xi(\vec{r}, \hat{p}', E', t')\rangle = \frac{1}{N_F}\delta(\vec{r} - \vec{r}')\delta(t - t')\delta(E - E')G(\hat{p}, \hat{p}', \vec{r}, E).$$
$$(D.14)$$

Using eqn (2.12), we can relate this back to the average distribution function by noting that for purely elastic scattering,

$$G(\hat{p}, \hat{p}') = \int d\hat{p}''[\delta(\hat{p} - \hat{p}') - \delta(\hat{p}' - \hat{p}'')][W(\hat{p}, \hat{p}'')\bar{f}(\hat{p})(1 - \bar{f}(\hat{p}''))$$
$$+W(\hat{p}'', \hat{p})\bar{f}(\hat{p}'')(1 - \bar{f}(\hat{p}))].$$
$$(D.15)$$

Here all functions are evaluated at the same position, the same time and with the same energy E.

In the diffusive limit we then get for the fluctuations of the local current density

$$\langle \delta j^s(\vec{r}, t)\delta j^s(\vec{r}', t')\rangle$$
$$= e^2 N_F^2 \ell_{el}^2 \int dE dE' \int d\hat{p}d\hat{p}'\hat{p}\hat{p}'\langle \xi(\vec{r}, \hat{p}, E, t)\xi(\vec{r}', \hat{p}', E', t')\rangle \tag{D.16}$$
$$= e^2 N_F^2 \ell_{el}^2 \delta(\vec{r} - \vec{r}')\delta(t - t')\int dE \int d\hat{p}d\hat{p}'\hat{p}\hat{p}'G(\hat{p}, \hat{p}', \vec{r}, E).$$

Substituting the expansion $f(\hat{p}) = f_0 + \vec{\delta f} \cdot \hat{p}$ we can perform the integrals over $\hat{p}$ and $\hat{p}'$ and get

$$\langle \delta j^s(\vec{r}, t)\delta j^s(\vec{r}', t')\rangle = 2\sigma\delta(\vec{r} - \vec{r}')\delta(t - t')\Lambda(\vec{r}) \tag{D.17}$$

with

$$\Lambda(\vec{r}) = \int dE \bar{f}_0(\vec{r}, E)[1 - \bar{f}_0(\vec{r}, E)], \tag{D.18}$$

specified via the average s-wave distribution function $\bar{f}_0$.

Fluctuations in a quasi-one-dimensional geometry

Combining the results (D.10), (D.17) and (D.18), we finally get for the full noise correlator

$$
\begin{aligned}
S(t,t') &\equiv 2\langle \delta I(t)\delta I(t')\rangle = 2\left\langle \int d\vec{y}\,\delta j(t,\vec{y}) \int d\vec{y}'\,\delta j(t',\vec{y}')\right\rangle \\
&= \frac{2}{L^2}\int_0^L dx \int_0^L dx' \int d\vec{y}\,d\vec{y}'\,\langle \delta j^s(x,\vec{y},t)\delta j^s(x',\vec{y}',t')\rangle \qquad \text{(D.19)} \\
&= \frac{4A\sigma}{L^2}\int_0^L dx\,\Lambda(x)\delta(t-t').
\end{aligned}
$$

Here we denoted the transverse coordinates by $\vec{y}$ and the cross-section of the sample by A. We thus find white noise with the power spectral density

$$
S(\omega \ll \omega^*) = \frac{4G_N}{L}\int_0^L dx \int dE\,\bar{f}_0(E,x)(1-\bar{f}_0(E,x)). \qquad \text{(D.20)}
$$

Here $G_N = A\sigma/L$ is the conductance of the wire. The frequency scale ω^* is related to the (inverse) time scales for particle diffusion through the sample or for charge relaxation, whichever is smaller. This analysis applies if the considered frequencies are much below these.

Reflection coefficient in electronic circuits

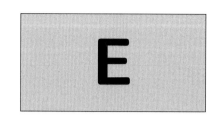

A popular means of measuring the impedance of nanoelectronic samples fast and accurately is to connect the sample as a termination impedance at the end of an electronic waveguide and measure the reflection[1] coefficient of waves sent to the waveguide. This appendix presents the calculation of the reflection coefficient for a waveguide terminated with impedance $Z(\omega)$.

A waveguide is characterized by an inductance L and a capacitance C per unit length. We may derive the equation of motion for the voltages and currents in the waveguide by first discretizing it into units of length h as in Fig. E.1 The currents $I(x_i) = I_i$ flowing through inductors and the stray currents $I_c(x_i) = I_{ci}$ flowing through the capacitors are related with the voltages $V(x_i) = V_i$ of the nodes of the circuit by

[1]Or the transmission coefficient, but here we only focus on the reflection coefficient.

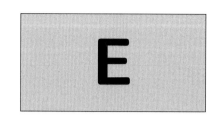

Fig. E.1 Discretization of a waveguide.

$$I_i = \frac{1}{hL} \int^t (V_i(t') - V_{i-1}(t'))dt' \tag{E.1a}$$

$$I_{ci} = Ch\dot{V}_i(t), \tag{E.1b}$$

where $\dot{V}_i(t)$ denotes the time derivative of the voltage. Current conservation into node i yields

$$I_i - I_{i+1} + I_{ci} = \frac{1}{hL} \int^t (2V_i(t') - V_{i+1}(t') - V_{i-1}(t'))dt' - Ch\dot{V}_i(t) = 0. \tag{E.2}$$

We can identify the first term as the discretization of the second derivative. Dividing both sides by h, taking the limit $h \to 0$ and differentiating once by time yields the wave equation

$$\partial_t^2 V(x,t) = \frac{1}{LC}\partial_x^2 V(x,t). \tag{E.3}$$

Here we identify the effective speed of light, $c = 1/\sqrt{LC}$. For a given voltage profile, the current can be obtained from eqn (E.1a) in the limit $h \to 0$,

$$I(x) = \frac{1}{L} \int^t \partial_x V(x,t')dt'. \tag{E.4}$$

Let us assume that the waveguide is terminated at $x = 0$ by an impedance $Z(\omega)$ as in Fig. E.2. There we obtain for the Fourier transformed voltages and currents $V(0,\omega) = Z(\omega)I(0,\omega)$. Combining this

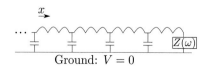

Fig. E.2 Waveguide terminated with impedance $Z(\omega)$.

with eqn (E.4) hence gives a boundary condition

$$Z(\omega)\partial_x V(x=0,\omega) = i\omega L V(x=0,\omega),\qquad\text{(E.5)}$$

where the integral over time is replaced by $1/i\omega$.

Now finding the reflection coefficient is rather straightforward. Assume we send a wave with amplitude V_0 and frequency ω along the waveguide towards the sample, and get a reflected wave with amplitude rV_0. The total wave is thus

$$V(x,t) = V_0 e^{i\omega t}(e^{i(\omega/c)x} + r e^{-i(\omega/c)x}),\qquad\text{(E.6)}$$

where the wave number has been chosen so that the wave equation (E.3) is satisfied. The boundary condition (E.5) yields for the Fourier amplitude at frequency ω,

$$Z(\omega)i\omega V_0(1-r) = i\omega L c V_0(1+r).\qquad\text{(E.7)}$$

Solving this gives the reflection coefficient

$$r = \frac{Z(\omega)-Lc}{Z(\omega)+Lc} = \frac{Z(\omega)-Z_0}{Z(\omega)+Z_0},\qquad\text{(E.8)}$$

where $Z_0 = Lc = \sqrt{L/C}$ is the characteristic impedance of the waveguide. Measuring the changes in the amplitude and/or the phase of r as a function of some control parameter, the sample impedance can be measured very accurately. For the coaxial cables typically used in microwave experiments, Z_0 is often of the order of 50 Ω.

References

Abrahams, E., Anderson, P. W., Licciardello, D. C., and Ramakrishnan, T. V. (1979). Scaling theory of localization: Absence of quantum diffusion in two dimensions. *Phys. Rev. Lett.*, **42**, 673–676. (Cited on page *69*).

Aharonov, Y. and Bohm, D. (1959). Significance of electromagnetic potentials in the quantum theory. *Phys. Rev.*, **115**(3), 485–491. (Cited on page *61*).

Albada, Meint P. Van and Lagendijk, Ad (1985). Observation of weak localization of light in a random medium. *Phys. Rev. Lett.*, **55**(24), 2692–2695. (Cited on page *70*).

Allain, P. and Fuchs, J. (2011). Klein tunneling in graphene: optics with massless electrons. *Eur. Phys. J. B*, **83**, 301–317. (Cited on page *194*).

Altland, Alexander and Zirnbauer, Martin R. (1997). Nonstandard symmetry classes in mesoscopic normal–superconducting hybrid structures. *Phys. Rev. B*, **55**, 1142–1161. (Cited on page *72*).

Altshuler, B. L. and Aronov, A. G. (1985). Electron–electron interaction in disordered conductors. In *Electron–Electron Interactions in Disordered Systems* (ed. A. L. Efros and M. Pollak), Chapter 1, pp. 1. Elsevier, Amsterdam. (Cited on pages *26, 113, and 114*).

Altshuler, B. L., Khmel'nitzkii, D., Larkin, A. I., and Lee, P. A. (1980). Magnetoresistance and Hall effect in a disordered two-dimensional electron gas. *Phys. Rev. B*, **22**, 5142–5153. (Cited on page *69*).

Al'tshuler, Boris L. and Lee, Patrick A. (1988). Disordered electronic systems. *Phys. Today*, **41**, 36. (Cited on page *76*).

Ambegaokar, Vinay and Baratoff, Alexis (1963). Tunneling between superconductors. *Phys. Rev. Lett.*, **10**(11), 486–489. (Cited on page *89*).

Anderson, P. W. (1958). Absence of diffusion in certain random lattices. *Phys. Rev.*, **109**(5), 1492–1505. (Cited on pages *63 and 64*).

Anderson, P. W. (1961). Localized magnetic states in metals. *Phys. Rev.*, **124**(1), 41–53. (Cited on page *142*).

Anderson, P. W., Thouless, D. J., Abrahams, E., and Fisher, D. S. (1980). New method for a scaling theory of localization. *Phys. Rev. B*, **22**, 3519–3526. (Cited on page *63*).

Ando, Tsuneya (2005). Theory of electronic states and transport in carbon nanotubes. *J. Phys. Soc. Jpn*, **74**(3), 777–817. (Cited on page *8*).

Andreev, A. F. (1964). The thermal conductivity of the intermediate state in superconductors. *Sov. Phys. JETP*, **19**, 1228. (Cited on page *86*).

Armour, A. D. (2004). Current noise of a single-electron transistor coupled to a nanomechanical resonator. *Phys. Rev. B*, **70**, 165315. (Cited on page *216*).

Ashcroft, Neil W. and Mermin, N. David (1976). *Solid State Physics*. Saunders College, Philadelphia. (Cited on pages *4, 35, and 52*).

Astafiev, O., Pashkin, Yu. A., Nakamura, Y., Yamamoto, T., and Tsai, J. S. (2004). Quantum noise in the Josephson charge qubit. *Phys. Rev. Lett.*, **93**(26), 267007. (Cited on page *112*).

Averin, D. V. and Likharev, K. K. (1986). Coulomb blockade of single-electron tunneling, and coherent oscillations in small tunnel junctions. *J. Low Temp. Phys.*, **62**, 345–373. (Cited on page *124*).

Averin, D. V. and Nazarov, Yu. V. (1990). Virtual electron diffusion during quantum tunneling of the electric charge. *Phys. Rev. Lett.*, **65**(19), 2446–2449. (Cited on page *130*).

Averin, Dmitri V. and Pekola, Jukka P. (2010). Violation of the fluctuation–dissipation theorem in time-dependent mesoscopic heat transport. *Phys. Rev. Lett.*, **104**, 220601. (Cited on page *117*).

Baibich, M. N., Broto, J. M., Fert, A., Van Dau, F. Nguyen, Petroff, F., Etienne, P., Creuzet, G., Friederich, A., and Chazelas, J. (1988). Giant magnetoresistance of (001)Fe/(001)Cr magnetic superlattices. *Phys. Rev. Lett.*, **61**, 2472–2475. (Cited on pages *32 and 33*).

Bao, W., Jing, L., Jr, J. Velasco, Lee, Y., Liu, G., Tran, D., Standley, B., Aykol, M., Cronin, S. B., D. Smirnov, M. Koshino, McCann, E., Bockrath, M., and Lau, C. N. (2011). Stacking-dependent band gap and quantum transport in trilayer graphene. *Nature Phys.*, **7**, 948. (Cited on page *191*).

Bardeen, J., Cooper, L. N., and Schrieffer, J. R. (1957). Theory of superconductivity. *Phys. Rev.*, **108**(5), 1175–1204. (Cited on page *78*).

Barone, A. and Paterno, G. (1982). *Physics and Applications of the Josephson Effect*. John Wiley & Sons. (Cited on pages *92 and 181*).

Baur, M., Fedorov, A., Steffen, L., Filipp, S., da Silva, M. P., and Wallraff, A. (2012). Benchmarking a quantum teleportation protocol in superconducting circuits using tomography and an entanglement witness. *Phys. Rev. Lett.*, **108**, 040502. (Cited on page *180*).

Beenakker, C. W. J. (1991*a*). Theory of Coulomb-blockade oscillations in the conductance of a quantum dot. *Phys. Rev. B*, **44**(4), 1646–1656. (Cited on pages *146 and 148*).

Beenakker, C. W. J. (1991*b*). Universal limit of critical-current fluctuations in mesoscopic Josephson junctions. *Phys. Rev. Lett.*, **67**(27), 3836–3839. (Cited on pages *90 and 91*).

Beenakker, C. W. J. (1992). Quantum transport in semiconductor–superconductor microjunctions. *Phys. Rev. B*, **46**(19), 12841–12844. (Cited on page *101*).

Beenakker, C. W. J. (1997). Random-matrix theory of quantum transport. *Rev. Mod. Phys.*, **69**(3), 731–808. (Cited on pages *51, 58, and 72*).

Belzig, Wolfgang (2003). Full counting statistics in quantum contacts. [arXiv:0312180]. (Cited on pages *104 and 116*).

Belzig, W., Wilhelm, F. K., Bruder, C., Schön, G., and Zaikin, A. D. (1999). Quasiclassical Green's function approach to mesoscopic superconductivity. *Superlatt. Microstruct.*, **25**, 1251. (Cited on page *92*).

Bergeret, F. S., Virtanen, P., Heikkilä, T. T., and Cuevas, J. C. (2010). Theory of microwave-assisted supercurrent in quantum point contacts. *Phys. Rev. Lett.*, **105**(11), 117001. (Cited on page *91*).

Bergmann, Gerd (1984). Weak localization in thin films: a time-of-flight experiment with conduction electrons. *Phys. Rep.*, **107**(1), 1 – 58. (Cited on page *69*).

Bernal, J.D. (1924). The structure of graphite. *Proc. Roy. Soc.*, **A160**, 749. (Cited on page *189*).

Betz, A. C., Vialla, F., Brunel, D., Voisin, C., Picher, M., Cavanna, A., Madouri, A., Fève, G., Berroir, J.-M., Plaçais, B., and Pallecchi, E. (2012). Hot electron cooling by acoustic phonons in graphene. *Phys. Rev. Lett.*, **109**, 056805. (Cited on page *105*).

Binasch, G., Grünberg, P., Saurenbach, F., and Zinn, W. (1989). Enhanced magnetoresistance in layered magnetic structures with antiferromagnetic interlayer exchange. *Phys. Rev. B*, **39**, 4828–4830. (Cited on pages *32 and 33*).

Bistritzer, R. and MacDonald, A. H. (2009). Electronic cooling in graphene. *Phys. Rev. Lett.*, **102**(20), 206410. (Cited on page *28*).

Blanter, Ya. and Büttiker, M. (2000). Shot noise in mesoscopic conductors. *Phys. Rep.*, **336**, 1. (Cited on pages *54, 55, 100, 104, and 118*).

Bleszynski-Jayich, A. C., Shanks, W. E., Peaudecerf, B., Ginossar, E., von Oppen, F., Glazman, L., and Harris, J. G. E. (2009). Persistent currents in normal metal rings. *Science*, **326**(5950), 272–275. (Cited on pages *74, 75, and 77*).

Blonder, G. E., Tinkham, M., and Klapwijk, T. M. (1982). Transition from metallic to tunneling regimes in superconducting microconstrictions: Excess current, charge imbalance, and supercurrent conversion. *Phys. Rev. B*, **25**, 4515–4532. (Cited on page *87*).

Bluhm, Hendrik, Koshnick, Nicholas C., Bert, Julie A., Huber, Martin E., and Moler, Kathryn A. (2009). Persistent currents in normal metal rings. *Phys. Rev. Lett.*, **102**, 136802. (Cited on page *75*).

Blumenthal, M.D., Kaestner, B., Li, L., Giblin, S., Janssen, T.J.B.M., Pepper, M., Anderson, D., Jones, G., and Ritchie, D.A. (2007). Gigahertz quantized charge pumping. *Nature Phys.*, **3**, 343–347. (Cited on page *138*).

Bolotin, K. I., Sikes, K. J., Hone, J., Stormer, H. L., and Kim, P. (2008). Temperature-dependent transport in suspended graphene. *Phys. Rev. Lett.*, **101**, 096802. (Cited on page *197*).

Bomze, Yu., Gershon, G., Shovkun, D., Levitov, L. S., and Reznikov, M. (2005). Measurement of counting statistics of electron transport in a tunnel junction. *Phys. Rev. Lett.*, **95**(17), 176601. (Cited on page *117*).

Börlin, J., Belzig, W., and Bruder, C. (2002). Full counting statistics of a superconducting beam splitter. *Phys. Rev. Lett.*, **88**(19), 197001. (Cited on page *107*).

Boukhvalov, D. W., Katsnelson, M. I., and Lichtenstein, A. I. (2008). Hydrogen on graphene: Electronic structure, total energy, structural distortions and magnetism from first-principles calculations. *Phys. Rev. B*, **77**, 035427. (Cited on page *186*).

Brataas, Arne, Nazarov, Yu. V., and Bauer, Gerrit E. W. (2000). Finite-element theory of transport in ferromagnet–normal metal systems. *Phys. Rev. Lett.*, **84**, 2481–2484. (Cited on page *57*).

Breit, G. and Wigner, E. (1936). Capture of slow neutrons. *Phys. Rev.*, **49**(7), 519–531. (Cited on page *54*).

Brey, L. and Fertig, H. A. (2006). Electronic states of graphene nanoribbons studied with the Dirac equation. *Phys. Rev. B*, **73**, 235411. (Cited on page *198*).

Brouwer, P. W. (1998). Scattering approach to parametric pumping. *Phys. Rev. B*, **58**, R10135–R10138. (Cited on page *56*).

Bruus, H. and Flensberg, K. (2004). *Many-Body Quantum Theory in Condensed Matter Physics*. Oxford University Press. (Cited on pages *5, 14, 67, 99, 143, 146, 152,* and *157*).

Buks, Eyal and Blencowe, M. P. (2006). Decoherence and recoherence in a vibrating rf SQUID. *Phys. Rev. B*, **74**, 174504. (Cited on page *214*).

Bunch, J.S., van der Zande, A.M., Verbridge, S.S., Frank, I.W., Tanenbaum, D.M., Parpia, J.M., Craighead, H.G., and McEuen, P.L. (2007). Electromechanical resonators from graphene sheets. *Science*, **315**(5811), 490–493. (Cited on pages *204, 208,* and *214*).

Büttiker, M. (1986). Four-terminal phase-coherent conductance. *Phys. Rev. Lett.*, **57**, 1761–1764. (Cited on page *37*).

Büttiker, M. (1990). Quantized transmission of a saddle-point constriction. *Phys. Rev. B*, **41**(11), 7906–7909. (Cited on page *44*).

Büttiker, M. (1992). Scattering theory of current and intensity noise correlations in conductors and wave guides. *Phys. Rev. B*, **46**(19), 12485–12507. (Cited on pages *105, 107,* and *243*).

Büttiker, M., Imry, Y., and Landauer, R. (1983). Josephson behavior in small normal one-dimensional rings. *Phys. Lett.*, **96A**(7), 365. (Cited on page *74*).

Büttiker, M., Thomas, H., and Prêtre, A. (1993). Mesoscopic capacitors. *Phys. Lett. A*, **180**(4–5), 364–369. (Cited on page *38*).

Büttiker, M., Thomas, H., and Prêtre, A. (1994). Current partition in multiprobe conductors in the presence of slowly oscillating external potentials. *Z. Phys. B*, **94**, 133. (Cited on page *56*).

Byers, N. and Yang, C. N. (1961). Theoretical considerations concerning quantized magnetic flux in superconducting cylinders. *Phys. Rev. Lett.*, **7**, 46–49. (Cited on page *75*).

Bylander, Jonas, Duty, Tim, and Delsing, Per (2005). Current measurement by real-time counting of single electrons. *Nature*, **434**, 361. (Cited on page *139*).

Callen, Herbert B. and Welton, Theodore A. (1951). Irreversibility and generalized noise. *Phys. Rev.*, **83**, 34–40. (Cited on page *96*).

Camley, R. E. and Barnaś, J. (1989). Theory of giant magnetoresistance effects in magnetic layered structures with antiferromagnetic coupling. *Phys. Rev. Lett.*, **63**, 664–667. (Cited on page *32*).

Castro Neto, A. H., Guinea, F., Peres, N. M. R., Novoselov, K. S., and Geim, A. K. (2009). The electronic properties of graphene. *Rev. Mod. Phys.*, **81**(1), 109–162. (Cited on pages *9, 76, 183,* and *201*).

Chakravarty, S. and Schmid, A. (1986). Weak localization: The quasiclassical theory of electrons in a random potential. *Phys. Rep.*, **140**, 193. (Cited on pages *65, 68,* and *76*).

Chaste, J., Eichler, A., Moser, J., Ceballos, G., Rurali, R., and Bachtold, A. (2012). A nanomechanical mass sensor with yoctogram resolution. *Nature Nanotech.*, **7**, 301. (Cited on page *203*).

Chaste, J., Pallecchi, E., Morfin, P., Fève, G., Kontos, T., Berroir, J.-M., Hakonen, P., and Plaçais, B. (2010). Thermal shot noise in top-gated single carbon nanotube field effect transistors. *Appl. Phys. Lett.*, **96**(19), 192103. (Cited on page *105*).

Cheianov, Vadim V., Fal'ko, Vladimir, and Altshuler, B. L. (2007). The focusing of electron flow and a Veselago lens in graphene p-n junctions. *Science*, **315**(5816), 1252–1255. (Cited on page *193*).

Cheianov, Vadim V. and Fal'ko, Vladimir I. (2006). Selective transmission of Dirac electrons and ballistic magnetoresistance of np junctions in graphene. *Phys. Rev. B*, **74**, 041403. (Cited on page *196*).

Chen, Y.-F., Dirks, T., Al-Zoubi, G., Birge, N.O., and Mason, N. (2009). Nonequilibrium tunneling spectroscopy in carbon nanotubes. *Phys. Rev. Lett.*, **102**(3), 036804. (Cited on pages *29 and 30*).

Cheung, Ho-Fai, Riedel, Eberhard K., and Gefen, Yuval (1989). Persistent currents in mesoscopic rings and cylinders. *Phys. Rev. Lett.*, **62**, 587–590. (Cited on page *74*).

Chiatti, O., Nicholls, J. T., Proskuryakov, Y. Y., Lumpkin, N., Farrer, I., and Ritchie, D. A. (2006). Quantum thermal conductance of electrons in a one-dimensional wire. *Phys. Rev. Lett.*, **97**(5), 056601. (Cited on page *44*).

Chiodi, F., Ferrier, M., Tikhonov, K., Virtanen, P., Heikkilä, T.T., Feigelman, M., Guéron, S., and Bouchiat, H. (2011). Probing the dynamics of Andreev states in a coherent normal/superconducting ring. *Sci. Rep.*, **1**, 3. (Cited on page *4*).

Christen, T. and Büttiker, M. (1996). Gauge-invariant nonlinear electric transport in mesoscopic conductors. *EPL (Europhys. Lett.)*, **35**(7), 523. (Cited on page *56*).

Chtchelkatchev, Nikolai M., Blatter, Gianni, Lesovik, Gordey B., and Martin, Thierry (2002). Bell inequalities and entanglement in solid-state devices. *Phys. Rev. B*, **66**, 161320. (Cited on page *108*).

Clark, A. M., Miller, N. A., Williams, A., Ruggiero, S. T., Hilton, G. C., Vale, L. R., Beall, J. A., Irwin, K. D., and Ullom, J. N. (2005). Cooling of bulk material by electron-tunneling refrigerators. *Appl. Phys. Lett.*, **86**, 173508. (Cited on page *165*).

Cleland, Andrew N. (2003). *Foundations of Nanomechanics, from Solid-State Theory to Device Applications.* Springer. (Cited on pages *204, 205, 214, and 227*).

Cleland, A. N. and Roukes, M. L. (1996). Fabrication of high frequency nanometer scale mechanical resonators from bulk Si crystals. *Appl. Phys. Lett.*, **69**(18), 2653–2655. (Cited on page *211*).

Clerk, A. A., Girvin, S. M., Nguyen, A. K., and Stone, A. D. (2002). Resonant Cooper-pair tunneling: Quantum noise and measurement characteristics. *Phys. Rev. Lett.*, **89**(17), 176804. (Cited on page *161*).

Crosser, M. S., Virtanen, P., Heikkilä, T.T., and Birge, N.O. (2006). Supercurrent-induced temperature gradient across a nonequilibrium SNS Josephson junction. *Phys. Rev. Lett.*, **96**(16), 167004. (Cited on page *30*).

Cuevas, Juan Carlos and Scheer, Elke (2010). *Molecular Electronics: An Introduction to Theory and Experiment.* World Scientific, Singapore. (Cited on page *157*).

Das Sarma, S., Adam, Shaffique, Hwang, E. H., and Rossi, Enrico (2011). Electronic transport in two-dimensional graphene. *Rev. Mod. Phys.*, **83**, 407–470. (Cited on pages *183, 197, and 201*).

Datta, S. (1995). *Electronic Transport in Mesoscopic Systems.* Cambridge University Press. (Cited on pages *12, 14, 52, 58, 65, and 76*).

Datta, Supriyo (2004). Electrical resistance: an atomistic view. *Nanotechnology*, **15**(7), S433. (Cited on page *144*).

Datta, Supriyo (2009). Nanohub: the NEGF approach to nano-device simulation. http://nanohub.org/topics/Negf. (Cited on page *53*).

Day, P.K., LeDuc, H.G., Mazin, B.A., Vayonakis, A., and Zmuidzinas, J. (2003). A broadband superconducting detector suitable for use in large arrays. *Nature*, **425**, 817. (Cited on page *176*).

de Gennes, P.G. (1999). *Superconductivity of Metals and Alloys*. Westview Press. (Cited on page *92*).

de Jong, M. J. M. and Beenakker, C. W. J. (1994). Doubled shot noise in disordered normal-metal–superconductor junctions. *Phys. Rev. B*, **49**, 16070–16073. (Cited on page *101*).

de Picciotto, R., Reznikov, M., Heiblum, M., Umansky, V., Bunin, G., and Mahalu, D. (1997). Direct observation of fractional charge. *Nature*, **389**, 162. (Cited on page *101*).

Dean, C. R., Young, A. F., Meric, I., Lee, C., Wang, L., Sorgenfrei, S., Watanabe, K., Taniguchi, T., Kim, P., and Hone, K. L. Shepardand J. (2010). Boron nitride substrates for high-quality graphene electronics. *Nature Nanotech.*, **5**, 722. (Cited on page *193*).

Della Rocca, M. L., Chauvin, M., Huard, B., Pothier, H., Esteve, D., and Urbina, C. (2007). Measurement of the current-phase relation of superconducting atomic contacts. *Phys. Rev. Lett.*, **99**(12), 127005. (Cited on page *89*).

Despósito, M. A. and Levy Yeyati, A. (2001). Controlled dephasing of Andreev states in superconducting quantum point contacts. *Phys. Rev. B*, **64**(14), 140511. (Cited on page *89*).

Devoret, Michel H., Martinis, John M., and Clarke, John (1985). Measurements of macroscopic quantum tunneling out of the zero-voltage state of a current-biased Josephson junction. *Phys. Rev. Lett.*, **55**, 1908–1911. (Cited on page *171*).

Devoret, M. H., Wallraff, A., and Martinis, J. M. (2004). Superconducting qubits: A short review. [arXiv:cond-mat/0411174]. (Cited on page *181*).

Dittrich, T., Hänggi, P., Ingold, G.-L., Kramer, B., Schön, G., and Zwerger, W. (1998). *Quantum Transport and Dissipation*. Wiley-VCH. (Cited on pages *14, 21, 64, 65, 67, 69, 76, 113, 138, 161, and 162*).

DiVincenzo, David P. (2000). The physical implementation of quantum computation. *Fortschr. Phys.*, **48**, 771. (Cited on page *177*).

Doh, Y.-J., van Dam, J.A., Roest, A.L., Bakkers, E.P.A.M., Kouwenhoven, L.P., and De Franceschi, S. (2005). Tunable supercurrent through semiconductor nanowires. *Science*, **309**, 272–275. (Cited on page *89*).

Dresselhaus, M. S. and Dresselhaus, G. (2002). Intercalation compounds of graphite. *Adv. Phys.*, **51**(1), 1–186. (Cited on page *188*).

Drude, P. (1900*a*). Zur Elektronentheorie der Metalle. *Ann. Phys. (Berlin)*, **306**, 566. (Cited on page *10*).

Drude, P. (1900*b*). Zur Elektronentheorie der Metalle; II. Teil. Galvanomagnetische und Thermomagnetische Effecte. *Ann. Phys. (Berlin)*, **308**, 369. (Cited on page *10*).

Eichler, A., Moser, J., Chaste, J., Zdrojek, M., Wilson-Rae, I., and Bachtold, A. (2011). Nonlinear damping in mechanical resonators made from carbon nanotubes and graphene. *Nature Nanotech.*, **6**, 339. (Cited on page *210*).

Ekinci, K. L. and Roukes, M. L. (2005). Nanoelectromechanical systems. *Rev. Sci. Instrum.*, **76**(6), 061101. (Cited on pages *210 and 227*).

Elzerman, J. M., Hanson, R., van Beveren, L. H. Willems, Witkamp, B., Vandersypen, L. M. K., and Kouwenhoven, L. P. (2004). Single-shot read-out of an individual electron spin in a quantum dot. *Nature*, **430**, 431. (Cited on page *158*).

Etaki, S, Poot, M, Mahboob, I, Onomitsu, K, Yamaguchi, H, and van der Zant, H S J (2008). Motion detection of a micromechanical resonator embedded in a d.c. SQUID. *Nature Phys.*, **4**(10), 785–788. (Cited on page *215*).

Fay, A., Danneau, R., Viljas, J. K., Wu, F., Tomi, M. Y., Wengler, J., Wiesner, M., and Hakonen, P. J. (2011). Shot noise and conductivity at high bias in bilayer graphene: Signatures of electron–optical phonon coupling. *Phys. Rev. B*, **84**, 245427. (Cited on page *105*).

Fetter, A. L. and Walecka, J.D. (2003). *Quantum Theory of Many-Particle Systems*. Dover. (Cited on page *99*).

Feynman, R., Leighton, R., and Sands, M. (2005). *The Feynman Lectures on Physics* (2nd edn), Volume 3. Addison Wesley. (Cited on page *84*).

Fisher, Daniel S. and Lee, Patrick A. (1981). Relation between conductivity and transmission matrix. *Phys. Rev. B*, **23**(12), 6851–6854. (Cited on page *52*).

Flensberg, Karsten (1993). Capacitance and conductance of mesoscopic systems connected by quantum point contacts. *Phys. Rev. B*, **48**(15), 11156–11166. (Cited on page *123*).

Flipse, J., Bakker, F.L., Slachter, A., Dejene, F.K., and van Wees, B.J. (2012). Direct observation of the spin-dependent Peltier effect. *Nature Nanotech.*, **7**, 166. (Cited on page *35*).

Geim, A.K., Grigorieva, I.V., Dubonos, S.V., Lok, J.G.S., Maan, J.C., Filippov, A.E., and Peeters, F.M. (1997). Phase transitions in individual sub-micrometre superconductors. *Nature*, **390**, 259. (Cited on page *80*).

Geim, A.K. and Novoselov, K.S. (2007). The rise of graphene. *Nature Mater.*, **6**, 183. (Cited on pages *9* and *183*).

Giazotto, Francesco, Heikkilä, Tero T., Luukanen, Arttu, Savin, Alexander M., and Pekola, Jukka P. (2006). Opportunities for mesoscopics in thermometry and refrigeration: Physics and applications. *Rev. Mod. Phys.*, **78**(1), 217–274. (Cited on pages *2, 15, 16, 35, 118, 135,* and *181*).

Glazman, L.I., Lesovik, G.B., Khmel'nitskii, D.E., and Shekter, R.I. (1988). Reflectionless quantum transport and fundamental ballistic-resistance steps in microscopic constrictions. *JETP Lett.*, **48**, 238. (Cited on page *44*).

Golubov, A. A., Kupriyanov, M. Yu., and Il'ichev, E. (2004). The current-phase relation in Josephson junctions. *Rev. Mod. Phys.*, **76**, 411. (Cited on page *92*).

Gorbachev, R. V., Tikhonenko, F. V., Mayorov, A. S., Horsell, D. W., and Savchenko, A. K. (2007). Weak localization in bilayer graphene. *Phys. Rev. Lett.*, **98**(17), 176805. (Cited on page *68*).

Gorelik, L. Y., Isacsson, A., Voinova, M. V., Kasemo, B., Shekhter, R. I., and Jonson, M. (1998). Shuttle mechanism for charge transfer in coulomb blockade nanostructures. *Phys. Rev. Lett.*, **80**, 4526–4529. (Cited on page *214*).

Gustavsson, S., Leturcq, R., Simovič, B., Schleser, R., Ihn, T., Studerus, P., Ensslin, K., Driscoll, D. C., and Gossard, A. C. (2006). Counting statistics of single electron transport in a quantum dot. *Phys. Rev. Lett.*, **96**(7), 076605. (Cited on page *117*).

Gustavsson, S., Studer, M., Leturcq, R., Ihn, T., Ensslin, K., Driscoll, D. C., and Gossard, A. C. (2007). Frequency-selective single-photon detection using a double quantum dot. *Phys. Rev. Lett.*, **99**(20), 206804. (Cited on page *7*).

Haldane, F. D. M. (1978). Scaling theory of the asymmetric Anderson model. *Phys. Rev. Lett.*, **40**(6), 416–419. (Cited on page *152*).

Han, Melinda Y., Özyilmaz, Barbaros, Zhang, Yuanbo, and Kim, Philip (2007). Energy band-gap engineering of graphene nanoribbons. *Phys. Rev. Lett.*, **98**, 206805. (Cited on page *198*).

Hanson, R., Kouwenhoven, L. P., Petta, J. R., Tarucha, S., and Vandersypen, L. M. K. (2007). Spins in few-electron quantum dots. *Rev. Mod. Phys.*, **79**, 1217–1265. (Cited on pages *151, 153, and 157*).

Harrabi, K., Yoshihara, F., Niskanen, A. O., Nakamura, Y., and Tsai, J. S. (2009). Engineered selection rules for tunable coupling in a superconducting quantum circuit. *Phys. Rev. B*, **79**, 020507. (Cited on page *180*).

Hecker, K., Hegger, H., Altland, A., and Fiegle, K. (1997). Conductance fluctuations in mesoscopic normal-metal/superconductor samples. *Phys. Rev. Lett.*, **79**(8), 1547–1550. (Cited on page *71*).

Heersche, Hubert B., Jarillo-Herrero, Pablo, Oostinga, Jeroen B., Vandersypen, Lieven M. K., and Morpurgo, Alberto F. (2007). Bipolar supercurrent in graphene. *Nature*, **446**, 56. (Cited on pages *14 and 89*).

Heikkilä, T.T., Virtanen, P., Johansson, G., and Wilhelm, F.K. (2004). Measuring non-Gaussian fluctuations through incoherent Cooper-pair current. *Phys. Rev. Lett.*, **93**(24), 247005. (Cited on page *133*).

Heikkilä, T. and Volovik, G. (2011). Dimensional crossover in topological matter: Evolution of the multiple Dirac point in the layered system to the flat band on the surface. *JETP Lett.*, **93**, 59–65. (Cited on page *191*).

Heikkilä, T. T., Hatami, Moosa, and Bauer, Gerrit E. W. (2010). Spin heat accumulation and its relaxation in spin valves. *Phys. Rev. B*, **81**, 100408. (Cited on page *30*).

Heikkilä, T. T. and Nazarov, Yuli V. (2009). Statistics of temperature fluctuations in an electron system out of equilibrium. *Phys. Rev. Lett.*, **102**(13), 130605. (Cited on page *117*).

Heikkilä, Tero T., Salomaa, Martti M., and Lambert, Colin J. (1999). Superconducting proximity effect and universal conductance fluctuations. *Phys. Rev. B*, **60**(13), 9291–9294. (Cited on page *64*).

Heikkilä, Tero T., Särkkä, Jani, and Wilhelm, Frank K. (2002). Supercurrent-carrying density of states in diffusive mesoscopic Josephson weak links. *Phys. Rev. B*, **66**, 184513. (Cited on pages *91 and 92*).

Hekking, F. W. J., Niskanen, A. O., and Pekola, J. P. (2008). Electron–phonon coupling and longitudinal mechanical-mode cooling in a metallic nanowire. *Phys. Rev. B*, **77**(3), 033401. (Cited on page *28*).

Henny, M., Oberholzer, S., Strunk, C., Heinzel, T., Ensslin, K., Holland, M., and Schönenberger, C. (1999). The fermionic Hanbury Brown and Twiss experiment. *Science*, **284**(5412), 296–298. (Cited on page *107*).

Hershfield, Selman, Davies, John H., and Wilkins, John W. (1991). Probing the Kondo resonance by resonant tunneling through an Anderson impurity. *Phys. Rev. Lett.*, **67**, 3720–3723. (Cited on page *152*).

Hirvi, K. P., Kauppinen, J. P., Korotkov, A. N., Paalanen, M. A., and Pekola, J. P. (1995). Arrays of normal metal tunnel junctions in weak Coulomb blockade regime. *Appl. Phys. Lett.*, **67**, 2096–2098. (Cited on page *136*).

Huard, B., Anthore, A., Pierre, F., Pothier, H., Birge, Norman O., and Esteve, D. (2004). Intensity of Coulomb interaction between quasiparticles in diffusive metallic wires. *Solid State Commun.*, **131**, 599. (Cited on page *27*).

Huard, B., Pothier, H., Birge, N.O., Esteve, D., Waintal, X., and Ankerhold, J. (2007). Josephson junctions as detectors for non-Gaussian noise. *Ann. Phys.*, **16**, 736. (Cited on page *171*).

Hüttel, Andreas K., Steele, Gary A., Witkamp, Benoit, Poot, Menno, Kouwenhoven, Leo P., and van der Zant, Herre S. J. (2009). Carbon nanotubes as ultrahigh quality factor mechanical resonators. *Nano Lett.*, **9**(7), 2547–2552. PMID: 19492820. (Cited on page *203*).

Imry, Y. (2002). *Introduction to Mesoscopic Physics* (2nd edn). Oxford University Press. (Cited on pages *12, 13, and 76*).

Ingold, G.-L. and Nazarov, Yu. V. (1992). Charge tunneling rates in ultrasmall junctions. In *Single Charge Tunnelling* (ed. H. Grabert and M. Devoret), Volume 294 of *Nato ASI Series B*, pp. 21–107. Plenum Press, New York. Also found at [cond-mat/0508728]. (Cited on pages *122, 134, and 138*).

Isacsson, A. and Nord, T. (2004). Low-frequency current noise of the single-electron shuttle. *EPL (Europhys. Lett.)*, **66**(5), 708. (Cited on page *216*).

Ivanchenko, Yu. M. and Zilberman, L.A. (1969). The Josephson effect in small tunnel contacts. *Sov. Phys. JETP*, **28**, 1272. (Cited on page *172*).

Jedema, F. J., Filip, A. T., and van Wees, B. J. (2001). Electrical spin injection and accumulation at room temperature in an all-metal mesoscopic spin valve. *Nature*, **410**, 345. (Cited on pages *32, 33, and 36*).

Jensen, H. Dalsgaard and Martinis, John M. (1992). Accuracy of the electron pump. *Phys. Rev. B*, **46**, 13407–13427. (Cited on page *130*).

Jensen, K., Kim, Kwanpyo, and Zettl, A. (2008). An atomic-resolution nanomechanical mass sensor. *Nature Nanotech.*, **3**, 533. (Cited on page *203*).

Ji, Yang, Chung, Yunchul, Sprinzak, D., Heiblum, M., Mahalu, D., and Shtrikman, Hadas (2003). An electronic Mach-Zehnder interferometer. *Nature*, **422**, 415–418. (Cited on page *107*).

Jia, X., Hofmann, M., Meunier, V., Sumpter, B.G., Campos-Delgado, J., Romo-Herrera, J.M., Son, H., Hsieh, Y.-P., Reina, A., Kong, J., Terrones, M., and Dresselhaus, M.S. (2009). Controlled formation of sharp zigzag and armchair edges in graphitic nanoribbons. *Science*, **323**(5922), 1701–1705. (Cited on page *198*).

Johnson, J. B. (1928). Thermal agitation of electricity in conductors. *Phys. Rev.*, **32**(1), 97. (Cited on page *97*).

Kafanov, S., Kemppinen, A., Pashkin, Yu. A., Meschke, M., Tsai, J. S., and Pekola, J. P. (2009). Single-electronic radio-frequency refrigerator. *Phys. Rev. Lett.*, **103**(12), 120801. (Cited on page *165*).

Kaminski, A. and Glazman, L. I. (2001). Electron energy relaxation in the presence of magnetic impurities. *Phys. Rev. Lett.*, **86**, 2400–2403. (Cited on page *27*).

Kasumov, A. Yu., Deblock, R., Kociak, M., Reulet, B., Bouchiat, H., Khodos, I. I., Gorbatov, Yu. B., Volkov, V. T., Journet, C., and Burghard, M. (1999). Supercurrents through single-walled carbon nanotubes. *Science*, **284**(5419), 1508–1511. (Cited on pages *89 and 93*).

Katsnelson, M.I., Novoselov, K.S., and Geim, A.K. (2006). Chiral tunnelling and the Klein paradox in graphene. *Nature Phys.*, **2**, 620. (Cited on page *194*).

Kawabata, Arisato (1989). Theory of ballistic transport through a constriction-quantization of conductance. *J. Phys. Soc. Jpn*, **58**(2), 372–375. (Cited on page *44*).

Keller, Mark W., Martinis, John M., Zimmerman, Neil M., and Steinbach, Andrew H. (1996). Accuracy of electron counting using a 7-junction electron pump. *Appl. Phys. Lett.*, **69**(12), 1804–1806. (Cited on page *138*).

Kharitonov, Maxim Yu. and Efetov, Konstantin B. (2008). Universal conductance fluctuations in graphene. *Phys. Rev. B*, **78**, 033404. (Cited on page *72*).

Kindermann, M. and Nazarov, Yu. V. (2003). Full counting statistics in electric circuits. In *Quantum Noise in Mesoscopic Physics* (ed. Y. V. Nazarov), pp. 403. Kluwer Academic Publishers, Dordrecht. Also at [arXiv:cond-mat/0303590]. (Cited on page *116*).

Kindermann, M. and Pilgram, S. (2004). Statistics of heat transfer in mesoscopic circuits. *Phys. Rev. B*, **69**, 155334. (Cited on page *117*).

Klein, O. (1929). Die Reflexion von Elektronen an einem Potentialsprung nach der relativistischen Dynamik von Dirac. *Z. Phys.*, **53**, 157. (Cited on page *194*).

Koch, Jens, Yu, Terri M., Gambetta, Jay, Houck, A. A., Schuster, D. I., Majer, J., Blais, Alexandre, Devoret, M. H., Girvin, S. M., and Schoelkopf, R. J. (2007). Charge-insensitive qubit design derived from the Cooper pair box. *Phys. Rev. A*, **76**(4), 042319. (Cited on page *179*).

Kogan, Sh. (1996). *Electronic noise and fluctuations in solids*. Cambridge University Press, Cambridge CB2 1RP. (Cited on page *118*).

Kogan, Sh. M. and Shulman, A. Ya. (1969). On the theory of fluctuations in a nonequilibrium electron gas. *Sov. Phys. JETP*, **29**, 467. (Cited on page *104*).

Kondo, Jun (1964). Resistance minimum in dilute magnetic alloys. *Prog. Theor. Phys.*, **32**, 37–49. (Cited on page *151*).

Kopnin, N. B., Heikkilä, T. T., and Volovik, G. E. (2011). High-temperature surface superconductivity in topological flat-band systems. *Phys. Rev. B*, **83**, 220503. (Cited on pages *190 and 192*).

Kopnin, N. B., Mel'nikov, A. S., and Vinokur, V. M. (2006). Resonance energy and charge pumping through quantum SINIS contacts. *Phys. Rev. Lett.*, **96**, 146802. (Cited on page *91*).

Koppens, F. H. L., Folk, J. A., Elzerman, J. M., Hanson, R., van Beveren, L. H. Willems, Vink, I. T., Tranitz, H. P., Wegscheider, W., Kouwenhoven, L. P., and Vandersypen, L. M. K. (2005). Control and detection of singlet-triplet mixing in a random nuclear field. *Science*, **309**, 1346–1350. (Cited on page *156*).

Koskinen, Pekka, Malola, Sami, and Häkkinen, Hannu (2008). Self-passivating edge reconstructions of graphene. *Phys. Rev. Lett.*, **101**, 115502. (Cited on page *198*).

Kouwenhoven, L. and Glazman, L. (2001). Revival of the Kondo effect. *Phys. World*, **14**, 33. Can also be found at [arXiv:cond-mat/0104100]. (Cited on page *151*).

Kouwenhoven, L P, Austing, D G, and Tarucha, S (2001). Few-electron quantum dots. *Rep. Prog. Phys.*, **64**(6), 701. (Cited on pages *140, 142, 150, and 157*).

Kovalev, Alexey A., Hayden, Lorien X., Bauer, Gerrit E. W., and Tserkovnyak, Yaroslav (2011). Macrospin tunneling and magnetopolaritons with nanomechanical interference. *Phys. Rev. Lett.*, **106**, 147203. (Cited on page *227*).

Kramer, B and MacKinnon, A (1993). Localization: theory and experiment. *Rep. Prog. Phys.*, **56**(12), 1469. (Cited on pages *53 and 70*).

Kubo, Ryogo (1957). Statistical-mechanical theory of irreversible processes. I. general theory and simple applications to magnetic and conduction problems. *J. Phys. Soc. Jpn*, **12**(6), 570–586. (Cited on pages *96 and 247*).

Kuemmeth, F., Ilani, S., Ralph, D.C., and McEuen, P.L. (2008). Coupling of spin and orbital motion of electrons in carbon nanotubes. *Nature*, **452**, 448. (Cited on pages *8 and 14*).

Kulik, I.O. and Shekter, R.I. (1975). Kinetic phenomena and charge discreteness effects in granulated media. *Sov. Phys. JETP*, **41**, 308. (Cited on page *124*).

Laakso, M. A., Heikkilä, T. T., and Nazarov, Yuli V. (2010). Fully overheated single-electron transistor. *Phys. Rev. Lett.*, **104**, 196805. (Cited on pages *117 and 127*).

Laakso, M. A., Heikkilä, T. T., and Nazarov, Yuli V. (2012). Manifestly non-Gaussian fluctuations in superconductor–normal metal tunnel nanostructures. *Phys. Rev. Lett.*, **108**, 067002. (Cited on page *117*).

Lafarge, P., Joyez, P., Esteve, D., Urbina, C., and Devoret, M. H. (1993). Measurement of the even–odd free-energy difference of an isolated superconductor. *Phys. Rev. Lett.*, **70**(7), 994–997. (Cited on page *162*).

Landau, L. D. and Lifshitz, E.M. (1985). *Statistical Physics* (3rd edn). Pergamon Press, Oxford. (Cited on pages *117, 204, and 205*).

Landauer, R. (1957). Spatial variation of currents and fields due to localized scatterers in metallic conduction. *IBM J. Re. Dev.*, **1**, 223. (Cited on page *37*).

Landauer, Rolf (1970). Electrical resistance of disordered one-dimensional lattices. *Philos. Mag.*, **21**(172), 863–867. (Cited on page *63*).

Landauer, Rolf (1981). Can a length of perfect conductor have a resistance? *Phys. Lett. A*, **85**(2), 91–93. (Cited on page *50*).

Latil, Sylvain and Henrard, Luc (2006). Charge carriers in few-layer graphene films. *Phys. Rev. Lett.*, **97**, 036803. (Cited on page *190*).

le Sueur, H., Joyez, P., Pothier, H., Urbina, C., and Esteve, D. (2008). Phase controlled superconducting proximity effect probed by tunneling spectroscopy. *Phys. Rev. Lett.*, **100**(19), 197002. (Cited on page *92*).

Leivo, M. M., Pekola, J. P., and Averin, D. V. (1996). Efficient Peltier refrigeration by a pair of normal metal/insulator/superconductor junctions. *Appl. Phys. Lett.*, **68**, 1996–1998. (Cited on page *164*).

Lemay, G., Janssen, J.W., van den Hout, M., Mooij, M., Bronikowski, M.J., Willis, P.A., Smalley, R.E., Kouwenhoven, L.P., and C.Dekker (2001). Two-dimensional imaging of electronic wavefunctions in carbon nanotubes. *Nature*, **412**, 617. (Cited on page *157*).

Levitov, L.S. and Lesovik, G.B. (1993). Charge distribution in quantum shot noise. *JETP Lett.*, **58**, 230. (Cited on page *115*).

Levitov, Leonid S., Lee, Hyunwoo, and Lesovik, Gordey B. (1996). Electron counting statistics and coherent states of electric current. *J. Math. Phys.*, **37**(10), 4845–4866. (Cited on page *115*).

Lévy, L. P., Dolan, G., Dunsmuir, J., and Bouchiat, H. (1990). Magnetization of mesoscopic copper rings: Evidence for persistent currents. *Phys. Rev. Lett.*, **64**, 2074–2077. (Cited on page *75*).

Likharev, K. K. (1979). Superconducting weak links. *Rev. Mod. Phys.*, **51**, 101–159. (Cited on page *92*).

Liu, R.C., Odom, B., Yamamoto, Y., and Tarucha, S. (1998). Quantum interference in electron collision. *Nature*, **391**, 264. (Cited on page *44*).

Loss, Daniel and DiVincenzo, David P. (1998). Quantum computation with quantum dots. *Phys. Rev. A*, **57**, 120–126. (Cited on pages *155 and 156*).

Mahan, G.D. (2000). *Many-Particle Physics* (3rd edn). Kluwer Academic/Plenum Publishers. (Cited on pages *4 and 152*).

Mailly, D., Chapelier, C., and Benoit, A. (1993). Experimental observation of persistent currents in GaAs–AlGaAs single loop. *Phys. Rev. Lett.*, **70**, 2020–2023. (Cited on page *75*).

Makhlin, Yuriy, Schön, Gerd, and Shnirman, Alexander (2001). Quantum-state engineering with Josephson-junction devices. *Rev. Mod. Phys.*, **73**(2), 357–400. (Cited on page *181*).

Manucharyan, Vladimir E., Koch, Jens, Glazman, Leonid I., and Devoret, Michel H. (2009). Fluxonium: Single Cooper-pair circuit free of charge offsets. *Science*, **326**(5949), 113–116. (Cited on page *179*).

Marquardt, Florian, Chen, Joe P., Clerk, A. A., and Girvin, S. M. (2007). Quantum theory of cavity-assisted sideband cooling of mechanical motion. *Phys. Rev. Lett.*, **99**, 093902. (Cited on pages *218 and 221*).

Marquardt, Florian and Girvin, Steve (2009). Optomechanics. *Physics*, **2**, 40. (Cited on pages *216 and 227*).

Martin, J., Akerman, N., Ulbricht, G., Lohmann, T., Smet, J. H., von Klitzing, K., and Yacoby, A. (2008). Observation of electron-hole puddles in graphene using a scanning single-electron transistor. *Nature Phys.*, **4**, 144. (Cited on page *193*).

Martin, Paul C. and Schwinger, Julian (1959). Theory of many-particle systems. I. *Phys. Rev.*, **115**(6), 1342–1373. (Cited on page *247*).

Martinis, John M., Nam, S., Aumentado, J., and Urbina, C. (2002). Rabi oscillations in a large Josephson-junction qubit. *Phys. Rev. Lett.*, **89**(11), 117901. (Cited on page *179*).

Massel, F., Heikkilä, T.T., Pirkkalainen, J.-M., Cho, S.U., Saloniemi, H., Hakonen, P.J., and Sillanpää, M.A. (2011). Microwave amplification with nanomechanical resonators. *Nature*, **480**, 351. (Cited on pages *216, 217, 218, 221, 223, and 228*).

Matveev, K. A. (1995). Coulomb blockade at almost perfect transmission. *Phys. Rev. B*, **51**(3), 1743–1751. (Cited on page *123*).

McClure, J.W. (1969). Electron energy band structure and electronic properties of rhombohedral graphite. *Carbon*, **7**, 425. (Cited on page *189*).

Mehta, M.L. (2004). *Random Matrices*. Academic Press. (Cited on page *58*).

Meir, Yigal and Wingreen, Ned S. (1992). Landauer formula for the current through an interacting electron region. *Phys. Rev. Lett.*, **68**(16), 2512–2515. (Cited on page *145*).

Meschke, M., Guichard, W., and Pekola, J.P. (2006). Single-mode heat conduction by photons. *Nature*, **444**, 187–190. (Cited on pages *17 and 24*).

Meschke, M., Pekola, J. P., Gay, F., Rapp, R. E., and Godfrin, H. (2004). Electron thermalization in metallic islands probed by Coulomb blockade thermometry. *J. Low Temp. Phys.*, **134**, 1119. (Cited on pages *135 and 136*).

Metzger, C. Höhberger and Karrai, Khalek (2004). Cavity cooling of a microlever. *Nature*, **432**, 1002. (Cited on page *218*).

Moskalets, M. and Büttiker, M. (2002). Floquet scattering theory of quantum pumps. *Phys. Rev. B*, **66**, 205320. (Cited on page *56*).

Moskalets, M. and Büttiker, M. (2008). Dynamic scattering channels of a double barrier structure. *Phys. Rev. B*, **78**, 035301. (Cited on page *56*).

Nagaev, K. E. (1995). Influence of electron–electron scattering on shot noise in diffusive contacts. *Phys. Rev. B*, **52**, 4740–4743. (Cited on page *35*).

Naik, A., Buu, O., LaHaye, M. D., Armour, A. D., Clerk, A. A., Blencowe, M. P., and Schwab, K. C. (2006). Cooling a nanomechanical resonator with quantum back-action. *Nature*, **443**, 193. (Cited on pages *161, 203, and 211*).

Nakahara, Mikio and Ohmi, Tetsuo (2008). *Quantum Computing: from Linear Algebra to Physical Realizations*. Taylor & Francis. (Cited on pages *177, 181, and 224*).

Nakamura, Y., Chen, C. D., and Tsai, J. S. (1996). Quantitative analysis of Josephson-quasiparticle current in superconducting single-electron transistors. *Phys. Rev. B*, **53**(13), 8234–8237. (Cited on page *161*).

Nakamura, Y., Pashkin, Yu. A., and Tsai, J.S. (1999). Coherent control of macroscopic quantum states in a single-Cooper-pair box. *Nature*, **398**, 786. (Cited on page *178*).

Nazarov, Yu. V. (1994). Limits of universality in disordered conductors. *Phys. Rev. Lett.*, **73**, 134. (Cited on page *51*).

Nazarov, Yuli V. (1999*a*). Coulomb blockade without tunnel junctions. *Phys. Rev. Lett.*, **82**(6), 1245–1248. (Cited on page *123*).

Nazarov, Yuli V. (1999*b*). Novel circuit theory of Andreev reflection. *Superlatt. Microstruct.*, **25**, 1221. (Cited on page *28*).

Nazarov, Yu. V. (ed.) (2003). *Quantum Noise in Mesoscopic Physics*. Kluwer Academic Publishers, Dordrecht. (Cited on page *118*).

Nazarov, Yu. V. and Blanter, Ya. M. (2009). *Quantum Transport: Introduction to Nanoscience*. Cambridge University Press. (Cited on pages *13 and 69*).

Neder, I., Ofek, N., Chung, Y., Heiblum, M., Mahalu, D., and Umansky, V. (2007). Interference between two indistinguishable electrons from independent sources. *Nature*, **448**, 333. (Cited on pages *105 and 108*).

Neeley, M., Bialczak, R.C., Lenander, M., Lucero, E., Mariantoni, M., O'Connell, A.D., Sank, D., Wang, H., Weides, M., Wenner, J., Yin, Y., Yamamoto, T., Cleland, A.N., and Martinis, J.M. (2010). Generation of three-qubit entangled states using superconducting phase qubits. *Nature*, **467**, 570. (Cited on page *180*).

Negele, J.W. and Orland, H. (1998). *Quantum Many-Particle Systems*. Westview Press. (Cited on page *99*).

Nielsen, Michael A. and Chuang, Isaac L. (2000). *Quantum Computation and Quantum Information*. Cambridge University Press, Cambridge. (Cited on pages *177, 181, and 224*).

NobelGraphene (2010). Advanced information on the Nobel prize on graphene, http://www.nobelprize.org/nobel_prizes/physics/laureates/2010/advanced.html. (Cited on page *201*).

Novoselov, K. S., Geim, A. K., Morozov, S. V., Jiang, D., Katsnelson, M. I., Grigorieva, I. V., Dubonos, S. V., and Firsov, A. A. (2005). Two-dimensional gas of massless Dirac fermions in graphene. *Nature*, **438**, 197. (Cited on page *183*).

Novoselov, K. S., Geim, A. K., Morozov, S. V., Jiang, D., Zhang, Y., Dubonos, S. V., Grigorieva, I. V., and Firsov, A. A. (2004). Electric field effect in atomically thin carbon films. *Science*, **306**(5696), 666–669. (Cited on pages *183 and 202*).

Nozières, P. (1974). A 'Fermi-liquid' description of the Kondo problem at low temperatures. *J. Low Temp. Phys.*, **17**, 31–42. (Cited on page *151*).

Nygård, J., Cobden, D.H., and Lindelof, P.E. (2000). Kondo physics in carbon nanotubes. *Nature*, **408**, 342. (Cited on page *153*).

Nyquist, H. (1928). Thermal agitation of electric charge in conductors. *Phys. Rev.*, **32**, 110–113. (Cited on page *97*).

Oberholzer, S., Bieri, E., Schönenberger, C., Giovannini, M., and Faist, J. (2006). Positive cross correlations in a normal-conducting fermionic beam splitter. *Phys. Rev. Lett.*, **96**(4), 046804. (Cited on page *107*).

O'Connell, A.D., Hofheinz, M., Ansmann, M., Bialczak, R.C., Lenander, M., Lucero, E., Neeley, M., Sand, D., Wang, H., Weides, M., Wenner, J., Martinis, J.M., and Cleland, A.N. (2010). Quantum ground state and single-phonon control of a mechanical resonator. *Nature*, **464**, 697. (Cited on pages *203, 211, 222, 223, 224, and 225*).

Ohta, Taisuke, Bostwick, Aaron, Seyller, Thomas, Horn, Karsten, and Rotenberg, Eli (2006). Controlling the electronic structure of bilayer graphene. *Science*, **313**(5789), 951–954. (Cited on page *187*).

Ojanen, T. and Heikkilä, T.T. (2006). Quantum transitions induced by the third cumulant of current fluctuations. *Phys. Rev. B*, **73**, 20501(R). (Cited on page *110*).

Oliver, William D., Kim, Jungsang, Liu, Robert C., and Yamamoto, Yoshihisa (1999). Hanbury Brown and Twiss-type experiment with electrons. *Science*, **284**(5412), 299–301. (Cited on page *107*).

Ono, K., Austing, D. G., Tokura, Y., and Tarucha, S. (2002). Current rectification by Pauli exclusion in a weakly coupled double quantum dot system. *Science*, **297**(5585), 1313–1317. (Cited on pages *155 and 156*).

Paila, A., Gunnarsson, D., Sarkar, J., Sillanpää, M. A., and Hakonen, P. J. (2009). Current-phase relation and Josephson inductance in a superconducting Cooper-pair transistor. *Phys. Rev. B*, **80**(14), 144520. (Cited on page *176*).

Pályi, András and Burkard, Guido (2009). Hyperfine-induced valley mixing and the spin-valley blockade in carbon-based quantum dots. *Phys. Rev. B*, **80**, 201404. (Cited on page *155*).

Partoens, B. and Peeters, F. M. (2006). From graphene to graphite: Electronic structure around the K point. *Phys. Rev. B*, **74**, 075404. (Cited on page *190*).

Pedersen, Morten Holm and Büttiker, Markus (1998). Scattering theory of photon-assisted electron transport. *Phys. Rev. B*, **58**(19), 12993–13006. (Cited on page *59*).

Pekola, Jukka P., Giazotto, Francesco, and Saira, Olli-Pentti (2007). Radio-frequency single-electron refrigerator. *Phys. Rev. Lett.*, **98**(3), 037201. (Cited on page *165*).

Pekola, J. P. and Hekking, F. W. J. (2007). Normal-metal–superconductor tunnel junction as a brownian refrigerator. *Phys. Rev. Lett.*, **98**(21), 210604. (Cited on page *164*).

Pekola, J. P., Hirvi, K. P., Kauppinen, J. P., and Paalanen, M. A. (1994). Thermometry by arrays of tunnel junctions. *Phys. Rev. Lett.*, **73**(21), 2903–2906. (Cited on page *135*).

Pekola, Jukka P., Holmqvist, Tommy, and Meschke, Matthias (2008). Primary tunnel junction thermometry. *Phys. Rev. Lett.*, **101**(20), 206801. (Cited on pages *4 and 135*).

Pekola, J. P., Nieminen, T. E., Meschke, M., Kivioja, J. M., Niskanen, A. O., and Vartiainen, J. J. (2005). Shot-noise-driven escape in hysteretic Josephson junctions. *Phys. Rev. Lett.*, **95**, 197004. (Cited on page *171*).

Pekola, Jukka P., Vartiainen, Juha J., Möttönen, Mikko, Saira, Olli-Pentti, Meschke, Matthias, and Averin, Dmitri V. (2008). Hybrid single-electron transistor as a source of quantized electric current. *Nature Phys.*, **4**, 120. (Cited on pages *138, 166, and 167*).

Peltonen, Joonas (2011). *Fluctuations, Relaxation and Proximity Effect in Superconducting Circuits*. Ph. D. thesis, Aalto University. (Cited on page *172*).

Petta, J. R., Johnson, A. C., Taylor, J. M., Laird, E. A., Yacoby, A., Lukin, M. D., Marcus, C. M., Hanson, M. P., and Gossard, A. C. (2005). Coherent manipulation of coupled electron spins in semiconductor quantum dots. *Science*, **309**(5744), 2180–2184. (Cited on page *156*).

Pierre, F. (2000). *Interactions Electron–Electron dans les Fils Mesoscopiques*. Ph. D. thesis, Universite Paris 6. http://www.pa.msu.edu/cmp/birge-group/pierre/. (Cited on page *24*).

Pillet, J-D., Quay, C.H.L., Morfin, P., Bena, C., Levy Yeyati, A., and Joyez, P. (2010). Andreev bound states in supercurrent-carrying carbon nanotubes revealed. *Nature Phys.*, **6**, 965. (Cited on page *89*).

Plischke, Michael and Bergersen, Birger (1994). *Equilibrium Statistical Physics* (2nd edn). World Scientific. (Cited on page *35*).

Ponomarenko, L. A., Schedin, F., Katsnelson, M. I., Yang, R., Hill, E. W., Novoselov, K. S., and Geim, A. K. (2008). Chaotic Dirac billiard in graphene quantum dots. *Science*, **320**, 356–358. (Cited on page *140*).

Pothier, H., Gueron, S., Birge, Norman O., Esteve, D., and Devoret, M. H. (1997). Energy distribution function of quasiparticles in mesoscopic wires. *Phys. Rev. Lett.*, **79**, 3490–3493. (Cited on page *29*).

Potok, R.M., Rau, I.G., Shtrikman, H., Oreg, Y., and Goldhaber-Gordon, D. (2007). Observation of the two-channel Kondo effect. *Nature*, **446**, 167. (Cited on page *157*).

Qi, Xiao-Liang and Zhang, Shou-Cheng (2011). Topological insulators and superconductors. *Rev. Mod. Phys.*, **83**, 1057–1110. (Cited on page *3*).

Qian, Dong, Wagner, Gregory J, Liu, Wing Kam, Yu, Min-Feng, and Ruoff, Rodney S (2002). Mechanics of carbon nanotubes. *Appl. Mech. Rev.*, **55**(6), 495–533. (Cited on page *208*).

Rammer, J. (1998). *Quantum transport theory.* Perseus books, Reading, Massachusetts. (Cited on pages *14 and 28*).

Rammer, J. (2007). *Quantum Field Theory of Non-Equilibrium States.* Cambridge University Press. (Cited on pages *17 and 24*).

Reichl, L.E. (2004). *A Modern Course in Statistical Physics* (2nd edn). Wiley-VCH. (Cited on page *210*).

Reulet, B., Senzier, J., and Prober, D. E. (2003). Environmental effects in the third moment of voltage fluctuations in a tunnel junction. *Phys. Rev. Lett.*, **91**(19), 196601. (Cited on page *117*).

Roati, Giacomo, D'Errico, Chiara, Fallani, Leonardo, Fattori, Marco, Fort, Chiara, Zaccanti, Matteo, Modugno, Giovanni, Modugno, Michele, and Inguscio, Massimo (2008). Anderson localization of a non-interacting Bose–Einstein condensate. *Nature*, **453**, 895. (Cited on page *70*).

Rocheleau, T., Ndukum, T., Macklin, C., Hertzberg, J. B., Clerk, A. A., and Schwab, K. C. (2009). Preparation and detection of a mechanical resonator near the ground state of motion. *Nature*, **463**, 72–75. (Cited on pages *203 and 211*).

Russo, Saverio, Oostinga, Jeroen B., Wehenkel, Dominique, Heersche, Hubert B., Sobhani, Samira Shams, Vandersypen, Lieven M. K., and Morpurgo, Alberto F. (2008). Observation of Aharonov–Bohm conductance oscillations in a graphene ring. *Phys. Rev. B*, **77**(8), 085413. (Cited on page *61*).

Rycerz, A., Tworzydło, J., and Beenakker, C.W.J. (2007). Valley filter and valley valve in graphene. *Nature Phys.*, **3**, 172. (Cited on page *200*).

Rychkov, V. and Büttiker, M. (2006). Mesoscopic versus macroscopic division of current fluctuations. *Phys. Rev. Lett.*, **96**(16), 166806. (Cited on page *107*).

Ryu, Shinsei and Hatsugai, Yasuhiro (2002). Topological origin of zero-energy edge states in particle-hole symmetric systems. *Phys. Rev. Lett.*, **89**, 077002. (Cited on page *199*).

Safavi-Naeini, Amir H., Chan, Jasper, Hill, Jeff T., Alegre, Thiago P. Mayer, Krause, Alex, and Painter, Oskar (2012). Observation of quantum motion of a nanomechanical resonator. *Phys. Rev. Lett.*, **108**, 033602. (Cited on page *222*).

Saira, Olli-Pentti, Meschke, Matthias, Giazotto, Francesco, Savin, Alexander M., Möttönen, Mikko, and Pekola, Jukka P. (2007). Heat transistor: Demonstration of gate-controlled electronic refrigeration. *Phys. Rev. Lett.*, **99**(2), 027203. (Cited on pages *165 and 166*).

Salo, J., Hekking, F. W. J., and Pekola, J. P. (2006). Frequency-dependent current correlation functions from scattering theory. *Phys. Rev. B*, **74**(12), 125427. (Cited on page *48*).

Samuelsson, P. and Büttiker, M. (2002). Chaotic dot-superconductor analog of the Hanbury Brown–Twiss effect. *Phys. Rev. Lett.*, **89**(4), 046601. (Cited on page *107*).

Samuelsson, P., Sukhorukov, E. V., and Büttiker, M. (2004). Two-particle Aharonov–Bohm effect and entanglement in the electronic Hanbury Brown–Twiss setup. *Phys. Rev. Lett.*, **92**(2), 026805. (Cited on pages *105 and 108*).

Sánchez, David and Büttiker, Markus (2004). Magnetic-field asymmetry of nonlinear mesoscopic transport. *Phys. Rev. Lett.*, **93**, 106802. (Cited on page *56*).

Sapmaz, S., Blanter, Ya. M., Gurevich, L., and van der Zant, H. S. J. (2003). Carbon nanotubes as nanoelectromechanical systems. *Phys. Rev. B*, **67**(23), 235414. (Cited on page *213*).

Sapmaz, S., Jarillo-Herrero, P., Kong, J., Dekker, C., Kouwenhoven, L. P., and van der Zant, H. S. J. (2005). Electronic excitation spectrum of metallic carbon nanotubes. *Phys. Rev. B*, **71**(15), 153402. (Cited on page *150*).

Sazonova, V., Yaish, Y., Üstünel, H., Roundy, D., Arias, T.A., and McEuen, P.L. (2004). A tunable carbon nanotube electromechanical oscillator. *Nature*, **431**, 284. (Cited on page *213*).

Scheer, E., Agraït, N., Cuevas, J.C., Yeyati, A. Levy, Ludoph, B., Martín-Rodero, A., Bollinger, G. Rubio, van Ruitenbeek, J., and Urbina, C. (1998). The signature of chemical valence in the electrical conduction through a single-atom contact. *Nature*, **394**, 154. (Cited on page *42*).

Scheer, E., Joyez, P., Esteve, D., Urbina, C., and Devoret, M. H. (1997). Conduction channel transmissions of atomic-size aluminum contacts. *Phys. Rev. Lett.*, **78**(18), 3535–3538. (Cited on page *100*).

Schep, Kees M. and Bauer, Gerrit E. W. (1997). Universality of transport through dirty interfaces. *Phys. Rev. Lett.*, **78**(15), 3015–3018. (Cited on page *101*).

Schmid, Albert (1983). Diffusion and localization in a dissipative quantum system. *Phys. Rev. Lett.*, **51**(17), 1506–1509. (Cited on page *176*).

Schoelkopf, R. J., Clerk, A. A., Girvin, S. M., Lehnert, K. W., and Devoret, M. H. (2003). Qubits as spectrometers of quantum noise. In *Quantum Noise in Mesoscopic Physics* (ed. Y. V. Nazarov), pp. 175. Kluwer Academic Publishers, Dordrecht. Also at [arXiv:cond-mat/0210247]. (Cited on page *112*).

Schoelkopf, R. J., Wahlgren, P., Kozhevnikov, A. A., Delsing, P., and Prober, D. E. (1998). The radio-frequency single-electron transistor (rf-SET): A fast and ultrasensitive electrometer. *Science*, **280**(5367), 1238–1242. (Cited on pages *136 and 137*).

Schoeller, Herbert and Schön, Gerd (1994). Mesoscopic quantum transport: Resonant tunneling in the presence of a strong Coulomb interaction. *Phys. Rev. B*, **50**(24), 18436–18452. (Cited on page *122*).

Schön, G. and Zaikin, A. D. (1994). Parity effects on electron tunnelling through small superconducting islands. *EPL (Europhys. Lett.)*, **26**(9), 695. (Cited on page *162*).

Schottky, W. (1918). Über spontane Stromschwankungen in verschiedenen Elektrizitätsleitern. *Ann. d. Phys.*, **57**, 541–567. (Cited on page *97*).

Shekter, R.I. (1973). Zero anomalies in the resistance of a tunnel junction containing metallic inclusions in the oxide layer. *Sov. Phys. JETP*, **36**, 747. (Cited on page *124*).

Sillanpää, Mika A., Khan, Raphaël, Heikkilä, Tero T., and Hakonen, Pertti J. (2011). Macroscopic quantum tunneling in nanoelectromechanical systems. *Phys. Rev. B*, **84**, 195433. (Cited on pages *214 and 222*).

Sillanpää, M. A., Lehtinen, T., Paila, A., Makhlin, Yu., Roschier, L., and Hakonen, P. J. (2005). Direct observation of Josephson capacitance. *Phys. Rev. Lett.*, **95**(20), 206806. (Cited on page *176*).

Smit, R. H. M., Noat, Y., Untiedt, C., Lang, N. D., van Hemert, M. C., and van Ruitenbeek, J. M. (2002). Measurement of the conductance of a hydrogen molecule. *Nature*, **419**, 906. (Cited on pages *8 and 14*).

Son, Young-Woo, Cohen, Marvin L., and Louie, Steven G. (2006). Half-metallic graphene nanoribbons. *Nature*, **444**, 347. (Cited on page *200*).

Spietz, Lafe, Lehnert, K. W., Siddiqi, I., and Schoelkopf, R. J. (2003). Primary electronic thermometry using the shot noise of a tunnel junction. *Science*, **300**(5627), 1929–1932. (Cited on pages *100 and 120*).

Spivak, B., Kravchenko, S. V., Kivelson, S. A., and Gao, X. P. A. (2010). *Colloquium*: Transport in strongly correlated two dimensional electron fluids. *Rev. Mod. Phys.*, **82**, 1743–1766. (Cited on page *69*).

Steele, G. A., Huttel, A. K., Witkamp, B., Poot, M., Meerwaldt, H. B., Kouwenhoven, L. P., and van der Zant, H. S. J. (2009). Strong coupling between single-electron tunneling and nanomechanical motion. *Science*, **325**(5944), 1103–1107. (Cited on pages *204, 210, and 215*).

Steinbach, Andrew H., Martinis, John M., and Devoret, Michel H. (1996). Observation of hot-electron shot noise in a metallic resistor. *Phys. Rev. Lett.*, **76**, 3806–3809. (Cited on page *105*).

Stolyarova, E., Rim, K.T., Ryu, S., Maultzsch, J., Kim, P., Brus, L.E., Heinz, T.F., Hybertsen, M.S., and Flynn, G.W. (2007). High-resolution scanning tunneling microscopy imaging of mesoscopic graphene sheets on an insulating surface. *PNAS*, **104**(22), 9209–9212. (Cited on page *9*).

Sukhorukov, Eugene V. and Loss, Daniel (1999). Noise in multiterminal diffusive conductors: Universality, nonlocality, and exchange effects. *Phys. Rev. B*, **59**(20), 13054–13066. (Cited on pages *18 and 35*).

Sulkko, J., Sillanpää, M.A., Häkkinen, P., Lechner, L., Helle, M., Fefferman, A., Parpia, J., and Hakonen, P.J. (2010). Strong gate coupling of high-q nanomechanical resonators. *Nano Lett.*, **10**(12), 4884–4889. (Cited on pages *204 and 208*).

Teufel, J. D., Donner, T., Li, Dale, Harlow, J. W., Allman, M. S., Cicak, K., Sirois, A. J., Whittaker, J. D., Lehnert, K. W., and Simmonds, R. W. (2011). Sideband cooling of micromechanical motion to the quantum ground state. *Nature*, **475**(5), 359–363. (Cited on pages *203, 204, 211, 216, 222, and 228*).

Thouless, D. J. (1977). Maximum metallic resistance in thin wires. *Phys. Rev. Lett.*, **39**(18), 1167–1169. (Cited on page *65*).

Tien, P. K. and Gordon, J. P. (1963). Multiphoton process observed in the interaction of microwave fields with the tunneling between superconductor films. *Phys. Rev.*, **129**(2), 647–651. (Cited on page *59*).

Timofeev, A. V., Meschke, M., Peltonen, J. T., Heikkilä, T. T., and Pekola, J. P. (2007). Wideband detection of the third moment of shot noise by a hysteretic Josephson junction. *Phys. Rev. Lett.*, **98**(20), 207001. (Cited on pages *117 and 171*).

Tinkham, M. (1996). *Introduction to superconductivity* (2nd edn). McGraw-Hill, Singapore. (Cited on pages *78, 80, 92, 168, and 181*).

Tombros, N., Jozsa, C., Popinciuc, M., Jonkman, H.T., and van Wees, B.J. (2007). Electronic spin transport and spin precession in single graphene layers at room temperature. *Nature*, **448**, 571. (Cited on pages *32 and 33*).

Tombros, N., Veligura, A., Junesch, J., aes, M.H.D. Guimar Vera-Marun, I.J., Jonkman, H.T., and van Wees, B.J. (2011). Quantized conductance of a suspended graphene nanoconstriction. *Nature Phys.*, **7**, 697. (Cited on page *59*).

Tsuneta, T., Virtanen, P., Wu, F., Wang, T., Heikkilä, T. T., and Hakonen, P. J. (2009). Local and non-local shot noise in multiwalled carbon nanotubes. *EPL (Europhys. Lett.)*, **85**(3), 37004. (Cited on page *55*).

Tuominen, M. T., Hergenrother, J. M., Tighe, T. S., and Tinkham, M. (1992). Experimental evidence for parity-based 2e periodicity in a superconducting single-electron tunneling transistor. *Phys. Rev. Lett.*, **69**(13), 1997–2000. (Cited on page *162*).

Tworzydło, J., Trauzettel, B., Titov, M., Rycerz, A., and Beenakker, C. W. J. (2006). Sub-Poissonian shot noise in graphene. *Phys. Rev. Lett.*, **96**, 246802. (Cited on page *196*).

Uchida, K., Takahashi, S., Harii, K., Ieda, J., Koshibae, W., Ando, K., Maekawa, S., and Saitoh, E. (2008). Observation of the spin Seebeck effect. *Nature*, **455**, 778. (Cited on page *35*).

Usmani, O., Blanter, Ya. M., and Nazarov, Yu. V. (2007). Strong feedback and current noise in nanoelectromechanical systems. *Phys. Rev. B*, **75**, 195312. (Cited on page *216*).

Valet, T. and Fert, A. (1993). Theory of the perpendicular magnetoresistance in magnetic multilayers. *Phys. Rev. B*, **48**, 7099–7113. (Cited on page *30*).

van Dam, Jorden A., Nazarov, Yuli V., Bakkers, Erik P.A.M., De Franceschi, Silvano, and Kouwenhoven, Leo P. (2006). Supercurrent reversal in quantum dots. *Nature*, **442**, 667. (Cited on page *89*).

van der Molen, Sense Jan and Liljeroth, Peter (2010). Charge transport through molecular switches. *J. Phys: Condens. Matter*, **22**(13), 133001. (Cited on page *7*).

van der Wiel, W. G., De Franceschi, S., Elzerman, J. M., Fujisawa, T., Tarucha, S., and Kouwenhoven, L. P. (2002). Electron transport through double quantum dots. *Rev. Mod. Phys.*, **75**(1), 1–22. (Cited on page *157*).

van Ostaay, J. A. M., Akhmerov, A. R., Beenakker, C. W. J., and Wimmer, M. (2011). Dirac boundary condition at the reconstructed zigzag edge of graphene. *Phys. Rev. B*, **84**, 195434. (Cited on pages *198 and 200*).

van Son, P. C., van Kempen, H., and Wyder, P. (1987). Boundary resistance of the ferromagnetic–nonferromagnetic metal interface. *Phys. Rev. Lett.*, **58**, 2271–2273. (Cited on page *31*).

van Wees, B. J., van Houten, H., Beenakker, C. W. J., Williamson, J. G., Kouwenhoven, L. P., van der Marel, D., and Foxon, C. T. (1988). Quantized conductance of point contacts in a two-dimensional electron gas. *Phys. Rev. Lett.*, **60**(9), 848–850. (Cited on page *44*).

Vandersypen, L. M. K., Elzerman, J. M., Schouten, R. N., van Beveren, L. H. Willems, Hanson, R., and Kouwenhoven, L. P. (2004). Real-time detection of single-electron tunneling using a quantum point contact. *Appl. Phys. Lett.*, **85**(19), 4394–4396. (Cited on page *44*).

Veselago, V.G. (1968). The electrodynamics of substances with simultaneously negative values of ϵ and μ. *Sov. Phys. Usp.*, **10**, 509. (Cited on page *194*).

Viljas, J. K. and Heikkilä, T. T. (2010). Electron–phonon heat transfer in monolayer and bilayer graphene. *Phys. Rev. B*, **81**(24), 245404. (Cited on page *28*).

Vion, D., Aassime, A., Cottet, A., Joyez, P., Pothier, H., Urbina, C., Esteve, D., and Devoret, M. H. (2002). Manipulating the quantum state of an electrical circuit. *Science*, **296**(5569), 886–889. (Cited on pages *172, 179, and 182*).

Virtanen, P. and Heikkilä, T.T. (2007). Thermoelectric effects in superconducting proximity structures. *Appl. Phys. A*, **89**, 625–637. (Cited on pages *16 and 92*).

Virtanen, Pauli and Heikkilä, Tero T. (2004). Thermopower induced by a supercurrent in superconductor–normal-metal structures. *Phys. Rev. Lett.*, **92**, 177004. (Cited on page *92*).

Voss, Richard F. and Webb, Richard A. (1981). Macroscopic quantum tunneling in 1-μm Nb Josephson junctions. *Phys. Rev. Lett.*, **47**(4), 265–268. (Cited on page *171*).

Waintal, Xavier (2011). KNIT project. http://inac.cea.fr/Pisp/xavier.waintal/KNIT.php. (Cited on page *53*).

Waintal, Xavier, Myers, Edward B., Brouwer, Piet W., and Ralph, D. C. (2000). Role of spin-dependent interface scattering in generating current-induced torques in magnetic multilayers. *Phys. Rev. B*, **62**, 12317–12327. (Cited on page *58*).

Wallace, P. R. (1947). The band theory of graphite. *Phys. Rev.*, **71**(9), 622–634. (Cited on pages *8 and 183*).

Wallraff, A., Schuster, D. I., Blais, A., Frunzio, L., Huang, R.-S., Majer, J., Kumar, S., Girvin, S. M., and Schoelkopf, R. J. (2004). Circuit quantum electrodynamics: Coherent coupling of a single photon to a Cooper pair box. *Nature*, **431**, 162. (Cited on page *181*).

Walls, D.F. and Milburn, G.J. (2010). *Quantum Optics.* Springer, Berlin. (Cited on page *220*).

Weiss, U. (1999). *Quantum Dissipative Systems* (2nd edn). World Scientific Publishing. (Cited on pages *108, 118, 170, and 181*).

Wendin, G. and Shumeiko, V. S. (1996). Giant Josephson current through a single bound state in a superconducting tunnel junction. *Phys. Rev. B*, **53**, R6006–R6009. (Cited on page *88*).

Wharam, D A, Thornton, T J, Newbury, R, Pepper, M, Ahmed, H, Frost, J E F, Hasko, D G, Peacock, D C, Ritchie, D A, and Jones, G A C (1988). One-dimensional transport and the quantisation of the ballistic resistance. *J. Phys. C*, **21**(8), L209. (Cited on page *44*).

Williams, J.R., Low, Tony, Lundstrom, M.S., and Marcus, C.M. (2011). Gate-controlled guiding of electrons in graphene. *Nature Nanotech.*, **6**, 222. (Cited on page *194*).

Wilson, Kenneth G. (1975). The renormalization group: Critical phenomena and the Kondo problem. *Rev. Mod. Phys.*, **47**(4), 773–840. (Cited on page *152*).

Xiao, Jiang, Bauer, Gerrit E. W., and Brataas, Arne (2008). Spin-transfer torque in magnetic tunnel junctions: Scattering theory. *Phys. Rev. B*, **77**, 224419. (Cited on page *58*).

Yacoby, A., Heiblum, M., Mahalu, D., and Shtrikman, Hadas (1995). Coherence and phase sensitive measurements in a quantum dot. *Phys. Rev. Lett.*, **74**, 4047–4050. (Cited on page *62*).

Young, Andrea F. and Kim, Philip (2009). Quantum interference and Klein tunnelling in graphene heterojunctions. *Nature Phys.*, **5**, 222. (Cited on page *196*).

Żak, Robert Andrzej, Maslov, Dmitrii L., and Loss, Daniel (2012). Ferromagnetic order of nuclear spins coupled to conduction electrons: A combined effect of electron–electron and spin–orbit interactions. *Phys. Rev. B*, **85**, 115424. (Cited on page *157*).

Zazunov, A., Shumeiko, V. S., Bratus', E. N., Lantz, J., and Wendin, G. (2003). Andreev level qubit. *Phys. Rev. Lett.*, **90**(8), 087003. (Cited on page *89*).

Zhang, Y., Tan, Y.-W., Stormer, H.L., and Kim, P. (2005). Experimental observation of the quantum Hall effect and Berry's phase in graphene. *Nature*, **438**, 201. (Cited on page *187*).

Zhang, Y., Tang, T.-T., Girit, C., Hao, Z., Martin, M.C., Zettl, A., Crommie, M.F., Ron Shen, Y., and Wang, F. (2009). Direct observation of a widely tunable bandgap in bilayer graphene. *Nature*, **459**, 820. (Cited on pages *187 and 189*).

Žutić, Igor, Fabian, Jaroslav, and Das Sarma, S. (2004). Spintronics: Fundamentals and applications. *Rev. Mod. Phys.*, **76**, 323–410. (Cited on pages *33 and 35*).

Index

In case of multiple page numbers for a single entry, the numbers in bold indicate pages with definitions or detailed explanations, whereas page numbers in roman type indicate textual references. The occurrences of the terms in exercises are in italics.